Springer Proceedings in Mathematics & Statistics

Volume 379

This book series features volumes composed of selected contributions from workshops and conferences in all areas of current research in mathematics and statistics, including operation research and optimization. In addition to an overall evaluation of the interest, scientific quality, and timeliness of each proposal at the hands of the publisher, individual contributions are all refereed to the high quality standards of leading journals in the field. Thus, this series provides the research community with well-edited, authoritative reports on developments in the most exciting areas of mathematical and statistical research today.

More information about this series at https://link.springer.com/bookseries/10533

Alexander Domoshnitsky · Alexander Rasin ·
Seshadev Padhi
Editors

Functional Differential Equations and Applications

FDEA-2019, Ariel, Israel, September 22–27

Springer

Editors
Alexander Domoshnitsky
Department of Mathematics
Ariel University
Ariel, Israel

Alexander Rasin
Department of Mathematics
Ariel University
Ariel, Israel

Seshadev Padhi
Department of Mathematics
Birla Institute of Technology, Mesra
Ranchi, Jharkhand, India

ISSN 2194-1009 ISSN 2194-1017 (electronic)
Springer Proceedings in Mathematics & Statistics
ISBN 978-981-16-6296-6 ISBN 978-981-16-6297-3 (eBook)
https://doi.org/10.1007/978-981-16-6297-3

Mathematics Subject Classification: 34K06, 34K10, 34K13, 00B25, 34A12, 34C10, 34C23, 34C25, 34D23, 35A01, 35A02, 39A21, 39A23, 39A28, 39A30

This Springer imprint is published by the registered company Springer Nature Singapore Pte Ltd.
The registered company address is: 152 Beach Road, #21-01/04 Gateway East, Singapore 189721, Singapore

Preface

Applications of functional differential equations have been the driving force for many researchers to study real-world problems. Given this context, the Department of Mathematics, Ariel University, Israel, has been organizing a series of international conferences on Functional Differential Equations and Applications.

This proceeding is the outcome of the Seventh International Conference on Functional Differential Equations and Applications, held at Ariel University, Ariel, Israel, from September 22–27, 2019. Researchers from Israel, the USA, Greece, Romania, Russia, Czech Republic, Poland, Slovakia, Ukraine, India, Georgia and Greece participated in the conference. The main focus on the conference was on the applications of functional differential equations, especially stability theory, positive solutions of differential equations, applications of boundary value problems, impulsive equations, integro-differential equations, feedback control and many other applications of functional differential equations.

A total of 84 researchers participated in the conference. Lectures were delivered on the oscillation and non-oscillation of solutions of ordinary and delay differential equations, and the stability of solutions of many mathematical models in engineering and medicines. A total of 20 articles were selected for publication in the proceedings. These are given chapter-wise as below.

The entire volume is divided into three parts. Part I addresses the dynamics of models in engineering. This part consists of six chapters. The chapters include topics on the dynamical behaviour of models in nanostructures, porous media, switched time-delay systems, ground robot path controlled by airborne autopilot with time-delay, diffusion–kinetic model of curing epoxy cancer and nonlinear equations of oscillations in modelling the magnetic separations. Part II addresses the dynamics of models in biology, medicine and ecology. This part consists of five chapters. This section includes topics on the dynamical behaviour of models of infectious disease, radiophysical sounding signals, bladder cancer treatment by using immunotherapy and a biological model with time-delay systems. Part III addresses the qualitative theory of differential equations. This section consists of nine chapters. The chapters included in this section are on the solutions of the modified Helmholtz equation, oscillation theory of differential equations, positive solutions of the Cantilever beam equation, bounded solutions to differential equations, periodic solutions of impulsive equations and many more. All the chapters are interesting and hope this will be useful to readers in their future research work.

We owe thanks to all participants of the conference and the authors of the chapters in this volume. We are thankful to the authorities of Ariel University for their support in organizing the conference. We should also mention here that it was a great pleasure to work with Shamim Ahmad, the Senior Editor of Springer Nature, and the production team, especially to Banu Dhayalan who took utmost care during the preparation of the volume.

Ariel, Israel Alexander Domoshnitsky
Ranchi, India Seshadev Padhi
Ariel, Israel Alexander Rasin

Contents

Dynamics of Models in Engineering

Dynamical Behaviour of Integro-Differential Equations Arising in Nano-Structures

Angela Slavova

Abstract Computational Nanotechnology has become an indispensable tool not only in predicting, but also in engineering the properties of multi-functional nano-structured materials. The presence of nano-heterogeneities in these materials affects or disturbs their elastic field at the local and the global scale and thus greatly influences their mechanical properties. In this paper we shall study dynamical behaviour of 2D dynamic coupled problem in multifunctional nano-heterogeneous piezoelectric composites. More in detail, we shall present first modeling of two-dimensional anti-plane (SH) wave propagation problem in piezoelectric anisotropic solids containing nano-holes or nano-inclusions. Nano-heterogeneities are considered in two aspects as wave scatters provoking scattered and diffraction wave fields and also as stress concentrators creating local stress concentrations in the considered solid.There are only few numerical results for dynamic behavior of bounded piezoelectric domain with heterogeneities under anti-plane load.

Keywords Integro differential equations · Nano structures · Nano heterogeneous piezoelectric composites · Cellular nano scale networks

1 Introduction

In the present work we propose, develop and validate for different mechanical models computational tools based on the application of the theory of integro-differential equations for solution of 2D dynamic coupled problems in multifunctional nano-heterogeneous piezoelectric composites. We study two-dimensional in-plane (P-SV) and anti-plane (SH) wave propagation problems in piezoelectric anisotropic solids containing nano-inhomogeneities. The model is based on the principles of elastodynamics, wave propagation theory and surface/interface elasticity theory. The obtained results and conclusions may be potentially useful for characterizing the mechanical

A. Slavova (✉)
Institute of Mathemtics and Informatics, Bulgarian Academy of Sciences, Sofia 1113, Bulgaria
e-mail: slavova@math.bas.bg

A. Domoshnitsky et al. (eds.), *Functional Differential Equations and Applications*, Springer Proceedings in Mathematics & Statistics 379,
https://doi.org/10.1007/978-981-16-6297-3_1

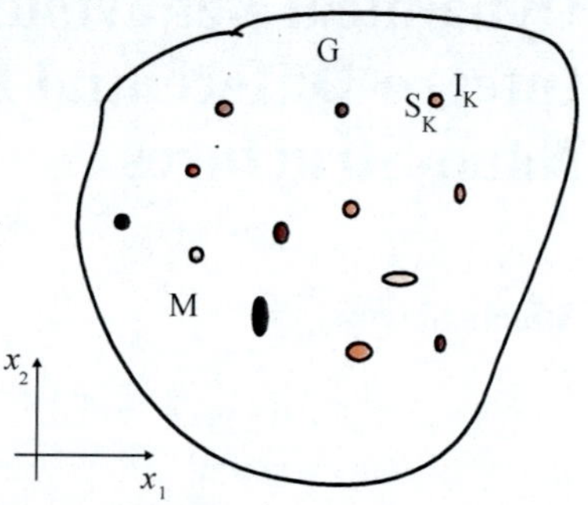

Fig. 1 The geometry: PEM inclusions in a bounded PEM matrix

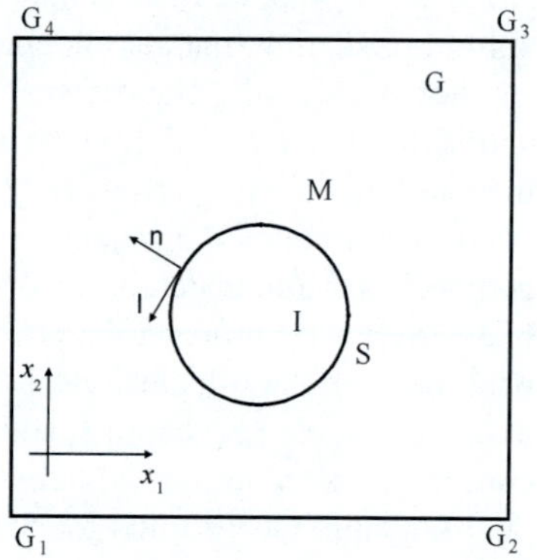

Fig. 2 Rectangular PEM matrix with circle inhomogeneity

stabilities of an array of nanowires or nano-tubes structures made by piezoelectric material under different type of dynamic loads.

Let $G \in R^2$ is a bounded piezoelectric domain (PEM) with a set of inhomogeneities $I = \cup I_k \in G$ (holes, inclusions, nano–holes, nano–inclusions) subjected to time–harmonic load on the boundary ∂G, see Fig. 1. Note that heterogeneities are of macro size if their diameter is greater than 10^{-6} m, while heterogeneities are of nano–size if their diameter is less than 10^{-7} m.

The aim is to find the field in every point of $M = G \backslash I$, and to evaluate stress concentration around the inhomogeneities.

Using the methods of continuum mechanics the problem can be formulated in terms of boundary value problem for a system of 2-nd order differential equations, see [3], Chap. 2. Let us for simplicity first formulate the problem in the case G is rectangular with a single circle inhomogeneity I, see Fig. 2.

There is a certain lack of work for solution of 2D in-plane and ant-plane dynamic problems for piezoelectric solids with nanoinclusions or nano-cavities [3–6]. The reason is that such a goal requires multidisciplinary knowledge and skills blending continuum mechanics, piezoelectricity, computational mechanics, material science, mathematical physics, and numerical method programming. Moreover, we shall apply Cellular Nanoscale Networks (CNN) [1, 8] in our investigations in order to obtain more accurate numerical results. In Sect. 2 we state the problem under consideration. We define the boundary conditions which play important role in the solutions. In Sect. 3 we propose CNN architecture which approximates the obtained integro-differential equation. We study the dynamics of CNN model via describing function method [7]. Section 4 deals with traveling wave solutions of the CNN model. In Sect. 5 we propose numerical simulations and validation for specific piezoelectric material. Feedback stabilization of the model is provided in Sect. 6.

2 Statement of the Problem

Following [5] let us define system of equations

$$\left| \begin{array}{l} c_{44}^N \Delta u_3^N + e_{15}^N \Delta u_4^N + \rho^N \omega^2 u_3 = 0, \\ e_{15}^N \Delta u_3^N - \varepsilon_{15}^N \Delta u_4^N = 0, \end{array} \right. \tag{1}$$

Here $x = (x_1, x_2)$, $\Delta = \frac{\partial^2}{\partial x_1^2} + \frac{\partial^2}{\partial x_2^2}$ is Laplace operator, $N = M$ for $x \in M$ and $N = I$ for $x \in I$; u_3^N is mechanical displacement, u_4^N is electric potential (the usual notation in mechanics is ϕ^N, but in order to have possibility for summation in formulas we use generalized notations u_J^N, $J = 3, 4$), ρ^N is the mass density, $c_{44}^N > 0$ is the shear stiffness, $e_{15}^N \neq 0$ is the piezoelectric constant and $\varepsilon_{11}^N > 0$ is the dielectric permittivity; ω is the frequency of the applied on ∂G load.

Let us define generalized stress σ_{iJ}, $i = 1, 2$; $J = 3, 4$ as

$$\left| \begin{array}{l} \sigma_{i3}^N = c_{44}^N \frac{\partial u_3^N}{\partial x_i} + e_{15}^N \frac{\partial u_4^N}{\partial x_i}, \\ \sigma_{i4}^N = e_{15}^N \frac{\partial u_3^N}{\partial x_i} - \varepsilon_{11}^N \frac{\partial u_4^N}{\partial x_i}, \end{array} \right. \tag{2}$$

Note that σ_{i3}^N is called mechanical stress, while σ_{i4}^N is called electrical displacement(the usual notation in mechanics is $D_i^N = \sigma_{i4}^N$, $i = 1, 2$).

Generalized traction at the point x on the line segment with normal vector $n = (n_1, n_2)$ is defined as

$$\left| \begin{array}{l} t_3^N = \sigma_{13}^N n_1 + \sigma_{23}^N n_2, \\ t_4^N = \sigma_{14}^N n_1 + \sigma_{24}^N n_2, \end{array} \right. \tag{3}$$

At every point $x \in S = \partial I$ we can define normal vector n and unit tangential vector l such that (l, n) forms right coordinate system.

On the exterior boundary ∂G boundary conditions are prescribed traction on the part of the boundary and prescribed displacement on the complemented part:

$$\left| \begin{array}{ll} t_J^{M0} & \text{on } \partial G_t, \\ u_J^{M0} & \text{on } \partial G_u = \partial G \backslash \partial G_t. \end{array} \right. \tag{4}$$

Here, the traction and the displacement vectors are defined as $t_J^{0M} = (t_3^{0M}, t_4^{0M})$ and $u_J^{0M} = (u_3^{0M}, u_4^{0M})$ respectively.

Boundary Conditions for Heterogeneities at Macro–Scale
(A) In the case I is a hole, formally we can consider that the constants in I are $c_{44}^I = 0$, $e_{15}^I = 0$, $\varepsilon_{11}^I = 0$ and boundary conditions on S are

$$t_J^M = 0 \quad \text{on } S, \tag{5}$$

Here, the traction vectors is defined as $t_J^M = (t_3^M, t_4^M)$. Then the boundary value problem (BVP) is: the equation (1) and boundary conditions (4), (5).
(B) In the case I is an inclusion, the constants in I are $c_{44}^I > 0$, $e_{15}^I \neq 0$, $\varepsilon_{11}^I > 0$; the constants in M are $c_{44}^M > 0$, $e_{15}^M \neq 0$, $\varepsilon_{11}^M > 0$ and boundary conditions on S are

$$\left| \begin{array}{l} u_J^M = u_J^I \quad \text{on } S, \\ t_J^I + t_J^M = 0 \quad \text{on } S, \end{array} \right. \tag{6}$$

The BVP is now: the Eq. (1) and boundary conditions (4), (6). Note, that $n_i^I = -n_i^M$, $i = 1, 2$, where n_i^I and n_i^M are the components of the outward normal for element along S considered as a boundary of the inclusion or matrix correspondingly. Additionally we have that $t_J^M = (t_3^M, t_4^M)$, $N = I, M$.

Boundary Conditions for Nano–Heterogeneities
Assume that the interface between the nano–inclusion I and its surrounding matrix M is regarded as a thin material surface S that possesses its own mechanical properties c_{44}^S, e_{15}^S, ε_{11}^S and surface tension τ^0.

More specifically, τ^0 is the residual surface tension under unstrained conditions that will induce an additional static deformation, but in dynamic analysis this is often ignored, i.e. $\tau^0 = 0$.
(C) In the case I is a nano–hole, formally we can consider that the constants in I are $c_{44}^I = 0$, $e_{15}^I = 0$, $\varepsilon_{11}^I = 0$ and boundary conditions on S are

$$t_J^M = \frac{\partial \sigma_{lJ}^S}{\partial l} \quad \text{on } S, \tag{7}$$

In this case BVP is: the Eq. (1) and boundary conditions (4), (7).

Boundary conditions (7) can be written in the following form for the mechanical and electrical part correspondingly:

$$t_3^M = \sigma_{n3}^M = \frac{\partial \sigma_{l3}^S}{\partial l}, \quad t_4^M = \sigma_{n4}^M = \frac{\partial \sigma_{l4}^S}{\partial l}$$

where t_3^M and t_4^M are the normal component of mechanical stress and electrical displacement (see Eq.(eq2)) in the matrix, while $\frac{\partial \sigma_{l3}^S}{\partial l}$ and $\frac{\partial \sigma_{l4}^S}{\partial l} = \frac{\partial D_l^S}{\partial l}$ are tangential derivatives of tangential components of stress σ_{l3}^S and tangential electrical displacement $\sigma_{l4}^S = D_l^S$ along the nano–hole boundary S.

(D) In the case I is a nano–inclusion, the constants in I are $c^I_{44} > 0, e^I_{15} \neq 0, \varepsilon^I_{11} > 0$; the constants in M are $c^M_{44} > 0$, $e^M_{15} \neq 0$, $\varepsilon^M_{11} > 0$.

On the heterogeneity boundary S where are defined constants $c^S_{44}, e^S_{15}, \varepsilon^S_{11}$ and with the notation for generalized displacement u^S along S the generalized tangential stress on S is defined as:

$$\left| \begin{array}{l} \sigma^S_{l3} = c^S_{44} \frac{\partial u^S_3}{\partial l} + e^S_{15} \frac{\partial u^S_4}{\partial l}, \\ \sigma^S_{l4} = e^S_{15} \frac{\partial u^S_3}{\partial l} - \varepsilon^S_{11} \frac{\partial u^S_4}{\partial l}, \end{array} \right. \tag{8}$$

Boundary conditions on S are

$$\left| \begin{array}{l} u^M_J = u^I_J \ \text{on} S, \\ t^I_J + t^M_J = \frac{\partial \sigma^S_{lJ}}{\partial l} \quad \text{on } S, \end{array} \right. \tag{9}$$

Then BVP is: the Eq. (1) and boundary conditions (4), (9).

Boundary conditions (9) can be written in the following form for the mechanical and electrical part correspondingly:

$$t^I_3 + t^M_3 = \frac{\partial \sigma^S_{l3}}{\partial l}, \quad t^I_4 + t^M_4 = \frac{\partial \sigma^S_{l4}}{\partial l}$$

where $t^N_3, t^N_4, N = I, M$ are the normal component of mechanical stress and electrical displacement (see Eq. (2)) in the inclusion and in the matrix, while $\frac{\partial \sigma^S_{l3}}{\partial l}$ and $\frac{\partial \sigma^S_{l4}}{\partial l} = \frac{\partial D^S_l}{\partial l}$ are tangential derivatives of tangential components of stress σ^S_{l3} and tangential electrical displacement $\sigma^S_{l4} = D^S_l$ along the interface boundary S. Here, it is take into consideration that $n^M_i = -n^I_i = -n_i, i = 1, 2$. Note that for the mechanical displacement u^N_3 and for the potential of the electric field $u^N_4 = \phi$ continuity conditions are satisfied, see first row of Eq. (9).

3 Integro-Differential CNN Model

Cellular Nonlinear/Nanoscale Networks (CNN) have been introduced in 1988 by Chua and Yang [1] as a new class of information processing systems which shows important potential applications (Fig. 3). The concept of CNN is based on some aspects of neurobiology and adapted to integrated circuits. CNN are defined as spatial arrangements of locally coupled dynamical systems, referred to as cells. The CNN dynamics are determined by a dynamic law of an isolated cell, by the coupling laws between the cells and by boundary and initial conditions. The cell coupling is confined to the local neighborhood of a cell within a defined sphere of influence. The dynamic law and the coupling laws of a cell are often combined and described by nonlinear ordinary differential- or difference equations (ODE), respectively, referred to as the state equations of cells. Thus a CNN is given by a system of coupled ODE

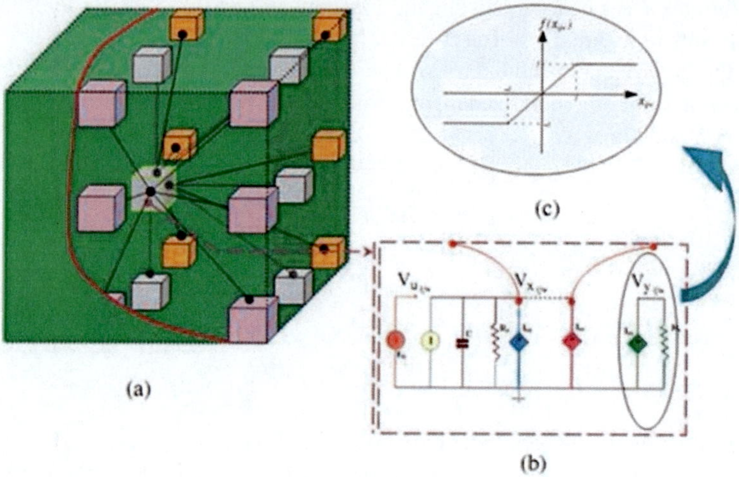

Fig. 3 **a** CNN architecture; **b** cell circuit; **c** output function of CNN

with a very compact representation in the case of translation invariant state equations. Despite of having a compact representation, CNN can show complex dynamics like chaotic behavior, self-organization, and pattern formation or nonlinear oscillation and wave propagation. Furthermore, Reaction-Diffusion Cellular Nonlinear/Nanoscale Networks (RD-CNN) have been applied for modeling complex systems [8].

Cellular Nanoscale Networks (CNN) [1, 8] are complex nonlinear dynamical systems, and therefore one can expect interesting phenomena like bifurcations and chaos to occur in such nets. It was shown that as the cell self-feedback coefficients are changed to a critical value, a CNN with opposite-sign template may change from stable to unstable. Namely speaking, this phenomenon arises as the loss of stability and the birth of a limit cycles.

We will give general definition of a CNN which follows the original one [1]:

Definition 1 An $M \times M$ cellular neural network is defined mathematically by four specifications:

(1) CNN cell dynamics;
(2) CNN synaptic law which represents the interactions (spatial coupling) within the neighbor cells;
(3) Boundary conditions;
(4) Initial conditions.

In terms of the definition we can present the dynamical systems describing CNN. For general CNN whose cells are made of time-invariant circuit elements, each cell $C(ij)$ is characterized by its CNN cell dynamics:

$$\dot{x}_{ij} = -g(x_{ij}, u_{ij}, I_{ij}^{s}),$$

where $x_{ij} \in \mathbf{R}^m$, u_{ij} is usually a scalar. In most cases, the interactions (spatial coupling) with the neighbor cell $C(i+k, j+l)$ are specified by a CNN synaptic law:

$$I^s_{ij} = A_{ij,kl}x_{i+k,j+l} + \tilde{A}_{ij,kl} * f_{kl}(x_{ij}, x_{i+k,j+l}) + \tilde{B}_{ij,kl} * u_{i+k,j+l}(t).$$

The first term $A_{ij,kl}x_{i+k,j+l}$ is simply a linear feedback of the states of the neighborhood nodes. The second term provides an arbitrary nonlinear coupling, and the third term accounts for the contributions from the external inputs of each neighbor cell that is located in the N_r neighborhood.

In [3] a system of integro-differential equations (IDE) is obtained for the unknowns u (displacement vectors) and τ (traction). The procedure is based on Gauss theorem [10] after finding the fundamental solutions of the boundary value problem formulated in the introduction.

Let us consider the following system of IDE, which is more general from the point of view of the applications in nano-technology:

$$\frac{\partial u(x)}{\partial \tau} = D\frac{\partial^2 u}{\partial x^2} - C_1 \int_S G(u(x))dx, \tag{10}$$

where C_1 is a constant depending on the ρ^M, $c^M_{44} > 0$, $e^M_{15} \neq 0$ and $\varepsilon^M_{11} > 0$, D is diffusion coefficient, $u = (u_3, u_4)$, function $G(x)$ is a function of the displacement vectors $u_{3,4}$ and the traction $\tau_{3,4}$.

It is known [1, 8] that some autonomous CNN represent an excellent approximation to nonlinear partial differential equations (PDEs). The intrinsic space distributed topology makes the CNN able to produce real-time solutions of nonlinear PDEs. There are several ways to approximate the Laplacian operator in discrete space by a CNN synaptic law with an appropriate A-template. In our case the CNN model of IDE (10) is:

$$\frac{du_{ij}}{dt} = DA_1 * u_{ij} - C_1 \int_S G(u_{ij}))dt, \; 1 \le i \le n, \; j = 3, 4, \tag{11}$$

where A_1 is 1-dimensional discretized Laplacian template [8] $A_1 : (1, -2, 1)$, $*$ is convolution operator, $n = M \times M$ is the number of cells of the CNN architecture.

Remark 1 Realized nano-scale CNN have been considered in a fast growing number of investigations dealing with image processing problems [1]. Despite of having a compact representation CNN can show very complex dynamics like chaotic behaviour, self organization and pattern formation or nonlinear oscillation and wave propagation. The future of CNN implementation is in nano-structure computer architecture. CNN not only represent a new paradigm for complexity but also establish novel approaches to information processing by nonlinear complex systems. Moreover, CNN have very impressive and promising applications in image processing and pattern recognition.

We develop the following algorithm for studying the dynamical behavior of CNN model (11) via describing function method [7]:

1. First, we apply double Fourier transform $F(s, z)$ to IDE CNN model (11)

$$F(s, z) = \sum_{k=-\infty}^{k=\infty} z^{-k} \int_{-\infty}^{\infty} f_k(t) exp(-st) dt. \tag{12}$$

from continuous time t and discrete space k to continuous temporal frequency ω, and continuous spatial frequency Ω, such that $z = exp(I\Omega)$, $s = I\omega$, I is the imaginary identity and therefore we obtain:

$$sU(s, z) = D[z^{-1}U(s, z) - 2U(s, z) + zU(s, z)] - C_1 s^{-1} G(U(s, z)). \tag{13}$$

2. We express $U(s, z)$ as a function of $G(U(s, z))$: $U(s, z) = \frac{C_1}{sD(z^{-1}-2+z)-s^2} G(U)$ and obtain the transfer function $H(s, z)$:

$$H(s, z) = \frac{C_1}{sD(z^{-1} - 2 + z) - s^2}. \tag{14}$$

According to the describing function technique [7], the transfer function can be expressed in terms of temporal frequency ω and spatial frequency Ω:

$$H_\Omega(\omega) = \frac{C_1}{I\omega D(2cos\ \Omega - 2) + \omega^2}. \tag{15}$$

3. We are looking for possible periodic solutions of our CNN model (11) in the form:

$$u_{ij}(t) = \xi(i\Omega + \omega t), 1 \leq i \leq n, j = 3, 4, \tag{16}$$

for some function $\xi : \mathbf{R} \to \mathbf{R}$ and for some spatial frequency $0 \leq \Omega \leq 2\pi$ and temporal frequency $\omega = \frac{2\pi}{T}$, where $T > 0$ is the minimal period.

4. According to the describing function technique [7] the following constraints hold:

$$\begin{aligned} \mathcal{R}(H_\Omega(\omega)) &= \frac{U_m}{Y_m}, \\ \mathcal{I}(H_\Omega(\omega)) &= 0. \end{aligned} \tag{17}$$

5. Thus (17) give us necessary set of equations for finding the unknowns U_m, Ω and ω. As we mentioned before we are looking for a periodic wave solution of (11), therefore U_m will determine approximate amplitude of the wave, an $T = \frac{2\pi}{\omega}$ will determine the wave speed. Now according to the describing function technique, if for a given value of Ω we can find the unknowns (U_m, ω), then we can predict the existence of a periodic solution of our CNN IDE (11) with an amplitude U_m and period of approximately $T = \frac{2\pi}{\omega}$.

Following the above algorithm the next theorem has been proved:

Theorem 1 *CNN IDE (11) of the BVP (1), (4) with circular array of $n = L \times L$ cells has periodic solutions $u_{ij}(t)$ with a finite set of spatial frequencies $\Omega = \frac{2\pi k}{n}$, $0 \leq k \leq n-1$ and a period $T = \frac{2\pi}{\omega}$.*

Remark 2 By applying the describing function technique we obtain a characterization of the periodic steady state solutions of our CNN model (11). In order to validate the accuracy of the achieved results we need to introduce a possible initial condition from which the network will reach, at steady state, a steady state solution characterized by the desired value of Ω. Therefore, we can take a initial condition $u_{ij}(0) = \sin(\tilde{\Omega} i)$, $1 \leq i \leq n$, $j = 3, 4$.

4 Travelling Wave Solutions of IDE CNN Model

We shall study traveling wave solutions of IDE CNN model (11) of the form:

$$u_i = \Phi(i - ct), \tag{18}$$

for some continuous function $\Phi : \mathbf{R}^1 \to \mathbf{R}^1$ and some unknow real number c. Let us denote $s = i - ct$. Let us substitute (18) in the IDE CNN model (11). Therefore $\Phi(s, c)$ and c satisfy the equation of the form:

$$-c\Phi^{'}(s, c) = \Phi(s-1, c) - 2\Phi(s, c) + \Phi(s+1, c) - C_1 \int_S f(\Phi(s, c)dt. \tag{19}$$

We consider solution of equation (19). The following theorem about travelling wave solution of our IDE CNN model hold:

Theorem 2 *Let $\Phi(s, c)$ is a solution of (19) and satisfies the following conditions:*

$$lim_{s \to -\infty} \Phi(s, c) = 0, \; lim_{s \to \infty} \Phi(s, c) = 1.$$

Then

(i) If $c = c^* < 0$, $\Phi(s, c)$ is a stable traveling wave solution of IDE CNN model.
(ii) If $c = c_* > 2$, $\Phi(s, c)$ is unstable traveling wave solution.

Remark 3 Our objective in this section is to study the structure of the travelling wave solutions of the CNN model (11). There has been studies on the travelling wave solutions of spatially discrete or both spatially and time discrete systems, but as far as we know there are no studies of periodic traveling wave solutions in CNN.

5 Numerical Simulations and Validation

Let us consider the square domain of piezoelectric solid $G_1G_2G_3G_4$ with a side a. For heterogeneities at nano–scale we have: the side of the square is $a = 10^{-7}$ m; material parameters inside I for hole are 0; material parameters on $S = \partial I$ for hole and for an inclusion are: $c_{44}^S = 0.1c_{44}^M$, $e_{15}^S = 0.1e_{15}^M$, $\varepsilon_{11}^S = 0.1\varepsilon_{11}^M$, $\rho^S = \rho^M$.

The characteristic that is of interest in nano-structures is normalized Stress Concentration Field (SCF) (σ/σ_0) and it is calculated by the following formula:

$$\sigma = -\sigma_{13}sin(\varphi) + \sigma_{23}cos(\varphi), \tag{20}$$

where φ is the polar angle of the observed point, σ_{ji} is the stress (2) near S.

Material parameters of the matrix are for transversely isotropic piezoelectric material PZT4 are:

- Elastic stiffness: $c_{44}^M = 2.56 \times 10^{10}\,\mathrm{N/m^2}$;
- Piezoelectric constant: $e_{15}^M = 12.7\,\mathrm{C/m^2}$;
- Dielectric constant: $\varepsilon_{11}^M = 64.6 \times 10^{-10}\,\mathrm{C/Vm}$;
- Density: $\rho^M = 7.5 \times 10^3\,\mathrm{kg/m^3}$.

The applied load is time harmonic uni-axial along vertical direction uniform mechanical traction with frequency ω and amplitude $\sigma_0 = 400 \times 10^6\,\mathrm{N/m^2}$and electrical displacement with amplitude $D_0 = k\frac{\varepsilon_{11}^M}{e_{15}^M}\sigma_0$. This means that the boundary conditions (4) are:

- on $G_1G_2 : t_3^M = -\sigma_0, t_4^M = -D_0$;
- on $G_2G_3 : t_3^M = t_4^M = 0$;
- on $G_3G_4 : t_3^M =_0, t_4^M = D_0$;
- on $G_4G_1 : t_3^M = t_4^M = 0$.

Then simulating our CNN IDE model (11) we obtain the following periodic wave solutions (see Fig. 4):

The simulations of IDE CNN model are obtained by simulation system MATCNN applying 4th- order Runge-Kutta integration. In order to minimize the computational complexity and to maximize the significance of the mean square error only outputs of 4 cells are taken into account.

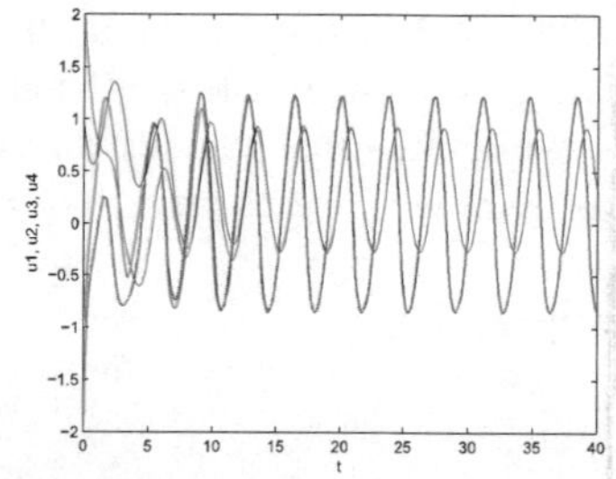

Fig. 4 Simulation of IDE CNN model with 4 cells

Remark 4 The CNN solution of integro-differential equations has four basic properties it is (i) continuous in time; (ii) continuous and bounded in value; (iii) continuous in interaction parameters; (iv) discrete in space. If we consider the output equation of CNN to be of integro-differential type the architecture becomes quite general. Analog CNN Chip hardware implementations have been developed and will further advance in the future. Miniaturized CNN based devices are used already commercially in real time applications e.g. in high speed image and video processing with an equivalent computational power of super computers. CNN in the form of Quantum Dot Cellular Automata appear to become a promising architecture for future nano-structured computers.

6 Stabilizing Feedback Control for IDE CNN Model

Let us extend the IDE CNN model (11) by adding to each cell the local linear feedback [9]:

$$\frac{du_{ij}}{dt} = D(u_{i-1j} - 2u_{ij} + u_{i+1j}) - C_1 \int_S G(u_{ij}))dt - ku_{ij}, \tag{21}$$

where k is the feedback controls coefficient, which is assumed to be equal for all cells. The problem is to prove that this simple and available for the implementation feedback can stabilize the IDE CNN model (11). In the following we present a proof of this statement and give sufficient condition on the feedback coefficient values which provide stability of the CNN nonlinear model (21). The following theorem holds:

Theorem 3 *Let the parameters of IDE CNN system and feedback coefficient k (21) have positive values. Then its linearized model is asymptotically stable for all $k > 0$.*

Proof Define the quadratic Lyapunov function candidate $L(z) = \frac{1}{2}z^T z$. Then its derivative along the linearized control IDE CNN is $\frac{dL(z)}{dt} = \frac{1}{2}z^T(J^T(k) + J(k))z = -z^T Q(k)z$. Therefore $\frac{dL(z)}{dt} < 0$ implies a positive definiteness of $Q(k)$. It can be shown that $Q(k)$ positive definiteness implies $k > 0$. For verification of the above statement the eigenvalues of $J(k)$ were calculated related on the values of feedback coefficient k. Stability of the linear system requires that the eigenvalues $\lambda^i_j, i = 1, \ldots, 4$ satisfy the inequality $max_i Re\lambda^i_j < 0$.

Simulations of the stabilized IDE CNN are given on Fig. 5:

Remark 5 The numerical solution procedure follows the algorithm developed in the previous section. The following steps of the numerical procedure are realized:

(a) solution of the algebraic system (17) for the unknowns U_m, Ω and ω;
(b) computation of the displacement and traction at any point in the PEM plane;

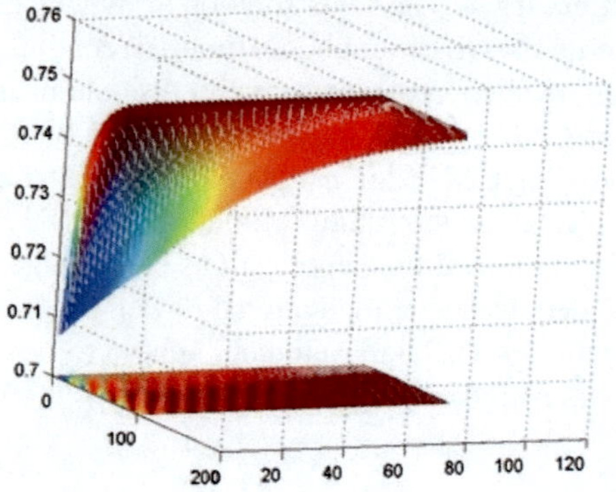

Fig. 5 Simulation of stabilized IDE CNN model

(c) SCF computation;
(d) creation of validated software based on Matlab.

Numerical simulations show that the stress concentration field near defects is strongly influenced by the type and the size of the defect(crack, hole or inclusion), the material anisotropy, the defect location and geometry, the dynamic load characteristics and the mutual interactions between defects and between them and the solid's boundary.

Acknowledgements This paper is supported by the bilateral project between Israel Academy of Sciences and Bulgarian Academy of Sciences.

References

1. Chua, L.O., Yang, L.: Cellular neural network: theory and applications. IEEE Trans. CAS. **35**, 1257–1272 (1988)
2. Chua, L.O., Yang, L.: CNN: applications. IEEE Trans. CAS. **35**, 1273–1299 (1988)
3. Dineva,P., Gross, D., Muller, R., Rangelov,T.: Dynamic fracture of piezoelectric materials. Solutions of Time-harmonic problems via BIEM, Solid Mechanics and its Applications. vol. 212. Springer International Publishing, Switzerland (2014)
4. Fang, X.Q., Liu, J.X., Dou, L.H., Chen, M.Z.: Dynamic strength arround two interactive piezoelectric nano-fibers with surfaces/interfaces in solid under electro-elastic wave. Thin Solid Films. **520**, 3587–3592 (2012)
5. Gurtin, M.E., Murdoch, A.I.: A continuum theory of elastic material surfaces. Arch. Ration. Mach. Anal. **57**, 291–323 (1975)
6. Jammes, M., Mogilevskaya, S.G., Crouch, S.L.: Multiple circular nano-inhomogeneities and/or nano-pores in one of two joined isotropic half-planes. Eng. Anal. Bound. Elem. **33**, 233–248 (2009)
7. Mees, A.I.: Dynamics of Feedback Systems. Wiley, London (1981)
8. Slavova, A.: Cellular Neural Networks: Dynamics and Modeling. Kluwer Academic Publishers (2003)
9. Vidyasagar, M.: Nonlinear Systems Analysis. Society for Industrial and Applied Mathematics, Philadelphia (2002)
10. Vladimirov, V.: Equations of Mathematical Physics. Marcel Dekker Inc, New York (1971)

Nonlinear Models of the Fluid Flow in Porous Media and Their Methods of Study

Jiří Benedikt, Petr Girg, and Lukáš Kotrla

Abstract We survey mathematical models of the fluid flow in porous media based on quasilinear parabolic partial differential equations. We focus on singular and/or degenerate parabolic equations, which are suitable for modeling of turbulent filtration such as groundwater flow trough gravel and/or fractured crystalline rocks and turbulent polytropic filtration of natural gas through rocks in standard deposits, on one hand, and isothermic nanoporous (slow) filtration of natural gas in shale formations, on the other hand. Since in the case of singular and/or degenerate parabolic equations, it is almost impossible to find explicit solutions, we survey some existence and regularity theory together with maximum and comparison principles. We apply this theory on some selected examples from practice.

Keywords Ground water · Drought · Flow in porous medium · Turbulence · Nonlinear Darcy law · Leibenson's equations · Natural gas · p-Laplacian · Doubly nonlinear equation · Comparison principles

1 Introduction

Climate change and shortage of natural freshwater resources are becoming very serious issues nowadays. There is a need for better management of existing resources, while looking for unconventional resources of this vital substance. Our aim is to contribute to these important issues by surveying several nonlinear mathematical

J. Benedikt · P. Girg (✉) · L. Kotrla
Department of Mathematics and NTIS, Faculty of Applied Sciences, University of West Bohemia, Univerzitní 8, CZ-301 00, Plzeň, Czech Republic
e-mail: pgirg@kma.zcu.cz

J. Benedikt
e-mail: benedikt@kma.zcu.cz

L. Kotrla
e-mail: kotrla@kma.zcu.cz

A. Domoshnitsky et al. (eds.), *Functional Differential Equations and Applications*, Springer Proceedings in Mathematics & Statistics 379,
https://doi.org/10.1007/978-981-16-6297-3_2

models of the fluid flow in porous media and their methods of study. We hope that people from practice may find them useful.

Long-lasting droughts become serious problem not only in traditionally arid and/or semi-arid areas, but newly also in countries with moderate climate such as countries in Central Europe. Indeed, several regions of Europe including those in Central and Northern Europe experienced severe drought conditions during June and July 2019, resulting from a combination of the 2018 drought, the heatwaves of 2019 and below-average precipitations in spring 2019, according to JRC European Drought Observatory report [39]. Moreover, below-average precipitations in 2018–2019 lead to lowering of the groundwater level which caused drying of wells in many places in the Czech Republic as it can be seen from the weekly observations of water table in shallow boreholes (ca. 2–15 m deep) conducted by the Czech Hydrometeorological Institute [13], where most of the observations are significantly below long-term average values (collected data since 1950s). The drought events of 2015–2019 also contributed to bark beetle calamity, see, e.g., [31, 32, 34, 55, 69], peaking in Central Europe in 2019. Of course, the problems of drought were not limited to Europe in 2019, significant problems were experienced also in many more areas worldwide, e.g., in Southeast Australia [40], Southern Africa [42, 43], and India [41] in 2019. According to [53], two-thirds of the global population live under conditions of severe water scarcity for at least 1 month of the year and half a billion people face severe water scarcity all year round. It was already in 2008, when Goldman Sachs [28] estimated that the annual consumption of freshwater approximately doubles every 20 years, claimed that water will be oil of the forthcoming century, and recommended to private investors to invest into infrastructure related with freshwater supply.

Most of the mathematical models of the groundwater flow used in practice are based on the linear Darcy (constitutive) law relating groundwater flux with piezometric head loss per length:

$$q = \text{const.}\,\frac{\Delta h}{\Delta L}\,, \tag{1.1}$$

where $h = \frac{P}{\rho g} + z$ is the piezometric head, P is hydrostatic pressure, ρ is density, g is acceleration due to gravity and z is vertical coordinate measured from arbitrary (but fixed) horizontal level, Δh stands for the piezometric head loss (difference of h), ΔL is distance, and q is flux. This law was established empirically by Henry Darcy [15] already in 1856 and it is sufficiently accurate in the case that the flow is laminar, that is, when the Reynolds number related to flux is not "too high" (to be clarified in Sect. 4). If, however, the Reynolds number of the flux is "too high" (see Sect. 4), the turbulence occurs and the linear Darcy law should be replaced by a nonlinear one such as the Smreker–Izbash–Missbach law

$$\frac{\Delta h}{\Delta L} = \text{const.}\, q^m \quad \text{or, equivalently,} \quad q = \text{const.}\left(\frac{\Delta h}{\Delta L}\right)^{\frac{1}{m}}, \tag{1.2}$$

or the Forchheimer law

$$\frac{\Delta h}{\Delta L} = aq + bq^2 , \tag{1.3}$$

where the positive multiplicative constants and the exponent $m \in (1, 2]$ are to be determined empirically. Note that the turbulence often occurs for reasonable and realistic fluxes in practice in the case of coarse porous materials such as gravel or fractured impermeable media with sufficiently wide fractures. A thorough historical survey of development constitutive laws and their history is presented in [8].

With increasing demand on water supply, crystalline rock (or hard rock) aquifers are gaining attention in the last decades [29, 60]. By crystalline rock (or hard rock), we mean impermeable rocks of igneous or metamorphic origin (of negligible permeability) such as, e.g., basalts, granites, or gneisses, where the groundwater flow occurs only in a system of cracks and fractures. Since the water is stored and flows only in cracks and fractures, wells and boreholes in the crystalline rock aquifers have significantly smaller yield as compared to those in porous sedimentary rocks or alluvial aquifers. Nevertheless, crystalline rocks of the Precambrian continental shields occupy approx. 20% of the land surface [29]. Hence, crystalline rock aquifers may become important source of freshwater in rural areas. More importantly, crystalline rocks are commonly found in semi-arid areas where they may represent important source of scarce freshwater. Indeed, continental shields occupy approx. 40% of the semi-arid areas of the sub-Saharan Africa [50, 77]. It is estimated that 40% of groundwater in Australia is stored in the crystalline aquifers [27]. Crystalline aquifers are intensively exploited by farming communities as a source of freshwater mostly used for irrigation in semi-arid southern India [57]. Thus, good understanding of the flow in crystalline aquifers can improve quality of life in these areas. The crystalline rocks are commonly found in continental shields and massifs also in areas which do not have lack of precipitations such as Brazil, Canada, and Scandinavia. On one hand, the crystalline rock aquifers are used for water supply to rural communities in these areas. On the other hand, there are also large underground construction projects such as tunnels, mines, nuclear-waste disposal sites, and similar, see [29]. Thus, understanding groundwater flow in hard rock aquifers is important not only from the point of view of water extraction, but also from the point of view of dewatering of these construction projects.

Hand in hand with climate change, global water cycle intensifies and hydrological extremes including floods may occur more frequently, see, e.g., [30, 33, 71, 76]. Thus, further research and development of effective drainage systems is needed. It appears that coarse porous media such as gravel or geosynthetic materials are suitable for this task, but it turns out that movement of water in these materials is again governed by the nonlinear Smreker–Izbash–Missbach or Forchheimer law [10, 23].

Recent serious drought events are closely related to ongoing climate changes, see, e.g., [14, 74, 75] and references therein. Although it may be the case that the CO_2 emissions are not the main reason of global warming, see, e.g., *pro et contra* arguments in [11, 46, 47, 61, 62, 70], preference for fossil fuels with lower CO_2 emissions will most likely not make the situation worse. Natural gas is a hydrocarbon gas mixture consisting primarily of methane (CH_4), and thus has the most favorable ratio between carbon and hydrogen in terms of emission reduction of all fossil fuels. For comparison, the amount of CO_2 produced by burning natural gas to get a unit of energy is a half that of black coal (117 lb CO_2 per 1 million Btu versus 205–228.6 lb CO_2 per 1 million Btu) according to the U.S. Energy Information Administration (see [72]). With geographically narrowly localized conventional gas fields, unconventional deposits (e.g., shale gas deposits) are now increasingly being opened worldwide to meet increasing demand. In order to better exploit valuable natural resources, one needs good mathematical models. Natural gas flow in the rock is a very complicated process which involves heat exchange with collector rock and may involve turbulence. One of the first to develop satisfactory mathematical models of non-stationary flow of natural gas in a collector rock of a conventional gas field was Leibenson [48].

It turns out that the archetypal parabolic partial differential equation

$$\frac{\partial v}{\partial t} - \operatorname{div}\left(|v|^l \, |\nabla v|^{p-2} \nabla v\right) = f(x, t) \tag{1.4}$$

is a suitable model for all above situations of the fluid flow in porous medium. Note that (1.4) becomes Leibenson's equation of filtration of a polytropic gas in a porous strata for $3/2 < p \le 2$, $l > 0$, see Sect. 3, and equation for the water table in an unconfined aquifer for $3/2 \le p \le 2$, $l = 1$, see Sect. 2. Note that the case $p = 2$ corresponds to *laminar flow* in both Leibenson's equation and the water table equation, while the case $p = 3/2$ corresponds to a *flow with fully developed turbulence*. Most importantly, for practical considerations, the intermediate case $3/2 < p < 2$ corresponds to a *flow with some effects of turbulence*. Moreover, (1.4) with $l = 1$ and $2 < p < 10$ is also model of fluid flow in nanoporous media (see [54]). Note that such type of gas filtration occurs in the shale deposits, whose importance in natural gas extraction has recently increased significantly.

2 Basic Terminology in Hydrology

2.1 Porous Medium

The attempt to formulate an exact definition of *porous medium* brings many pitfalls, see Bear [6, Sects. 1.2 and 1.3]. We adopt the conceptual model presented in [6, Sects. 1.3 and 4.5.2]. Moreover, we restrict ourselves to the case where a portion of space (domain from mathematical point of view) is occupied by two homogeneous kinds of matter. Solid phase (say rock) forms a rigid container for fluid phase. The space occupied by solid phase is called *solid matrix* and the space filled by fluid phase is called *pore space*. Porous medium contains solid matrix and pore space in any sufficiently large subdomain (but still much smaller than the whole domain). In fact, the pore space includes many relatively narrow channels or tubes of various length, cross-section, and orientation. We call a *junction* the part of void space where at least three channels meet each other. The channels and the junctions have more or less uniform spatial distribution.

In the case of fluid flow, we can assume that any two points in pore space may be connected by a curve that lies completely within it since there is no flow in isolated *pores* (subsets of pore space). Consequently, the isolated pores are considered as the part of solid matrix, see [6, Sect. 1.2]. The remaining pore space (interconnected by channels) is usually called effective pore space. We will assume that the pore space includes only effective pore space for simplicity and hence we will omit the term "effective".

2.2 Groundwater

Typical porous media considered in hydrogeology are soils, sands, gravels, porous rocks such as sandstones, and fractured crystalline rocks such as basalts, granites, and gneisses. In general, the pore space of these porous media can be filled by air, vapor, and liquid phase of water. Part of the porous medium where all pores are filled by the water in liquid phase is called *saturated zone* and the part where the pores contain gaseous phase (air and vapor) and also liquid phase (of water) for at least part of the time is called *aeration zone*. For the purposes of this paper, the term *groundwater* is limited to the water present beneath Earth's surface in the saturated zone. Mathematical models presented in this paper are restricted to the motion of water in saturated zone.

2.3 Aquifer

Note that the saturated zone can be either overlain by an impermeable layer (of rock or clay) or it can have a free upper surface, which is called water table. The water table is characterized as a surface where the pressure is equal to the atmospheric pressure. An *aquifer* is such saturated zone which allows groundwater flow. The aquifer with the free upper surface is called *unconfined aquifer* while the aquifer enclosed between two impermeable layers is called *confined aquifer*. In the presented paper, we are interested in groundwater flow through unconfined aquifer.

2.4 Velocities and Flux

The real velocity of the groundwater in the porous medium is highly and unpredictably fluctuating in space and time due to irregularity of the channels and their

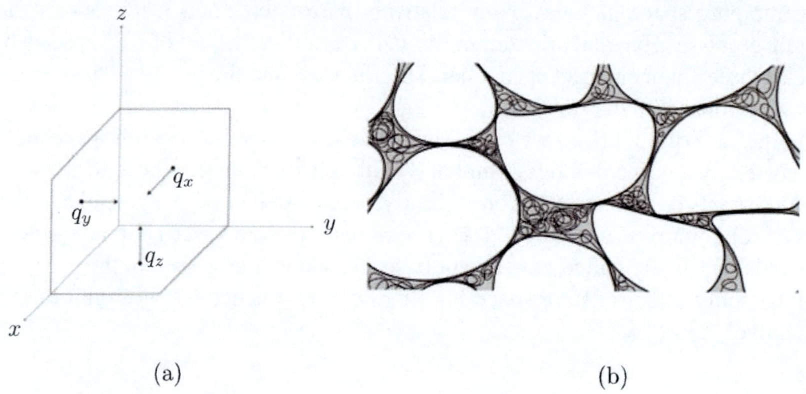

Fig. 1 Flow through porous medium. (a) specific discharge (or Darcy velocity) *versus* (b) streamlines of the real velocity field of the fluid

joints (and due to turbulence for high values of the Reynolds number). Thus, the real velocity is useless for the practical purposes. Instead, *average velocity* (which can be measured in practice) is used. Let us choose Cartesian coordinate system xyz, with z being the vertical axis. Now, let us consider cross-sectional area A_x perpendicular to x-axis. Let Q_x be the volume of water that passes through A_x per unit of time. The sign of Q_x is positive, if the water (in bulk) passes through A_x in the direction of axis x and negative otherwise. Then

$$q_x \stackrel{\text{def}}{=} \frac{Q_x}{A_x}.$$

In analogous way, we define q_y and q_z. Then, $\vec{q}=(q_x, q_y, q_z)$ and $q=\sqrt{q_x^2+q_y^2+q_z^2}$. The quantity $\vec{q}$ is called *specific discharge* or *Darcy velocity*. We also define *average velocity* $\vec{v}=\vec{q}/n$, where n is porosity. Similarly, $v \stackrel{\text{def}}{=} |\vec{v}| = q/n$. This approach works for any incompressible fluid (Fig. 1).

2.5 Groundwater Energy and Piezometric Head

The total mechanical energy of a unit volume of groundwater (or any other incompressible fluid) is the sum of gravitational potential energy, pressure energy, and kinetic energy

$$E_T = z\varrho g + P + \frac{1}{2}\varrho v^2,$$

see, e.g., [59]. Here, v stands for the magnitude of average velocity of the flow, see above. Groundwater is losing total energy while flowing due to friction with porous medium. Thus, its total energy decreases in the direction of the flow. The *total head* h_T is the height of the fictive column of groundwater with the gravitational potential energy equal to E_T, i.e.,

$$h_T = z + \frac{P}{\varrho g} + \frac{1}{2\,g} v^2 \,.$$

Since the average velocity of the groundwater flow in real situations is maximally of the order of a meter per day (that is, of the order 0.00001 m/s), the term corresponding to kinetic energy is negligible and can be dropped. In this way, we obtain *piezometric head*

$$h = z + \frac{P}{\varrho g} \,,$$

which is the state variable in the mathematical models of underground movement. Constitutive relations between specific discharge and piezometric head were observed by in-field observations [67] as well as experimentally established in laboratory conditions [15]. In general (for isotropic medium), these relations can be written as

$$q = \Phi \left(\frac{\Delta h}{\Delta L} \right) \,,$$

where Φ is some nondecreasing function such that $\Phi(0) = 0$. Note that the constitutive laws are inferred from experiments for one-dimensional flow. However, groundwater flow in the real world is three dimensional. The properties of the isotropic porous medium are the same in all directions. Thus, in this case, the three-dimensional constitutive law can be inferred from the one-dimensional one in a straightforward manner, taking into account that the specific discharge takes the opposite direction of the gradient of the piezometric head and no flow occurs if the gradient of the piezometric head is zero, i.e.,

$$\vec{q} = \begin{cases} \vec{0} & \text{for } \nabla h = \vec{0} \,, \\ -\Phi(|\nabla h|) \dfrac{\nabla h}{|\nabla h|} & \text{for } \nabla h \neq \vec{0} \,. \end{cases} \tag{2.1}$$

In particular, we obtain the linear Darcy law (1.1) for

$$\Phi(r) = k\,r \,, \qquad r > 0 \,, \tag{2.2}$$

the Smreker–Izbash–Missbach power law (1.2) for

$$\Phi(r) = c r^{\frac{1}{m}} \,, \qquad r > 0 \,, \tag{2.3}$$

$1 < m < 2$, inverse Forchheimer law (inverse formula to (1.3)) for

$$\Phi(r) = \frac{\sqrt{a^2 + 4br} - a}{2b} = \frac{2r}{\sqrt{a^2 + 4br} + a}, \qquad r > 0. \tag{2.4}$$

It has been observed by King [45] that the flow of water in low-permeable clays obeys (2.3) with $0 < m < 1$. The work [68, p. 239] contains an overview of values of m for various materials where m ranges from 0.27 to 0.89 (note that $1/m = n$, exponent n taken from [68, Table (A), Appendix I, p. 239]). Recently, it has been found that very slow filtration (i.e., $0 < m < 1$) occurs in petroleum and gas extraction from tight shales reservoirs. For laboratory experiments with real fluids and media, see [25, 63].

2.6 Problem of the Free Surface, the Dupuit–Forchheimer Assumption, and Simplified Problem

In the case of unconfined aquifers, the free surface of the groundwater is the upper boundary of the aquifer. Thus, we need to solve a partial differential equation for both an unknown $h = h(x, y, z, t)$ and an unknown bounded domain $\Theta \equiv \Theta(t)$ in $\mathbb{R}^3$ that represents the aquifer.

In 1863, Dupuit [21] simplified the problem of unknown boundary by observing that the maximal piezometric head loss per length $\Delta h/\Delta L$ is between 0.001 and 0.01 in typical unconfined aquifers and unconfined aquifer is bounded from below by horizontal impermeable layer. Based on these observations, he formulated the following assumptions on the flow:

(DF1) groundwater flows horizontally (and thus piezometric head is constant in vertical direction z) and

(DF2) the Darcy law (1.1) applies to this flow (Dupuit assumed that the groundwater flow is slow enough at these values of piezometric head loss per length so that the nonlinear effects can be neglected).

We will derive a simplified model of groundwater flow in unconfined aquifer using the assumption (DF1). We also assume that the lower boundary of the aquifer formed by impermeable layer is the xy-plane. We choose Ω a bounded domain in $\mathbb{R}^2$ such that orthogonal projection to $\Theta(t)$ to the xy-plain is contained in Ω for every $t \in [0, T]$. We remind that for a fixed x, y, t, the point (x, y, z) belongs to the water table if and only if $h(x, y, z, t) = z$. In case there is no water above $(x, y) \in \Omega$ at t, we extend the definition of the water table to contain the point $(x, y, 0)$. We assume that the water table is the graph of a function of x, y, and t, that is, there exists nonnegative and sufficiently smooth function $H : \mathbb{R}^2 \times [0, T] \to \mathbb{R}$ such that $h(x, y, H(x, y, t), t) = H(x, y, t)$. Then the mass of the water column stacked above arbitrary two-dimensional disk $A \subset \Omega$ at time t is

$$m_A(t) = \int_A n\, \varrho_{\text{water}}\, H(x, y, t) \mathrm{d}x\mathrm{d}y\,.$$

Hence, the integral form of mass conservation law for water (or any other incompressible fluid) has the following form:

$$m_A(t_2) - m_A(t_1) + \int_{t_1}^{t_2} \int_{\partial A} H(x, y, t)\, \vec{j}(x, y, t) \cdot \vec{n}\, \mathrm{d}s = \int_{t_1}^{t_2} \int_A f(x, y, t)\, \mathrm{d}x\mathrm{d}y\mathrm{d}t\,, \tag{2.5}$$

where $\vec{j} = n\, \varrho_{\text{water}}\, \vec{v} = \varrho_{\text{water}}\, \vec{q}$ is the mass flow, $\vec{n}$ is normal vector of ∂A, and f quantifies the sources or absorption in column over the point (x, y) at time t. Let us recall that $\vec{j}(x, y, z, t) = \vec{j}(x, y, t)$ by the assumption (DF1).

Using $m_A(t_2) - m_A(t_1) = \int_{t_1}^{t_2} m'_A(t)\, \mathrm{d}t = \int_A \int_{t_1}^{t_2} n\, \varrho_{\text{water}} \frac{\partial H}{\partial t}(x, y, t)\, \mathrm{d}t\mathrm{d}x\mathrm{d}y$, and the divergence theorem on the second term in (2.5), we arrive at

$$\int_A \int_{t_1}^{t_2} n\, \varrho_{\text{water}} \frac{\partial H}{\partial t}(x, y, t)\, \mathrm{d}t\mathrm{d}x\mathrm{d}y - \int_{t_1}^{t_2} \int_A \mathrm{div}\,(H(x, y, t)\varrho_{\text{water}}\, \vec{q})\, \mathrm{d}x\mathrm{d}y\mathrm{d}t = \int_{t_1}^{t_2} \int_A f(x, y, t)\, \mathrm{d}x\mathrm{d}y\mathrm{d}t\,.$$

Since the integral identity is valid for any test disk $A \subset \Omega$ and any interval $[t_1, t_2] \subset [0, T]$, we infer the local form of the mass conservation law

$$n \frac{\partial H}{\partial t}(x, y, t) - \mathrm{div}\,(H(x, y, t)\, \vec{q}) = \frac{f(x, y, t)}{\varrho_{\text{water}}}$$

a.e. in $\Omega \times [0, T]$. Since $h(x, y, \cdot, t) \equiv \text{const.}$ by the assumption (DF1) and $h(x, y, H(x, y, t), t) = H(x, y, t)$ on the water table, $h(x, y, z, t) \equiv h(x, y, t) = H(x, y, t)$ and

$$n \frac{\partial h}{\partial t}(x, y, t) - \mathrm{div}\,(h(x, y, t)\, \vec{q}) = \frac{f(x, y, t)}{\varrho_{\text{water}}}\,. \tag{2.6}$$

By (DF2), we apply the Darcy law, i.e., (2.1) with (2.2) to conclude

$$\frac{\partial h}{\partial t} - \frac{k}{n} \mathrm{div}\,(|h|\, \nabla h) = f(x, y, t) \tag{2.7}$$

with a little bit of abuse of notation ("hiding" multiplicative constants into f). Note that we can assume that $k/n = 1$ since we can get rid of this multiplicative constant by a linear substitution in the time variable.

Based on numerous experiments and in-field observations summarized in [26], Ph. Forchheimer [26, see p. 1782 and "Anhang," pp. 1787–1788] pointed out that the assumption (DF2) (i.e., the Darcy law (1.1)) is not accurate enough for piezometric head loss per length greater than 0.0005 for certain porous media (sands) and thus (1.3) has to be used instead while the assumption (DF1) is still applicable. Following Forchheimer, we apply Forchheimer law, i.e., (2.1) with (2.4) to conclude

$$\frac{\partial h}{\partial t} - \frac{1}{n} \mathrm{div} \left(\frac{2\,|h|\, \nabla h}{\sqrt{a^2 + 4b|\nabla h|} + a} \right) = f(x, y, t)\,. \tag{2.8}$$

Or alternatively, we apply Smreker–Izbash–Missbach law, i.e., (2.1) with (2.3) to conclude

$$\frac{\partial h}{\partial t} - \frac{c}{n} \operatorname{div}\left(|h|\,|\nabla h|^{p-2}\,\nabla h\right) = f(x, y, t)\,, \tag{2.9}$$

where $p = 1 + 1/m$. It turns out that the equation (2.9) is easier to handle both theoretically and computationally and thus it is preferred in the literature.

3 Leibenson's Equation and Flow of the Natural Gas

Following Leibenson [49], we assume that the porous medium is nondeformable, isotropic, and homogeneous at macroscopic scale with constant porosity n and the gas is a homogeneous mixture. The condition on the gas ensures that its density depends on the pressure only. We also suppose that the examined thermodynamic process is *polytropic*, i.e., it obeys the following relation:

$$\frac{P}{\varrho^{\gamma}} = \beta^{\gamma}\,. \tag{3.1}$$

Here, $x \in \mathbb{R}^3$, $\varrho = \varrho(t, x)$ is the density, $P = P(t, x)$ is the pressure, $\gamma > 1$ is the *polytropic index* of the process, and $\beta > 0$ is a constant. The flow of the gas (as of any fluid) in the porous medium is governed by continuity equation in the form

$$n\frac{\partial \varrho}{\partial t} + \operatorname{div}(\varrho\vec{q}) = 0 \tag{3.2}$$

and an appropriate constitutive law which relates specific discharge $\vec{q} = n\vec{v}$ and pressure gradient ∇P. Specific discharge is volumetric flux per unit area and the term $\varrho\vec{q}$ represents mass flux per unit area. We refer to [6, Sect. 6.2] for derivation of (3.2) for homogeneous mixture.

For compressible fluid, the specific discharge $\vec{q}$ does not provide relevant information and mass flux must be used instead. In this way, a similar power law for compressible gas subjected to polytropic process,

$$\varrho\vec{q} = -C\,|\nabla P_1|^{s-1}\,\nabla P_1\,, \qquad \frac{1}{2} < s < 1\,, \tag{3.3}$$

was experimentally established, where $P_1 = P^{(\gamma+1)/\gamma}$ (see Leibenson [49]).

Plugging (3.3) into (3.2), we obtain

$$n\frac{\partial}{\partial t}\left(\frac{P_1^{\frac{1}{\gamma+1}}}{\beta}\right) - C\operatorname{div}\left(|\nabla P_1|^{s-1}\nabla P_1\right) = 0$$

by (3.1). This equation is often called the equation of turbulent polytropic filtration of gas in porous medium and it has attracted attention of many researches, see, e.g., [4, 17, 19, 20, 24, 36].

4 Turbulence in Porous Medium and Real-World Observations

The turbulence in porous medium was probably first *conjectured* from the experimentally established deviations from the Darcy law by Pavlovskii [56], who proposed to use the Reynolds number for the distinction of the validity range of the linear Darcy law from the validity range of nonlinear laws. He also observed that formula for the Reynolds number in the porous medium must be different than the one for a pipe. He proposed a definition suitable for grained porous media (e.g., sand or gravel) formed of grains of approximately the same diameter. His formula reads

$$\mathrm{Re} = \frac{6.5\, q\, d\, \rho}{\mu\,(0.75\, n + 0.23)}\,,$$

where d is effective diameter of the grain, ρ is density of the incompressible fluid (water), and μ is its dynamic viscosity. For this definition of Reynolds number, it follows from the experiments that the Darcy law (2.2) is valid if the value of Re is approximately below 50 to 60 (the boundary between the two cases is somewhat blurred) and, for higher values of Re, the Smreker–Izbash–Missbach law (2.3) with $1 < m < 2$ or the Forchheimer law (2.4) must be used instead. According to V. I. Aravin and S. N. Numerov [1, p. 4 and p. 33], this was the first time in [56] when such specification of ranges of the Reynolds number appears in the literature. As pointed out in [1, p. 33], the value of the Reynolds number when the Darcy law becomes inaccurate does not have to be the same as the critical value of the Reynolds number when the turbulence in the flow occurs. Nowadays, it is known that there are at least three ranges of Reynolds number with three different laws:

- pre-Darcy law (2.3) with $0 < m < 1$ for very low values of the Reynolds number;
- Darcy law (2.2) for moderate values of the Reynolds number;
- post-Darcy law (2.3) with $1 < m < 2$ or (2.4) for high values of the Reynolds number.

To get the picture complete, experimental study of flow through porous media over the complete flow regime is presented in, e.g., [2, 3, 66, 68].

Since the constitutive law can become nonlinear even in the laminar regime (as pointed out by [1, p. 33]), we are often asked at conferences if the turbulence in the flow through the porous medium was indeed observed in the laboratory. The modern laboratory techniques can indeed capture the structure of turbulent vortexes, see the recent paper [78].

5 Functional Framework

In this section we survey relevant existence, uniqueness and regularity results concerning generalized solutions of doubly nonlinear parabolic equations. There are several approaches to generalized solutions of (singular/degenerate) doubly nonlinear parabolic equations, see, e.g., [16, 18, 22, 36, 52]. For our purposes, we chose the least technical approach presented in the survey paper by Ivanov [36] (for the complete proofs of results surveyed in [36], see [35, 37] for $p > 1$ and [38] for $p > 2$).

Let Ω be a bounded domain in $\mathbb{R}^N$, $N \geq 1$, and $T > 0$. We assume that Ω has $C^{0,1}$-boundary (i.e., Lipschitz boundary) $\partial\Omega$, see [58]. Then $Q_T \overset{\text{def}}{=} \Omega \times (0, T]$, $S_T \overset{\text{def}}{=} \partial\Omega \times [0, T]$ and $\Gamma_T \overset{\text{def}}{=} S_T \cup (\Omega \times \{0\})$. We will use standard function spaces for parabolic problems and, for the convenience of the reader, their traditional notation, which is often different from the notation used in [36]. By $C\left([0, T] \to L^q(\Omega)\right)$, for $q \geq 1$, we denote the space of all measurable functions v on Q_T such that the mapping $t \mapsto v(\cdot, t)$ acting from $[0, T]$ to $L^q(\Omega)$ is continuous, *i.e.*,

$$\lim_{n\to+\infty} |t_n - t| = 0 \Rightarrow \lim_{n\to+\infty} \|v(\cdot, t_n) - v(\cdot, t)\|_{L^q(\Omega)} = 0$$

for every sequence $\{t_n\}_{n=1}^{+\infty}$, $t_n \in [0, T]$ and $t \in [0, T]$.

By $L^p\left([0, T] \to W^{1,p}(\Omega)\right)$, we mean a space of all measurable functions on Q_T such that $v(\cdot, t) \in W^{1,p}(\Omega)$ for a.e. $t \in [0, T]$ and

$$\|v\|_{L^p([0,T]\to W^{1,p}(\Omega))} \overset{\text{def}}{=} \left(\int_0^T \|v(\cdot, t)\|^p_{W^{1,p}(\Omega)} \mathrm{d}t\right)^{1/p} < +\infty\,.$$

Note that if $v \in L^p\left([0, T] \to W^{1,p}(\Omega)\right)$ then the trace of $v(\cdot, t)$ on $\partial\Omega$ is defined for a.e. $t \in [0, T]$.

Finally, by $C^{\lambda,\lambda/p}(\overline{Q_T})$ we mean a space of all continuous functions v on $\overline{Q_T}$ such that

$$\|v\|_{C^{\lambda,\lambda/p}(\overline{Q_T})} = \max_{(x,t)\in\overline{Q_T}} |v(x, t)| + \sup_{\substack{(x,t),(y,s)\in\overline{Q_T}\\(x,t)\neq(y,s)}} \frac{|u(x, t) - u(y, s)|}{|x - y|^\lambda + |t - s|^{\lambda/p}} < +\infty\,.$$

We consider the prototype initial-boundary-value problem

$$\begin{cases} \dfrac{\partial v}{\partial t} - \operatorname{div} \vec{a}(x, t, v, \nabla v) = f(x, t) & \text{in } Q_T\,; \\ v = \psi & \text{on } \Gamma_T\,, \end{cases} \tag{5.1}$$

where $f \in L^\infty(Q_T)$, and $\psi \in C^{\lambda,\lambda/p}(\overline{Q_T})$ such that $\hat{\psi} \overset{\text{def}}{=} \psi^{1/(p-1)+1} \in L^p\left([0, T] \to W^{1,p}(\Omega)\right)$, are given functions.

The following *structural hypotheses* on the Carathéodory function $\vec{a}$ are assumed for a.e. $(x, t) \in Q_T$ and any $s \in \mathbb{R}$ and any $\vec{r} \in R^N$:

$$\vec{a}(x, t, s, \vec{r}) \cdot \vec{r} \geq \nu_0 |s|^l |\vec{r}|^p - \mu_0 \left(1 + |s|^\delta\right), \tag{5.2}$$

$$|\vec{a}(x, t, s, \vec{r})| \leq \nu_1 |s|^l |\vec{r}|^{p-1} + \mu_1 |s|^{l/p}. \tag{5.3}$$

Here $p > 1, l \geq 0, \nu_0, \nu_1 > 0$, and $\mu_0, \mu_1 \geq 0$ are certain given constants. Moreover, $0 \leq \delta < l + p$ is given constant for $l + p > 2$ and $\delta = 2$ for $1 < l + p \leq 2$. Note that these structural assumptions are satisfied in the particular case of the equation (1.4).

Note that Ω with Lipschitz boundary satisfies the following *structural hypothesis* from [36] (so-called *property of positive geometric density*) on the boundary $\partial\Omega$:

$$\exists \alpha_* \in (0, 1) \;\; \exists \varrho_* > 0 \;\; \forall x_0 \in \partial\Omega \;\; \forall \rho \in (0, \varrho_*]: \; \text{meas}\left(\Omega \cap B_\varrho (x_0)\right) \leq (1 - \alpha_*)\, \text{meas}\left(B_\rho (x_0)\right). \tag{5.4}$$

From [36, Def. 1.1 and Def. 2.1], we adapt the following notion of weak solution.

Definition 1 A nonnegative function $v \in L^\infty(Q_T)$ is a *weak solution* (*supersolution, subsolution*) if

(a) $v \in C\left([0, T] \to L^1(\Omega)\right), \partial v^{\sigma+1}/\partial x_i \in L^p(Q_T)$ for $\sigma \stackrel{\text{def}}{=} l/(p-1), i = 1, \ldots, N$, and $\hat{v} \stackrel{\text{def}}{=} v^{\sigma+1} \in L^p\left([0, T] \to W^{1,p}(\Omega)\right)$.
(b) for any $\phi \in C_0^1(Q_T)$ and any $t_1, t_2 \in [0, T]$,

$$\int_\Omega v\, \phi\, \mathrm{d}x \Big|_{t_1}^{t_2} + \int_{t_1}^{t_2} \int_\Omega \left\{ -v \frac{\partial \phi}{\partial t} + \vec{a}(x, t, v, \vec{v}_x) \cdot \nabla\phi - f\, \phi \right\} \mathrm{d}x\mathrm{d}t = 0, \tag{5.5}$$

($\phi \geq 0$, for supersolution: ≥ 0, for subsolution: ≤ 0), where $\vec{v}_x \stackrel{\text{def}}{=} (v_{x_1}, v_{x_2}, \ldots, v_{x_N})$ and

$$v_{x_i} \stackrel{\text{def}}{=} \begin{cases} (1+\sigma)^{-1} v^{-\sigma} \dfrac{\partial \hat{v}}{\partial x_i} & \text{in} \{(x, t) \in Q_T : v > 0\}, \\ 0 & \text{in} \{(x, t) \in Q_T : v = 0\}. \end{cases} \tag{5.6}$$

(c) v coincide with ψ on Γ_T, that is,

$$v = \psi \quad \text{on } S_T \quad \text{in the sense of } v^{\sigma+1} \text{trace}; \tag{5.7}$$

$$\lim_{t \to 0+} \|v(\cdot, t) - \psi(\cdot, 0)\|_{L^1(\Omega)} = 0. \tag{5.8}$$

This definition makes sense, cf. [36, p. 24], in the general case $p > 1, l > 1 - p$. However, we limit ourselves to $p > 1$ and $l \geq 0$, which are values that appear in our models. Note that the conditions (5.7) and (5.8) do not appear in [36, Definition 2.1] explicitly, however, they are mentioned in previous works by the same author, see, e.g., [38, Definition 1.2].

The following result is a basic weak comparison principle for the weak sub- and supersolutions of the doubly nonlinear equation from the initial-boundary-value problem (5.1).

Proposition 1 cc(see [36, Prop. 4.1]) *Let the assumptions* (5.2), (5.3) *be fulfilled. Assume that* $v_1 \in L^p\left([0,T] \to W^{1,p}(\Omega)\right)$ *is a subsolution of the equation*

$$\frac{\partial v}{\partial t} - \operatorname{div} \vec{a}(x,t,v,\nabla v) = f_1(x,t) \quad \text{in } Q_T\,, \tag{5.9}$$

and $v_2 \in L^p\left([0,T] \to W^{1,p}(\Omega)\right)$ *is a supersolution of the equation*

$$\frac{\partial v}{\partial t} - \operatorname{div} \vec{a}(x,t,v,\nabla v) = f_2(x,t) \quad \text{in } Q_T\,, \tag{5.10}$$

where $f_1, f_2 \in L^\infty(Q_T)$. *If*

$$v_1 \le v_2 \quad \text{on } \Gamma_T \text{ (in the sense of traces)}\,, \qquad \text{and } f_1 \le f_2 \quad \text{in } Q_T \tag{5.11}$$

then, for any $\tau \in (0,T]$, *we have*

$$\int_\Omega (v_1(x,\tau) - v_2(x,\tau))^+ \,\mathrm{d}x \le \int_\Omega (v_1(x,0) - v_2(x,0))^+ \,\mathrm{d}x\,. \tag{5.12}$$

From this proposition, we easily obtain uniqueness of weak solutions in the class $L^p\left([0,T] \to W^{1,p}(\Omega)\right)$.

Proposition 2 (see [36, Prop. 4.2]) *Let assumptions* (5.2), (5.3) *be fulfilled. Then there is at most one weak solution of the initial-boundary-value problem* (5.1) *belonging to* $L^p\left([0,T] \to W^{1,p}(\Omega)\right)$.

Note that in the case of the doubly nonlinear equation, a weak solution v is assumed to satisfy $v^{\sigma+1} \in L^p\left([0,T] \to W^{1,p}(\Omega)\right)$ for $\sigma = l/(p-1)$, which reduces to $v \in L^p\left([0,T] \to W^{1,p}(\Omega)\right)$ provided $l = 0$. For $l \neq 0$, weak solutions to (5.1) do not need to be of class $L^p\left([0,T] \to W^{1,p}(\Omega)\right)$, in general. Note that if $\inf_{Q_T} v > 0$ then $v^{\sigma+1} \in L^p\left([0,T] \to W^{1,p}(\Omega)\right)$ implies $v \in L^p\left([0,T] \to W^{1,p}(\Omega)\right)$.

Proposition 3 (see [36, Theorems 5.3 and 6.1]) *Let* $p > 1$ *and either*

$$l \ge 0\,, \quad p + l \ge 2$$

or

$$1 < p + l \le 2\,.$$

Moreover, assume

$\Omega \subset \mathbb{R}^N$ *is a bounded domain satisfying (20),* (5.13)

$f \in L^\infty(Q_T), \quad f \geq 0$ *a.e. in* Q_T, (5.14)

$\psi \in L^p\left([0,T] \to W_0^{1,p}(\Omega)\right) \cap C^{\beta,\beta/p}(\overline{Q_T})$, $\psi \geq 0$, *for some* $\beta \in (0,1)$. (5.15)

Then there exists exactly one quasi-strong solution of the Cauchy–Dirichlet problem

$$\begin{cases} \dfrac{\partial v}{\partial t} - \operatorname{div}\left(|v|^l \, |\nabla v|^{p-2} \nabla v\right) = f(x,t) & \text{in } Q_T\,; \\ v = \psi & \text{on } \Gamma_T\,, \end{cases}$$

which is Hölder continuous on $\overline{Q_T}$.

Moreover, $\nabla\left(v^{\alpha+1}\right) \in L^p(Q_T)$, *with* $\alpha = l/p$, *and*

$$\sup_{(x,t),(x',t') \in Q_T} \frac{|v(x,t) - v(x',t')|}{|x-x'|^\lambda + |t-t'|^{\lambda/p}} \leq K$$

with some $\lambda \in (0,1)$, $K > 0$ *depending only on* N, p, l, $\|f\|_{L^\infty(Q_T)}$, meas Ω, T, α^*, *and* ρ^* *(from condition (5.4)),* $\|\psi\|_{L^p\left([0,T] \to W_0^{1,p}(\Omega)\right)}$, $\|\psi\|_{C^{\beta,\beta/p}(\overline{Q_T})}$, *and* $\beta \in (0,1)$.

Proposition 4 (see [36, Theorem 3.1]) *Let* $1 < p < 2$, $p + l \geq 2$ *and assume that the structural conditions (5.2) and (5.3) are satisfied. Moreover, suppose that*

(a) for a.e. $(x,t) \in Q_T$ *and any* $s \in \mathbb{R}$ *there exist* $\nu_1 > 0$ *and* $\vec{b} = \vec{b}(x,t,s) \in \mathbb{R}^N$, $|\vec{b}(x,t,s)| < +\infty$, *such that for a.e.* $(x,t) \in Q_T$ *and all* $s \in \mathbb{R}$ *and* $\vec{r}_1, \vec{r}_2 \in \mathbb{R}^N$

$$[\vec{a}(x,t,s,\vec{r}_1) - \vec{a}(x,t,s,\vec{r}_2)] \cdot (\vec{r}_1 - \vec{r}_2) \geq \nu_1 |s|^l |\vec{r}_1 - \vec{r}_2|^2 \left\{|\vec{r}_1 - \vec{b}|^p + |\vec{r}_2 - \vec{b}|^p\right\}^{1-\frac{2}{p}}$$

holds.

(b) for a.e. $(x,t) \in Q_T$ *and any* $\vec{r} \in \mathbb{R}^N$, *the functions* $s^{-\alpha} a_i(x,t,s,\vec{r})$ *and* $s^{-\alpha} a_i(x,t,s,s^{-\alpha}\vec{r})$ *are continuous on* $\mathbb{R}$ *with respect to* s. *Here* $\alpha = 1/p$.

(c) $\psi(x,t)$ *is nonnegative in* Q_T, $\psi \in L^p\left([0,T] \to W_0^{1,p}(\Omega)\right) \cap L^\infty(Q_T)$, *and we have the Hölder condition*

$$\sup_{(x,t),(x',t') \in \overline{Q}_T} \frac{|\psi(x,t) - \psi(x',t')|}{(|x-x'|^p + |t-t'|)^{\gamma_0/p}} \leq K_0$$

for some $K_0 > 0$ *and* $\gamma_0 \in (0,1)$.

Then there exists a weak solution v *of the Cauchy–Dirichlet problem (5.1) which is Hölder continuous in* $\overline{Q}_T$. *Moreover,* $\nabla(v^{\alpha+1}) \in L^p(Q_T)$, $\alpha = 1/p$, *and the estimate*

$$\sup_{(x,t),(x',t') \in Q_T} \frac{|v(x,t) - v(x',t')|}{(|x-x'|^p + |t-t'|)^{\gamma/p}} \leq K$$

holds with constants $K > 0$ and $\gamma \in (0, 1)$ dependent only on the dimension N, the known parameters from (5.2), (5.3), a)–b), *the constants α_* and ϱ_*, $|\Omega|$, T, $\|\psi\|_{W^{1,p}(Q_T)}$, $\sup_{Q_T}(\psi)$, γ_0, and K_0.*

The following result stated in [38] guarantees the existence of a solution of (5.1) with time-dependent boundary conditions. Let us emphasize that the result is valid only for $p > 2$. As far as we know, a similar result has not been proved for $1 < p < 2$ yet. In Proposition 5, we use Einstein's summation convention as in [38].

Proposition 5 (see [38, Theorem 1.1]) *Let $p > 2$ and assume that the structural conditions* (5.2) *and* (5.3) *are satisfied. Moreover, suppose that*

(a) for any $s \in \mathbb{R}$, $\vec{r}_1, \vec{r}_2 \in \mathbb{R}^N$ and a.e. $(x, t) \in Q_T$,

$$(a_i(x, t, s, \vec{r}_1) - a_i(x, t, s, \vec{r}_2)) \cdot \left(r_{1,i} - r_{2,i}\right) \geq \nu_1 |s|^l |\vec{r}_1 - \vec{r}_2|^p$$

with $\nu_1 = \text{const.} > 0$.

(b) for a.e. $(x, t) \in Q_T$ and all $\vec{r} \in \mathbb{R}^N$, the limit

$$\lim_{s\to 0+} s^{-\alpha} a_i\left(x, t, s, s^{-\alpha}\vec{r}\right), \qquad \alpha = \frac{l}{p},$$

exists.

(c) for any $\vec{r} \in \mathbb{R}^N$ and a.e. $(x, t) \in Q_T$,

$$a_i(x, t, s, \vec{r})r_i - f(x, t)s > -c_1 s^2, \qquad c_1 = \text{const.} > 0,$$

for all $s < 0$.

(d) concerning the function $\psi(x, t)$, $(x, t) \in \overline{Q}_T$, defining the boundary condition in (5.1), *$\psi(x, t) \geq 0$, $\psi \in W^{1,p}(Q_T) \cap L^\infty(Q_T)$ and we have the Hölder condition*

$$\sup_{(x,t),(x',t')\in \overline{Q}_T} \frac{|\psi(x, t) - \psi(x', t')|}{(|x - x'|^p + |t - t'|)^{\gamma_0/p}} \leq K_0$$

for some $K_0 > 0$ and $\gamma_0 \in (0, 1)$.

Then problem (5.1) *has at least one nonnegative weak solution $v(x, t)$ for which*

$$\sup_{(x,t),(x',t')\in Q_T} \frac{|v(x, t) - v(x', t')|}{(|x - x'|^p + |t - t'|)^{\gamma/p}} \leq K$$

holds with $K > 0$ and $\gamma \in (0, 1)$.

6 Maximum and Comparison Principles

In case of singular and/or degenerate parabolic equations, it is impossible to find explicit solutions except for very rare cases, thus we heavily rely on qualitative methods of their study combined with numerical computations. *Maximum and comparison principles* play a prominent role among the qualitative methods. To remind what maximum and comparison principles are, let us start with the well-known *elliptic* Dirichlet Laplacian problem. Let $u_i \in W^{1,2}(\Omega)$, $i = 1, 2$, be the weak solutions of

$$-\Delta u_i = f_i(x) \quad \text{in } \Omega \,,$$

$f_i \in L^\infty(\Omega)$, in a bounded domain $\Omega \subset \mathbb{R}^N$. The *weak* comparison principle states that if $f_1 \le f_2$ in Ω and $u_1 \le u_2$ on $\partial\Omega$ (in the sense of traces) then $u_1 \le u_2$ in Ω. The *strong* comparison principle states that if, moreover, $f_1 \not\equiv f_2$ in Ω or $u_1 \not\equiv u_2$ on $\partial\Omega$ then $u_1 < u_2$ in Ω. In particular, the strong comparison principle says that $f_1 < f_2$ in a small part of Ω of positive measure (and $f_1 \equiv f_2$ elsewhere) is sufficient to have $u_1 < u_2$ everywhere in Ω.

Similar principles hold for the *parabolic* Cauchy–Dirichlet Laplacian problem. Let $u_i \in L^2\left([0, T] \to W^{1,2}(\Omega)\right)$, $i = 1, 2$, be the weak solutions of

$$\frac{\partial u_i}{\partial t} - \Delta u_i = f_i(x, t) \quad \text{in } Q_T \,,$$

$f_i \in L^\infty(Q_T)$. Notice that this equation is a special case of (5.1) with $\vec{a}(x, t, s, \vec{r}) = \vec{r}$ satisfying both (5.2) and (5.3) with $p = 2$. If $f_1 \le f_2$ in Q_T and $u_1 \le u_2$ on Γ_T (in the sense of traces) then $u_1 \le u_2$ in Q_T (weak comparison principle, cf. Proposition 1). If, moreover, at least one of the following three conditions holds:

- $f_1 \not\equiv f_2$ in $\Omega \times (0, t_0)$ whenever $0 < t_0 \le T$,
- $u_1 \not\equiv u_2$ on $\Omega \times \{0\}$ (in the sense of traces),
- $u_1 \not\equiv u_2$ on $\partial\Omega \times (0, t_0)$ (in the sense of traces) whenever $0 < t_0 \le T$,

then $u_1 < u_2$ in Q_T (strong comparison principle).

For the linear case $p = 2$, it is usual to prove the *maximum* principles first since the comparison principles come forth as a consequence. Let $u \in L^2\left([0, T] \to W^{1,2}(\Omega)\right)$ be the weak solution of

$$\frac{\partial u}{\partial t} - \Delta u = f(x, t) \quad \text{in } Q_T \,,$$

$f \in L^\infty(\Omega)$. The *weak* maximum principle states that if $f \ge 0$ in Q_T then $u \ge M \overset{\text{def}}{=} \operatorname*{ess\,inf}_{\Gamma_T} u$ (in the sense of traces) in Q_T. We note that although it would make more sense to call this statement a *minimum* principle and to call a *maximum principle* that $f \le 0$ implies $u \le \operatorname*{ess\,sup}_{\Gamma_T} u$, these two are equivalent (we get one from the other replacing u by $-u$) and thus we use only the term *maximum* principle. The *strong* maximum principle states that if, moreover, at least one of the following three conditions holds:

- $f \not\equiv 0$ in $\Omega \times (0, t_0)$ whenever $0 < t_0 \le T$,
- $u \not\equiv M$ on $\Omega \times \{0\}$ (in the sense of traces),
- $u \not\equiv M$ on $\partial\Omega \times (0, t_0)$ (in the sense of traces) whenever $0 < t_0 \le T$,

then $u > M$ in Q_T.

Once the maximum principle (weak or strong) is proved, the comparison principle (weak or strong, respectively) is easily obtained choosing $u = u_2 - u_1$ (thus $M \ge 0$) and $f = f_2 - f_1$. Notice that the linearity of the left-hand side of the equation is used. Conversely, if we have the comparison principle in our hands, the respective maximum principle can be derived choosing $u_1 \equiv M$, $f_1 \equiv 0$, $u_2 = u$ and $f_2 = f$ (no linearity is used here).

Let us now replace the Laplacian by the p-Laplace operator

$$\Delta_p u \stackrel{\text{def}}{=} \operatorname{div}\left(|\nabla u|^{p-2}\nabla u\right) = \operatorname{div} \vec{a}(x, t, u, \nabla u),$$

$p > 1$, where $\vec{a}(x, t, s, \vec{r}) = |\vec{r}|^{p-2}\vec{r}$ satisfies both (5.2) and (5.3). Similarly as above, the comparison principle implies the respective maximum principle. But since the operator is nonlinear, the maximum principle does not imply the comparison principle. In other words, the maximum principle is weaker because it is only a comparison with the constant solution. Moreover, the uniqueness of the weak solution is a consequence of the weak comparison principle (cf. Proposition 2) but not a consequence of the maximum principle.

As for the *elliptic* Dirichlet p-Laplacian problem for

$$-\Delta_p u = f(x) \quad \text{in } \Omega,$$

both the *weak* maximum and the *weak* comparison principle can be proved in a standard way choosing an appropriate test function. Basically, the *weak* comparison principle states that the p-Laplacian is a monotone operator. The *strong* maximum principle was proved by Vázquez in 1984 [73]. The *strong* comparison principle was proved by Cuesta and Takáč in 1998 [12] provided $0 \le f_1 \le f_2$, $f_1 \not\equiv f_2$ and $u \equiv 0$ on $\partial\Omega$ (they focus on the influence of the right-hand side rather than the boundary data).

While the *weak* maximum and the *weak* comparison principle for the *parabolic* Cauchy–Dirichlet p-Laplacian problem for

$$\frac{\partial u}{\partial t} - \Delta_p u = f(x, t) \quad \text{in } Q_T \tag{6.1}$$

is still standard (see Proposition 1), the *strong* maximum and comparison principle is much more involved when $p \ne 2$. It follows from Barenblatt [4] that we cannot expect the strong maximum principle in the degenerate case $p > 2$ (weak diffusion) even locally in time. Indeed, an explicit radially symmetric solution $u(x, t) \equiv \varrho(|x|, t) =$

$\varrho(r,t)$, $r = |x|$, of (6.1) with $f \equiv 0$, is obtained from the well-known Barenblatt solution of [4, Eq. (1.3)]:

$$c\frac{\partial \varrho}{\partial t} = \frac{1}{r^{N-1}}\frac{\partial}{\partial r}\left[r^{N-1}\left(\frac{\partial}{\partial r}\varrho^k\right)\left|\frac{\partial}{\partial r}\varrho^k\right|^{m-1}\right] \tag{6.2}$$

with $m = p - 1$, $k = 1$, and $c > 0$ a constant. Hence, the case $p > 2$ corresponds to $k > 1/m$. The support of such u (see [4, Fig. 1]) at each particular time is a compact ball with the radius starting from 0 at $t = 0$ (the initial condition is the Dirac distribution located at the origin) and increasing in time at finite speed. Consequently, if we choose Ω a ball in $\mathbb{R}^N$ and an initial time in which the support of the solution is a smaller ball (replacing t by $t + \varepsilon$ with an $\varepsilon > 0$ small enough in [4]), then $u \not\equiv M = 0$ on $\Omega \times \{0\}$ and $u \not> 0$ in Q_T since $u = 0$ in a part of Ω (spherical shell) for positive times until the support of the solution hits $\partial\Omega$. Another counterexample to the strong comparison principle in one spatial dimension where $u_1 \equiv u_2$ on Γ_T, $f_1 \le f_2$, $f_1 \not\equiv f_2$ but $u_1 \not< u_2$ is presented in [9]. On the other hand, a certain stronger condition on the separation of f_1 and f_2 that guarantee the strong comparison principle is formulated in [9].

Even in the singular case $1 < p < 2$ (strong diffusion) the strong maximum principle cannot hold for arbitrarily large T. It follows from the extinction in finite time (see DiBenedetto [19, Sect. 2 of Chap. VII.]) which implies that if $u > 0$ on $\Omega \times \{0\}$, $u \equiv 0$ on $\partial\Omega \times (0,T)$ and $f \equiv 0$ in Γ_T then $u(\cdot,t)$ vanishes in Ω for t large enough. Hence, the strong maximum principle $u > M = 0$ does not hold globally in time. A time-local version of the strong maximum principle was proved in [7] for even more general doubly nonlinear equation

$$\frac{\partial}{\partial t}b(u(x,t)) - \Delta_p u = f(x,t) \quad \text{in } Q_T, \tag{6.3}$$

where $b\colon \mathbb{R}_+ \to \mathbb{R}_+$ is a continuous function, $b(0) = 0$, and $b \in C^1(0,+\infty)$ with $b' > 0$ in $(0,+\infty)$. Notice that if $b(s) \equiv s$ then (6.3) reduces to (6.1).

Theorem 1 (see [7, Theorem 1.1]) *Let* $1 < p < 2$ *and*

$$\lim_{s\to 0+}\frac{s^{2-p}\,b'(s)}{|\log s|^{p-1}} = 0\,. \tag{6.4}$$

Assume that $u\colon \overline{\Omega} \times [0,T) \to \mathbb{R}_+$ *is a continuous, nonnegative, weak solution of* (6.3). *Then, for any fixed* $t_0 \in (0,T)$, *the solution* $u(\cdot,t_0)$ *is either positive everywhere on* Ω *or else identically zero on* Ω.

In particular, if $u(\xi,0) > 0$ *for some* $\xi \in \Omega$, *then there exists* $\tau \in (0,T]$ *such that* $u(x,t) > 0$ *for all* $(x,t) \in \Omega \times (0,\tau)$, *i.e., the strong maximum principle is valid in the* $(N+1)$*-dimensional space-time cylinder* $\Omega \times (0,\tau)$. *The number* $\tau \in (0,T)$ *can be estimated as*

$$\tau = \sup\{T' \in (0, T] \colon u(\xi, t) > 0 \, for\ all\ t \in [0, T')\} > 0\,.$$

Notice that $u(x, t) \equiv \varrho^k(|x|, t)$ where ϱ is the Barenblatt solution of (6.2) is a solution of (6.3) where $b(s) = s^{1/k}$, $p = m + 1$, and $f \equiv 0$. If $k \le 1/m$, i.e., $k \le 1/(p-1)$, then the Barenblatt solution is positive everywhere in $\mathbb{R}^N$ for any positive time. In other words, the speed of propagation is infinite, and it is reasonable to expect at least the time-local strong maximum principle to hold in this case. Indeed, for $b(s) = s^{1/k}$ the condition (6.4) reads as

$$\lim_{s \to 0+} \frac{s^{1-p+1/k}}{k|\log s|^{p-1}} = 0$$

which is satisfied if and only if $1 - p + 1/k \ge 0$, i.e., $k \le 1/(p-1)$. Obviously, condition (6.4) is natural and matches the Barenblatt result perfectly.

7 Basic Models

7.1 Parallel Ditches

Our first model is related to irrigation and drainage. Irrigation is especially important in agriculture while drainage is very important in building and construction. We assume that aquifer is homogeneous, isotropic, and resting on a horizontal impermeable layer. Bottom of all ditches reaches the impermeable layer and the water levels in all ditches are at equal elevation. In our first model, we will consider two infinite parallel ditches and we will study transient groundwater flow between them with the possible recharge due to rain. For the sketch of the problem, see the vertical cross-section perpendicular to the ditches in Fig. 2, where we place the axis x to be perpendicular to the ditches and $x = 0$ is set to be exactly in the middle between two ditches. Such problem has been intensively studied in [5, 51, 64, 65] (and others, see references therein). In the aforementioned works, Darcy law is used as constitutive law. Following Forchheimer's observations from [26], we use nonlinear Smreker–Izbash–Missbach law instead. Thus, the governing equation is (2.9), i.e.,

$$\frac{\partial h}{\partial t} - \operatorname{div}\left(|h|\,|\nabla h|^{p-2}\,\nabla h\right) = f(x, y, t) \ge 0\,. \tag{7.1}$$

We suppose that the problem is translation invariant with respect to y-axis, i.e., a possible recharge is described by $f(x, y, t) \equiv f(x, t) \ge 0$. Thus, $h(x, y, t) \equiv h(x, t)$ and Equation (2.9) reduces to

$$\frac{\partial h}{\partial t} - \frac{\partial}{\partial x}\left(|h|\left|\frac{\partial h}{\partial x}\right|^{p-2}\frac{\partial h}{\partial x}\right) = f(x, t)\,. \tag{7.2}$$

The level of water in the ditches is supposed to be a constant equal to H. This enforces the Dirichlet boundary conditions $h(\pm L/2, t) = H$. As an initial condition, we can consider any function $h_0(x)$ such that $h_0(\pm L/2) = H$ and it satisfies some reasonable additional conditions to be specified later. It will turn out that our assumptions on the initial condition are more general than those in [5, 51, 64, 65]. We distinguish two cases, $H = 0$ (dry ditches) and $H > 0$ (flooded ditches). Function $h_0(x) - H$ can be thought of as a sudden recharge at $t = 0$.

Case $H = 0$. We may directly apply Proposition 3 with $l = 1$ and

$$\text{(fully developed turbulent flow) } \frac{3}{2} < p < 2 \text{ (laminar flow)}$$

to obtain the existence and uniqueness of the solution of the Cauchy–Dirichlet problem (7.2) in $Q_T \stackrel{\text{def}}{=} (-L/2, L/2) \times (0, T]$ with $h(\pm L/2, t) = 0$ and $h_0(x)$ such that $h_0(\pm L/2) = 0$ whenever there exists an extension ψ of h_0 on $\overline{Q}_T$ such that (5.15) is satisfied. Note that such extension exists, e.g., in the case of h_0 with $h_0(\pm L/2) = 0$ being Lipschitz function by the McShane–Whitney extension theorem.

If $f \equiv 0$ and h_0 is Lipschitz continuous with support $[x_0 - \delta, x_0 + \delta] \subset (-L/2, L/2)$ and $h_0(x_0) > 0$, then we will show that there exists $\tau \in (0, T)$ such that $\operatorname{supp} h(\cdot, t) \subset (-L/2, L/2)$ for $0 < t < \tau$. Hence, the solution profile possesses the finite speed of propagation in the sense of Kalašnikov [44]. We wish to use some comparison principle. Unfortunately, the quasi-strong solutions obtained from Proposition 3 do not have to be from $L^p\left([0, T] \to W^{1,p}(\Omega)\right)$ and thus Proposition 1 is not applicable. The situation becomes somewhat intricate and different framework of weak solutions and corresponding weak comparison principle must be used (see Díaz [16, Theorem 9, p. 329]). The following function is used as supersolution:

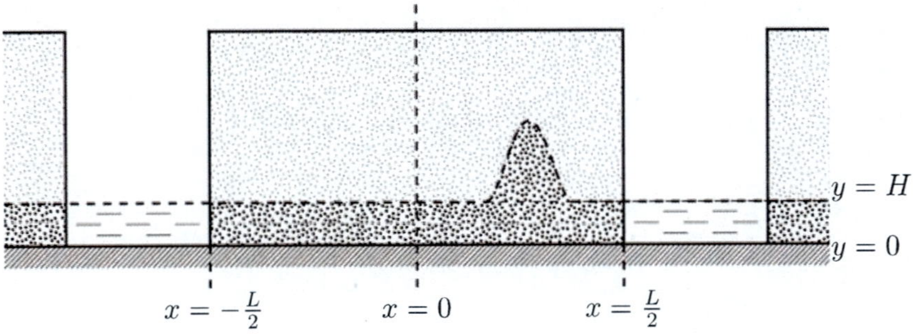

Fig. 2 Perturbed water level

$$U(x,t) = \frac{1}{(t+\tau)^{\lambda}} \left[C - k \frac{|x-x_0|^{p'}}{(t+\tau)^{\frac{p'}{2p-1}}} \right]_+^{p'}$$

with

$$p' = \frac{p}{p-1}, \quad \lambda = \frac{p}{2p^2-3p+1}, \quad k = \left(\frac{p-1}{p}\right)^2 \left(\frac{1}{2p-1}\right)^{\frac{p}{p-1}},$$

and some $\tau > 0$ and $C > 0$ such that $h_0(x) \le U(x,0)$ and

$$[x_0-\delta, x_0+\delta] \subset \operatorname{supp} U(\cdot\,,0) \subsetneq \left(-\frac{L}{2}, \frac{L}{2}\right).$$

Then

$$\operatorname{supp} h(\cdot\,,t) \subset \left(x_0 - \left(\frac{C}{k}\right)^{\frac{1}{p'}} (t+\tau)^{\frac{1}{2p-1}}, x_0 + \left(\frac{C}{k}\right)^{\frac{1}{p'}} (t+\tau)^{\frac{1}{2p-1}} \right).$$

This means that the water from the localized sudden recharge $h_0(x)$ does not reach any of the shores immediately.

Case $H > 0$. In contrast, if both the water level in the ditches and the water table are at constant level $H > 0$, then the localized sudden recharge $h_0 - H \ge 0$ with $\operatorname{supp}(h_0 - H) = [x_0-\delta, x_0+\delta] \subset (-L/2, L/2)$ and $h_0(x_0) > 0$ will cause immediate rise of the water table in the whole aquifer between the ditches. In order to apply theory from Sects. 5 and 6, we introduce a substitution $v(x,t) = h(x,t) - H$ and we arrive at

$$\begin{cases} \dfrac{\partial v}{\partial t} - \dfrac{\partial}{\partial x}\left(|v+H| \left|\dfrac{\partial v}{\partial x}\right|^{p-2} \dfrac{\partial v}{\partial x}\right) = f(x,t) & \text{for } (x,t) \in \left(-\dfrac{L}{2}, \dfrac{L}{2}\right) \times (0,T), \\ v\left(\pm\dfrac{L}{2}, t\right) = 0 & \text{for } t \in (0,T), \\ v(x,0) = h_0(x) - H & \text{for } x \in \left(-\dfrac{L}{2}, \dfrac{L}{2}\right). \end{cases} \tag{7.3}$$

For any $(x,t) \in Q_T$, we set

$$a(x,t,s,r) \stackrel{\text{def}}{=} \begin{cases} H\,|r|^{p-2} r & \text{for } s < 0, \\ (s+H)\,|r|^{p-2} r & \text{for } s \in [0,M], \\ (M+H)\,|r|^{p-2} r & \text{for } s > M, \end{cases} \tag{7.4}$$

where $M = \|h_0 - H\|_{L^\infty(\Omega)} + T\|f\|_{L^\infty(Q_T)}$. Then $a(x,t,s,r)$ given by (7.4) satisfies the assumptions of Proposition 4 with $l = 0$ and $\frac{3}{2} < p < 2$. Then by Proposition 4, Cauchy–Dirichlet problem (5.1) possesses the unique quasi-strong solution $v \in L^p([0, T] \to W^{1,p}(-L/2, L/2))$ for $a(x, t, s, r)$ given by (7.4), $\psi \in L^p([0, T] \to W_0^{1,p}(-L/2, L/2))$ is an extension of $h_0 - H$. Since v is a subsolution of (5.9) with $f_1 = f$ and $\overline{v} = \|h_0 - H\|_{L^\infty(\Omega)} + t\|f\|_{L^\infty(Q_T)}$ is a supersolution of (5.10) with $f_2 = \|f\|_\infty$. Thus, by Proposition 1, $(0 \leq)v \leq \overline{v} \leq M$ on Q_T. It follows that v is also the weak solution of (7.3). Since

$$\frac{\partial v}{\partial x} = \frac{\partial (v+H)}{\partial x} \qquad \text{and} \qquad (v+H)^{\frac{1}{p-1}} \frac{\partial (v+H)}{\partial x} = \frac{p-1}{p} \frac{\partial (v+H)^{\frac{p}{p-1}}}{\partial x}$$

we may rewrite PDE in (7.3) as

$$\frac{\partial v}{\partial t} - \left(\frac{p-1}{p}\right)^{p-1} \frac{\partial}{\partial x} \left(\left| \frac{\partial (v+H)^{\frac{p}{p-1}}}{\partial x} \right|^{p-2} \frac{\partial (v+H)^{\frac{p}{p-1}}}{\partial x} \right) = f(x,t).$$

Introducing another substitution $u = (v+H)^{p/(p-1)} - H^{p/(p-1)}$, we arrive at

$$\frac{\partial \left(\left(u + H^{\frac{p}{p-1}}\right)^{\frac{p-1}{p}} - H \right)}{\partial t} - \left(\frac{p-1}{p}\right)^{p-1} \frac{\partial}{\partial x} \left(\left|\frac{\partial u}{\partial x}\right|^{p-2} \frac{\partial u}{\partial x} \right) = f(x,t), \quad (7.5)$$

which is in fact (6.3) with $b(s) = \left(s + H^{\frac{p}{p-1}}\right)^{\frac{p-1}{p}} - H$. Note that $b(s)$ is a continuous function, $b(0) = 0$, $b \in C^1(0, +\infty)$ with $b' > 0$ in $(0, +\infty)$, and it satisfies (6.4) from Theorem 1. Since $v \in C^{\gamma,\gamma/p}(\overline{Q_T})$, $u = (v+H)^{p/(p-1)} - H^{p/(p-1)}$ is a continuous weak solution of (7.5) (and (6.3)), Theorem 1 guarantees the existence of $\tau \in (0, T)$ such that $u(x,t) > 0$ for all $(x,t) \in \Omega \times (0, \tau)$. In particular, for $f = 0$ this means that localized sudden recharge $h_0 - H$ causes the immediate water table rise in the whole aquifer between the ditches.

Conclusion. *In the case of dry ditches* $(H = 0)$, *the water from a localized sudden recharge does not reach the shores of the ditches immediately and the boundaries of the water mound expand toward the ditches with finite speed. In contrast, for the flooded ditches* $(H > 0)$, *the localized sudden recharge causes the immediate water table rise in whole aquifer between the ditches. In the real world, all movements take place at finite speeds. Thus, the above results should be interpreted as follows: for* $H > 0$, *the water mound expands toward the ditches much faster than for* $H = 0$.

7.2 Isothermic Nanoporous Filtration of Natural Gas

The shales are increasingly gaining importance in natural gas extraction due to their abundance in the world in comparison with classical gas reservoirs. The size of pores and channels in shales is of order of several nanometers which leads to extremely low permeability and the standard mathematical models fail in this situation, see [54] for more details. Thus, in [54], the following mathematical model of isothermic nanoporous filtration of natural gas was proposed (we slightly change their notation in order not to interfere with ours)

$$\begin{cases} \dfrac{\partial P}{\partial t} = \dfrac{\partial}{\partial x}\left(|P(x)|\left|\dfrac{\partial P}{\partial x}\right|^{p-2}\dfrac{\partial P}{\partial x}\right) & \text{for } (x,t) \in (0,+\infty)\times(0,T), \\ P(0,t) = P_0 & \text{for } t \in (0,T), \\ \lim\limits_{x\to\infty} P(x,t) = \overline{P} & \text{for } t \in (0,T), \\ P(x,0) = P_0 & \text{for } x \in (0,+\infty), \end{cases}$$

for $2 < p < 10$, $P_0 \geq 0$, $\overline{P} > 0$ being given constants. This model was analyzed using self-similarity of solutions in [54]. Using the methods of Sect. 5, we can analyze the problem in situations which are not self-similar including time-varying boundary conditions, but only on a bounded interval for x. Note that, e.g., if we assume $(x,t) \in Q_T \stackrel{\text{def}}{=} (0,L)\times(0,T]$, for some $L > 0$, $P(0,t) = P_0(t) \geq 0$, $P(L,t) = P_L(t) \geq 0$ for $t \in [0,T]$ and $P(x,0) = P_{\text{init}}(x) \geq 0$ for $x \in [0,L]$ are Lipschitz functions such that $P_0(0) = P_{\text{init}}(0)$, $P_L(0) = P_{\text{init}}(L)$. Then Proposition 5 guarantees existence of at least one weak solution on $Q_T = (0,L)\times(0,T]$ together with a priori bounds on its Hölder norm.

Acknowledgements The research of Jiří Benedikt, Petr Girg, and Lukáš Kotrla was partially supported by the Grant Agency of the Czech Republic (GAČR) under Project No. 18-03253S. Lukáš Kotrla was partially supported also by the Ministry of Education, Youth, and Sports (MŠMT, Czech Republic) under the program NPU I, Project No. LO1506 (PU-NTIS).

References

1. Aravin, V.I., Numerov, S.N.: Teoriya dvizheniya zhidkostei i gazov v nedeformiruemoi poristoi srede. Gosudarstv. Izdat. Tehn.-Teor. Lit., Moscow, 1953. English Transl. by A. Moscona: "Theory of Fluid Flow in Undeformable Porous Media", Israel Program for Scientific Translations, Jerusalem (1965)
2. Bağcı, Ö., Dukhan, N., Özdemir, M.: Flow regimes in packed beds of spheres from pre-darcy to turbulent. Transp. Porous Media **104**(3), 501–520 (2014)
3. Banerjee, A., Pasupuleti, S., Singh, M.K., Dutta, S.C., Kumar, G.N.P.: Modelling of flow through porous media over the complete flow regime. Transp. Porous Media **129**(1), 1–23 (2019)
4. Barenblatt, G.I.: On some unsteady motions of a liquid and gas in a porous medium. Akad. Nauk SSSR. Prikl. Mat. Meh. **16**, 67–78 (1952). In Russian
5. Basak, P.: An analytical solution for the transient ditch drainage problem. J. Hydrol. **41**(3), 377–382 (1979)
6. Bear, J.: Dynamics of Fluids in Porous Media. Dover Civil and Mechanical Engineering Series, Dover Publications Inc, New York (2014)

7. Benedikt, J., Girg, P., Kotrla, L., Takáč, P.: The strong maximum principle in parabolic problems with the p-Laplacian in a domain. Appl. Math. Lett. **63**, 95–101 (2017)
8. Benedikt, J., Girg, P., Kotrla, L., Takáč, P.: Origin of the p-Laplacian and A. Missbach. Electron. J. Differ. Equ. Paper No. 16, 17 pp (2018)
9. Benedikt, J., Girg, P., Kotrla, L., Takáč, P.: The strong comparison principle in parabolic problems with the p-Laplacian in a domain. Appl. Math. Lett. **98**, 365–373 (2019)
10. Bordier, C., Zimmer, D.: Drainage equations and non-darcian modelling in coarse porous media or geosynthetic materials. J. Hydrol. **228**(3), 174–187 (2000)
11. Bressler, S., Shaviv, G., Shaviv, N.: The sensitivity of the greenhouse effect to changes in the concentration of gases in planetary atmospheres. Acta Polytech. **53**(SUPPL. 1), 832–838 (2013)
12. Cuesta, M., Takáč, P.: A strong comparison principle for the Dirichlet p-Laplacian. In Reaction diffusion systems. Lecture Notes in Pure and Applied Mathematics, vol. 194. Dekker, Trieste, New York **1998**, pp. 79–87 (1995)
13. Czech Hydrometeorological Institute. Current Observations of Watertable in Shalow Boreholes [Data] (2020). http://hydro.chmi.cz/hpps/hpps_pzv_list.php?&objtyp[]=p&objtyp[]=m&objtyp[]=h# [Online; accessed 04-February-2020]. In Czech
14. Dai, A.: Drought under global warming: a review. Wiley Interdiscip. Rev.: Clim. Change **2**(1), 45–65 (2011)
15. Darcy, H.: Les fontaines publiques de la ville de Dijon. Victor Dalmont, Paris (1856)
16. Díaz, J.I.: Qualitative study of nonlinear parabolic equations: an introduction. Extracta Math. **16**(3), 303–341 (2001)
17. Diaz, J.I., de Thélin, F.: On a nonlinear parabolic problem arising in some models related to turbulent flows. SIAM J. Math. Anal. **25**(4), 1085–1111 (1994)
18. Díaz, J.I., Padial, J.F.: Uniqueness and existence of solutions in the $\mathrm{BV}_t(Q)$ space to a doubly nonlinear parabolic problem. Publ. Mat. **40**(2), 527–560 (1996)
19. DiBenedetto, E.: Degenerate Parabolic Equations. Universitext. Springer, New York (1993)
20. DiBenedetto, E., Gianazza, U., Vespri, V.: Harnack's Inequality for Degenerate and Singular Parabolic Equations. Springer Monographs in Mathematics, Springer, New York (2012)
21. Dupuit, J.: Études théoriques et pratiques sur le mouvement des eaux dans les canaux découverts et à travers les terrains perméables. Dunod, Paris (1863)
22. Ebmeyer, C., Urbano, J.M.: The smoothing property for a class of doubly nonlinear parabolic equations. Trans. Am. Math. Soc. **357**(8), 3239–3253 (2005)
23. Eck, B., Barrett, M., Charbeneau, R.: Forchheimer flow in gently sloping layers: Application to drainage of porous asphalt. Water Resour. Res. **48**, 1 (2012)
24. Esteban, J.R., Vázquez, J.L.: On the equation of turbulent filtration in one-dimensional porous media. Nonlinear Anal. **10**(11), 1303–1325 (1986)
25. Farmani, Z., Azin, R., Fatehi, R., Escrochi, M.: Analysis of pre-darcy flow for different liquids and gases. J. Petrol. Sci. Eng. **168**, 17–31 (2018)
26. Forchheimer, P.: Wasserbewegung durch boden. Zeit. Ver. Deutsch. Ing. **45**, 1736–1741 and 1781–1788 (1901)
27. Geoscience Australia: (2014). http://www.ga.gov.au/scientific-topics/water/groundwater/groundwater-in-australia/fractured-rocks, Accessed 21-February-2020
28. Goldman Sachs: The Essentials of Investing in the Water Sector, version 2.0 (2008). http://venturecenter.co.in/water/pdf/2008-goldman-sachs-water-primer.pdf. [Online; accessed 04-February-2020]
29. Gustafson, G., Krásný, J.: Crystalline rock aquifers: their occurrence, use and importance. Appl. Hydrogeol. **2**(2), 64–75 (1994)
30. Hall, J., Arheimer, B., Borga, M., Brázdil, R., Claps, P., Kiss, A., Kjeldsen, T.R., Kriaučiūnienė, J., Kundzewicz, Z.W., Lang, M., Llasat, M.C., Macdonald, N., McIntyre, N., Mediero, L., Merz, B., Merz, R., Molnar, P., Montanari, A., Neuhold, C., Parajka, J., Perdigão, R.A.P., Plavcová, L., Rogger, M., Salinas, J.L., Sauquet, E., Schär, C., Szolgay, J., Viglione, A., Blöschl, G.: Understanding flood regime changes in Europe: a state-of-the-art assessment. Hydrol. Earth Syst. Sci. **18**(7), 2735–2772 (2014)

31. Hinze, J., John, R.: Effects of heat on the dispersal performance of ips typographus. J. Appl. Entomol. **144**(1–2), 144–151 (2020)
32. Huang, J., Kautz, M., Trowbridge, A., Hammerbacher, A., Raffa, K., Adams, H., Goodsman, D., Xu, C., Meddens, A., Kandasamy, D., Gershenzon, J., Seidl, R., Hartmann, H.: Tree defence and bark beetles in a drying world: carbon partitioning, functioning and modelling. New Phytol. **225**(1), 26–36 (2020)
33. Huntington, T.G.: Evidence for intensification of the global water cycle: review and synthesis. J. Hydrol. **319**(1), 83–95 (2006)
34. Hussain, A., Classens, G., Guevara-Rozo, S., Cale, J.A., Rajabzadeh, R., Peters, B.R., Erbilgin, N.: Spatial variation in soil available water holding capacity alters carbon mobilization and allocation to chemical defenses along jack pine stems. Environ. Exp. Bot. **171**, 103902 (2020)
35. Ivanov, A.V.: The classes $\mathcal{B}_{m,l}$ and Hölder estimates for quasilinear parabolic equations that admit double degeneration. *Zap. Nauchn. Sem. S.-Peterburg. Otdel. Mat. Inst. Steklov. (POMI) 197*, Kraev. Zadachi Mat. Fiz. Smezh. Voprosy Teor. Funktsiĭ. **23**, 42–70, 179–180 (1992)
36. Ivanov, A.V.: Regularity for doubly nonlinear parabolic equations. Zap. Nauchn. Sem. S.-Peterburg. Otdel. Mat. Inst. Steklov. (POMI) 209, Voprosy Kvant. Teor. Polya i Statist. Fiz. **12**, 37–59, 261 (1994)
37. Ivanov, A.V.: Existence and uniqueness of a regular solution of the Cauchy-Dirichlet problem for doubly nonlinear parabolic equations. Z. Anal. Anwendungen **14**(4), 751–777 (1995)
38. Ivanov, A.V., Mkrtychyan, P.Z.: On the regularity up to the boundary of generalized solutions of the first initial-boundary value problem for quasilinear parabolic equations that admit double degeneration. Zap. Nauchn. Sem. Leningrad. Otdel. Mat. Inst. Steklov. (LOMI) **196**, Modul. Funktsii Kvadrat. Formy. **2**, 83–98, 173–174 (1991). Translated in J. Math. Sci. **70**(6), 2112–2122 (1994), 35K65 (35D10)
39. JRC European Drought Observatory: Drought in Europe (2019). https://edo.jrc.ec.europa.eu/documents/news/EDODroughtNews201908_Europe.pdf. Accessed 04-February-2020
40. JRC Global Drought Observatory (GDO): Drought in New South Wales (Australia), (2019). https://edo.jrc.ec.europa.eu/documents/news/GDODroughtNews201910_South_East_Australia.pdf. Accessed 04-February-2020
41. JRC Global Drought Observatory (GDO) and ERCC Analytical Team : Drought in India (2019). https://edo.jrc.ec.europa.eu/documents/news/GDODroughtNews201906_India.pdf. Accessed 04-February-2020
42. JRC Global Drought Observatory (GDO) and ERCC Analytical Team : Drought in Southern Africa (2019). https://edo.jrc.ec.europa.eu/documents/news/GDODroughtNews201901_SouthernAfrica.pdf. Accessed 04-February-2020
43. JRC Global Drought Observatory (GDO) and ERCC Analytical Team: Drought in Southern Africa (2019). https://edo.jrc.ec.europa.eu/documents/news/GDODroughtNews201912_Southern_Africa.pdf. Accessed 04-February-2020
44. Kalašnikov, A.S.: On the concept of finite rate of propagation of perturbations. Uspekhi Mat. Nauk **34**, 2(206), 199–200 (1979)
45. King, F.: Principles and conditions of the movements of ground water. Nineteenth Ann. Kept. U. S. Geol. Survey pt. **2**, 9–12 (1898), 209–215
46. Lanci, L., Galeotti, S., Grimani, C., Huber, M.: Evidence against a long-term control on earth climate by galactic cosmic ray flux. Global Planet. Change **185** (2020)
47. Laut, P.: Solar activity and terrestrial climate: an analysis of some purported correlations. J. Atmos. Solar Terr. Phys. **65**(7), 801–812 (2003)
48. Leibenson, L.S.: Turbulent movement of gas in a porous medium. *Bull. Acad. Sci. USSR. Sér. Géograph. Géophys. [Izvestia Akad. Nauk SSSR] 9*, 3–6 (1945). In Russian. Reprinted in Ref. [49], 499–502
49. Leibenson, L.S.: Sobranie trudov, Chast' II: Podzemnaya gidrodinamika [Collected Works, Vol. II: Underground Hydrodynamics]. Izdat'elstvo Akademii Nauk S.S.S.R., Moscow, U.S.S.R. (1953). In Russian
50. Macdonald, A., Davies, J., Calow, R.: African hydrogeology and rural water supply. Appl. Groundwater Stud. Afr., 127–148 (2008)

51. Marino, M.: Rise and decline of the water table induced by vertical recharge. J. Hydrol. **23**(3–4), 289–298 (1974)
52. Matas, A., Merker, J.: Existence of weak solutions to doubly degenerate diffusion equations. Appl. Math. **57**(1), 43–69 (2012)
53. Mekonnen, M.M., Hoekstra, A.Y.: Four billion people facing severe water scarcity. Sci. Adv. **2**, 2 (2016)
54. Monteiroa, P., Rycroftc, C., Barenblatt, G.: A mathematical model of fluid and gas flow in nanoporous media. Proc. Natl. Acad. Sci. U.S.A. **109**(50), 20309–20313 (2012)
55. Netherer, S., Matthews, B., Katzensteiner, K., Blackwell, E., Henschke, P., Hietz, P., Pennerstorfer, J., Rosner, S., Kikuta, S., Schume, H., Schopf, A.: Do water-limiting conditions predispose norway spruce to bark beetle attack? New Phytol. **205**(3), 1128–1141 (2015)
56. Pavlovskii, N.N.: The theory of movement of ground water under hydraulic structures and its main applications. Scientific Amelioration Institute, St. Petersburg, lecture notes. Lithographic (1922). In Russian
57. Perrin, J., Ahmed, S., Hunkeler, D.: The effects of geological heterogeneities and piezometric fluctuations on groundwater flow and chemistry in a hard-rock aquifer, southern india. Hydrogeol. J. **19**(6), 1189–1201 (2011)
58. Pick, L., Kufner, A., John, O., Fučík, S.: Function spaces. Vol. 1, extended ed., *De Gruyter Series in Nonlinear Analysis and Applications*, vol. 14. Walter de Gruyter & Co., Berlin (2013)
59. Sen, Z.: Applied Hydrogeology for Scientists and Engineers. CRC Press, Boca Raton (1995)
60. Shapiro, A.M.: Fractured-rock aquifers understanding an increasingly important source of water (2002). https://toxics.usgs.gov/pubs/FS-112-02/fs-112-02.pdf. Accessed 21-February-2020
61. Shaviv, N., Veizer, J.: Celestial driver of phanerozoic climate? GSA Today **13**(7), 4–10 (2003)
62. Shaviv, N.J.: On climate response to changes in the cosmic ray flux and radiative budget. J. Geophys. Res. Space Phys. **110**, A8 (2005)
63. Siddiqui, F., Soliman, M., House, W., Ibragimov, A.: Pre-darcy flow revisited under experimental investigation. J. Anal. Sci. Technol. **7**, 1 (2016)
64. Singh, R., Rai, S.: On subsurface drainage of transient recharge. J. Hydrol. **48**(3–4), 303–311 (1980)
65. Singh, R., Rai, S.: A solution of the nonlinear boussinesq equation for phreatic flow using an integral balance approach. J. Hydrol. **109**(3–4), 319–323 (1989)
66. Sivanesapillai, R., Steeb, H., Hartmaier, A.: Transition of effective hydraulic properties from low to high reynolds number flow in porous media. Geophys. Res. Lett. **41**(14), 4920–4928 (2014)
67. Smreker, O.: Entwicklung eines Gesetzes für den Widerstand bei derBewegung des Grundwassers. Zeitschr. des Vereines deutscher Ing., **22**, 117–128 and 193–204 4 and 5 (1878)
68. Soni, J., Islam, N., Basak, P.: An experimental evaluation of non-darcian flow in porous media. J. Hydrol. **38**(3–4), 231–241 (1978)
69. Stereńczak, K., Mielcarek, M., Kamińska, A., Kraszewski, B., Żaneta Piasecka, Miścicki, S., Heurich, M.: Influence of selected habitat and stand factors on bark beetle Ips typographus (L.) outbreak in the Białowiża forest. Forest Ecol. Manag. **459**, 117826 (2020)
70. Svensmark, H., Friis-Christensen, E.: Variation of cosmic ray flux and global cloud coverage - a missing link in solar-climate relationships. J. Atmos. Solar Terr. Phys. **59**(11), 1225–1232 (1997)
71. Trenberth, K.: Changes in precipitation with climate change. Climate Res. **47**(1–2), 123–138 (2011)
72. U.S.: Energy Information Administration. Carbon dioxide emissions coefficients (2016). https://www.eia.gov/environment/emissions/co2_vol_mass.php. Accessed 13-March-2020
73. Vázquez, J.L.: A strong maximum principle for some quasilinear elliptic equations. Appl. Math. Optim. **12**(3), 191–202 (1984)
74. Vicente-Serrano, S., McVicar, T., Miralles, D., Yang, Y., Tomas-Burguera, M.: Unraveling the influence of atmospheric evaporative demand on drought and its response to climate change. Climate Change, Wiley Interdisciplinary Reviews (2019)

75. Vicente-Serrano, S., Quiring, S., Peña Gallardo, M., Yuan, S., Domínguez-Castro, F.: A review of environmental droughts: Increased risk under global warming? Earth-Sci. Rev. **201** (2020)
76. Westra, S., Fowler, H., Evans, J., Alexander, L., Berg, P., Johnson, F., Kendon, E., Lenderink, G., Roberts, N.: Future changes to the intensity and frequency of short-duration extreme rainfall. Rev. Geophys. **52**(3), 522–555 (2014)
77. Wright, E.: The hydrogeology of crystalline basement aquifers in Africa. Geol. Soc. Spec. Pub. **66**, 1–27 (1992)
78. Ziazi, R.M., Liburdy, J.A.: Vortical structure characteristics of transitional flow through porous media. In: Volume 1: Fluid Mechanics Fluids Engineering Division Summer Meeting. V001T01A070 (2019)

Research on Solutions Stability for Dynamic Switched Time-Delayed Systems

Denys Khusainov and Oleksii Bychkov

Abstract In this paper, we study the stability for switched systems using linear differential subsystems with time delays. We have used Lyapunov functions to study our results. The results are new in the literature.

Keywords Common Lyapunov function, Uniformly asymptotically stability · Delay systems · Switched systems

1 Introduction

We understand switched time-delayed systems as systems described by a set of differential equations with constant time delay that function on the finite time intervals, switching while maintaining continuity and again by differential equations with time delay [1–4].

Functioning of that system is described by the set of equations

$$\dot{x}(t) = f_k(x(t), x(t-\tau), t), k \in K, t_k < t < t_{k+1}, x(t_k+0) = g_k(x(t_k-0), t_k-0).$$

Let us assume that the value of delay is lower than the functioning time of the subsystem of this kind, so the switched systems has solutions conformant to the uniqueness and continuously dependent on initial conditions.

Definition 1.1 Zero solution of the switched system is called stable by Lyapunov if for an arbitrary solution $x(t)$ for any $\varepsilon > 0$ there exists $\delta(\varepsilon) > 0$ such that for an arbitrary solution $x(t)$, the equation $|x(t)| < \varepsilon, t > t_0$, is correct whenever $|x(t_0)| < \delta(\varepsilon)$.

Definition 1.2 Zero solution for switched system is called asymptotically stable if it is stable by Lyapunov and $\lim\limits_{t\to+\infty} |x(t)| = 0$.

D. Khusainov (✉) · O. Bychkov
Taras Shevchenko National University of Kyiv, Kyiv, Ukraine
e-mail: d.y.khusainov@gmail.com

A. Domoshnitsky et al. (eds.), *Functional Differential Equations and Applications*, Springer Proceedings in Mathematics & Statistics 379,
https://doi.org/10.1007/978-981-16-6297-3_3

2 Stability of the Solutions for Switched Systems with Linear Differential Subsystems with Time Delay

Let us review the usage of the Lyapunov common function method during the investigation into the stability of the switched systems described by the differential subsystems preserving the continuity in the switching points.

Let $S(A, B) = \{S_i(A_i, B_i), i \in N\}$ be the set of the dynamical subsystems $S_i(A_i, B_i)$, which are systems of linear differential equations [5–7]

$$\dot{x}_i(t) = A_i x(t) + B_i x(t-\tau), i \in N_i,$$

functioning over the time intervals $t \in T_i$, $T_i : t_{i-1} \leq t < t_i$. At the moment $t = t_i$, the switching to $i+1$ subsystem occurs

$$\dot{x}(t) = A_{i+1} x(t) + B_{i+1}(t-\tau), i \in N_1.$$

And the functioning of this subsystem while preserving the continuity condition occurs on the interval $t_i \leq t < t_{i+1}$. Further dynamic process occurs likewise.

It is said that the solution $x(t) \equiv 0$ of a differential switched system $S(A, B)$ is stable if for an arbitrarily set systems $S_i(A, B)$, $i = 0, 1, 2, \ldots$ and time intervals T_s : $t_s \leq t < t_{s+1}$, $s = 0, 1, 2, \ldots$ for an arbitrary $\varepsilon > 0$, there exists $\delta(\varepsilon) > 0$, such that for any solution of the system $S(A, B)$ with initial conditions $|x(s)| < \delta(\varepsilon)$, $-\tau \leq s \leq 0$ with $t > 0$, the condition $|x(t)| < \varepsilon$ is met. Furthermore, if $\lim_{t \to +\infty} |x(t)| = 0$, then zero solution will be asymptotically stable.

Let us define the conditions for zero solution of a system $S(A, B)$ be asymptotically stable. Investigation into the stability will be carried out by the method of the Lyapunov function and the Razumikhin condition. There is the following result.

Theorem 2.1 *For a zero solution of a differential switched system $S(A, B)$ to be uniformly asymptotically stable, it is enough for all its subsystems $S_i(A_i, B_i)$ that the common Lyapunov function should exist.*

Let us see the switched systems described only by subtractional equations. Let $S(C) = \{S_i(C), i \in N\}$ be a system consisting of a set of subtractional subsystems

$$S_i(C) : x(k+1) = C_i x(k),$$

which function over the integer intervals $T_i = [k_{i-1}+1, k_{i-1}+2, \ldots, k_i]$. Subtractional switched system $S(C)$ is a dynamic system which is composed of a system of subtractional equations functioning over the intervals T_i, $i = 0, 1, 2, \ldots N_2$.

We can say that the solution $x(k) \equiv 0$ of a subtractional switched system $S(C)$ is stable on switchings, if for an arbitrary preset subsystems $S_i(C)$, time intervals T_i and arbitrary $\varepsilon > 0$, there exists $\delta(\varepsilon) > 0$, such that for an arbitrary solution of a $S(C)$ system with starting conditions $|x(0)| < \delta(\varepsilon)$, while $k = 1, 2, 3, \ldots$, the condition

$|x(k)| < \varepsilon$ will be met. If, additionally, $\lim_{k \to +\infty} |x(k)| = 0$, then zero solution will be asymptotically stable.

Theorem 2.2 *For a zero solution $x(k) \equiv 0$ of a switched system $S(C)$ to be uniformly asymptotically stable, it is enough that for all its subsystems $S_i(C)$, the common Lyapunov function should exist.*

Finally, let us consider the system $S(A, B, C) = \{S_i(A, B), S_j(C), i \in N_1, j \in N_2\}$, which consists of a set of subsystems $S_i(A, B)$, which are systems of linear differential equations with time delay [4],

$$\dot{x}(t) = A_j x(t) + B_i x(t - \tau), t \in N_j,$$

functioning over time intervals $T_i : t_i \le t < t_{i+1}, i = 0, 1, 2, \ldots$, and system of subtractional equations

$$x\left(t_j + 0\right) = B_j x\left(t_j - 0\right), j \in N_2.$$

At moments of time $t = t_j$, the switchings occur due to subtractional subsystems principles.

The stability conditions for a zero solution of a combined system $S(A, B) = \{S_i(A), S_j(B), i \in N_1, j \in N_2\}$ have a more strict form.

Theorem 2.3 *For a zero solution of subtractional switched system $S(A, B, C)$ to be asymptotically stable it is enough that for its differential and subtractional subsystems, a common Lyapunov functions should exist $V_{dif}(x)$, $V_{ras}(x)$. It is also necessary to have monotonic decrease at the break points*

$$V_{dif}(x(t_k - 0)) > V_{ris}(x(t_k - 0)) > V_{ris}(x(t_k + 0)) > V_{dif}(x(t_k + 0)).$$

Notice 2.1 It is very difficult to verify the condition formulated in Theorem 2.3. Therefore, we should formulate another, more strict but easier to verify stability condition.

Theorem 2.4 *For a zero solution of a switched system $S(A, B, C)$ to be asymptotically stable, it is enough that for its differential and subtractional subsystems, a common Lyapunov function $V_{obsh}(x)$ should exist.*

Let us consider the constructive conditions of a time-delayed switched system stability. It is known [5] that for a time-delayed linear systems

$$\dot{x}(t) = Ax(t) + Bx(t - \tau), \tag{1}$$

there are the following stability conditions.

Theorem 2.5 *Let $A + B$ is an asymptoticalle stable matrix and there exists positively defined matrix H such that the following is true*

$$\lambda_{\min}(C) - 2|HB|\left(1 + \sqrt{\varphi(H)}\right) > 0, \varphi(H) = \lambda_{\max}(H)/\lambda_{\min}(H),$$

where $\lambda_{\max}(H)$, $\lambda_{\min}(H)$*—extremal eigenvalues of a symmetrical positively defined matrix* H*, which is a solution for the Lyapunov matrix equation*

$$(A + B)^T H + H(A + B) = -C, \tag{2}$$

for any positively defined matrix C*. Then zero solution for time-delayed system* (2) *is asymptotically stable for any time delay* $\tau > 0$.

Theorem 2.6 *Let* $A + B$ *be asymptotically stable matrix. Then for* $\tau < \tau_0$,

$$\tau_0 = \frac{\lambda_{\min}(C)}{2|HB|(|A| + |B|)\sqrt{1 + \varphi(H)}}$$

time-delayed system (2) *will also be asymptotically stable.*

On the grounds of the aforementioned auxiliary statements, we shall formulate the stability conditions for switched systems, whose differential part is described by the linear time-delayed systems in the form of (2).

Theorem 2.7 *Let there be symmetrical positively defined matrix* H*, for which matrices* $C_i = -(A_i + B_i)^T H + H(A_i + B_i)$, $i \in N_i$*, are also positively defined and the following equation is true*

$$\lambda_{\min}(C_i) - 2|HB_i|\left(1 + \sqrt{\varphi(H)}\right) > 0, i \in N_i.$$

Then the switched system $S(A, B)$ *will be asymptotically stable with any time delay* $\tau > 0$.

We shall get the stability conditions dependent on time delay.

Theorem 2.8 *Let there be symmetrical positively defined matrix* H*, for which matrices* $C_i = -(A_i + B_i)^T H + H(A_i + B_i)$, $i \in N_i$*, are also positively defined. Then the switched system* $S(A, B)$ *will be asymptotically stable with time delay* $\tau < \tau_0$,

$$\tau_0 = \min\left\{\frac{\lambda_{\min}(C_i)}{2|HB_i|(|A_i| + |B_i|)\sqrt{1 + \sqrt{\varphi(H)}}}\right\}, i \in N_i.$$

References

1. Azbelev N.V., Maksimov, Rakhmatullina L.F. Introduction to the theory of functional-differential equations. – M.: Nauka, 277 s
2. Samojlenko A.N., Perestryuk N.A. Differential equations with impulse action. – K., Vy'shhya shkola, 1987. – 286 s

3. Branicky M.S., Hebbar R. Fast marching for hybrid control. In: Proceedings of the 38th IEEE Conference on Decision and Control (1999)
4. Berezansky, L., Idels, L., Troib, L.: Global dynamics of the class on nonlinear nonautonomous systems with time-varying delays. Nonlinear Anal. **74**(18), 7499–7512 (2011)
5. Khusainov D.Ya., Shaty'rko A.V.: Lyapunov's method of functions in the study of stability of differential functional systems. - Kiev, Izd.-vo Kievskogo universiteta (1997). - 236 s
6. Hale, J.K., Verduyn Lunel Sjoerd M. Introduction to Functional Differential Equations. Springer, New-York-Berlin-Ytidelberg-London-Paris (1991)
7. Kolmanovski V., Myshkis A.: Applied Theory of Funktional Differential Equations. Kluver Academic Publishers, Dordrecht-Boston-London (1992)

A Method for Stabilization of Ground Robot Path Controlled by Airborne Autopilot with Time Delay

Alexander Domoshnitsky, Oleg Kupervasser, Hennadii Kutomanov, and Roman Yavich

Abstract The paper addresses the problem of visual navigation of ground robots using a camera positioned at a certain elevation above the confined area. Also, the methods of the stability theory of delay differential equations are used in the study of an actual engineering problem of a ground robot autonomous path. We give a description of autopilot for the stabilization of the ground robot autonomous motion according to desirable path. Indeed, large time delay exists in obtaining by autopilot current information about robot position and orientation, because of big data processing by vision-based (visual) navigation system. Despite this fact, we can prove that autopilot can guarantee a stable desirable path. We demonstrate how to create an appropriate controlling signal for the described information time delay and calculate control parameters for case of polygonal chain path. This path consists of linear motion along with line segments and rotations in vertices.

Keywords Time delay · Differential equations · Stability · Ground robots · Airborne control · Tethered platform · Autopilot · Vision-based navigation · Visual navigation

1 Introduction

The methods developed in the stability theory for differential equations with time delays [1–6] are used in the paper for resolution of the important engineering problem. It is a problem of a ground robot path stabilization using airborne control carried out by autopilot with time delay. The similar application was made by us recently for drone autonomous stable flight controlled by autopilot with time delay [7]. However, in [7], we give only one numerical example of the solution. Here, we find a solution

A. Domoshnitsky · O. Kupervasser (✉) · H. Kutomanov · R. Yavich
Department of Mathematics, Ariel University, Ariel, Israel
e-mail: olegku@ariel.ac.il

O. Kupervasser
TRANSIST VIDEO LLC, Skolkovo, Russia

A. Domoshnitsky et al. (eds.), *Functional Differential Equations and Applications*, Springer Proceedings in Mathematics & Statistics 379,
https://doi.org/10.1007/978-981-16-6297-3_4

as function of equation's parameters. Moreover, equations of motion in [7] are completely different from equations here and describe flighting robot (not ground one). The main new results of this paper are the following:

1. An example of adaptation of the mathematical theory (which was during long time developed without any connection to physics or engineering) in solving actual engineering problem.
2. The approach, proposed here for solving the stabilization problem of ground robot autonomous path, is much better than previously used ones in approximate engineering solution.
3. The adaptation of this mathematical theory for ground robot path stabilization is a difficult problem, since the mathematical theory cannot be applied directly and explicitly for the system describing the ground robot motion. Indeed, we need to apply some nontrivial mathematical transform to the physical differential equations to make such use possible.
4. Even after getting from the mathematical theory constraints for controlled parameters defined by autopilot (which are necessary for stabilization ground robot path), it is also nontrivial problem to find the solution for these parameters.

This paper applies the mathematical stability theory of differential equations with delays [1–6] to the important engineering problem. Despite the fact that engineers prefer to use engineering approximate solutions instead of stability theory of differential equations with time delays, this theory develops very intensively. Indeed, we can see the publication of hundreds of papers on this theory every year.

Stability analysis is the necessary component in most of papers on robotics. The authors of these papers do not consider the time delay even though they understand the fact of existence of such time delay in their mathematical models. The method of Lyapunov's functions (initially described in papers of N. Krasovskii in 1950s) is usually applied, however, it is not appropriate frequently for stabilization by feedback control with time delay.

Now, the main engineering technique for describing a system with the time delay is replacing a system with time delays to a system without time delay and applying the usual stability theory (the method of Lyapunov's functions for nonlinear equations and characteristic equations for linear equations). It can be made by using two ideas or some their combinations [8–10]:

(1) extrapolation motion of a robot forward during time of delay.

(2) finding error estimation for a current state appearing as a result of the time delay and to use the error propagation methods for future analysis.

Using the method (2) results in the obvious decrease in the accuracy of control and its effectiveness.

The application of method (1) is possible in the case of an inert system where the control does not have a strong effect during the time of delay. However, a complex algorithm is necessary even in this case. As a result, we get the increase of the calculation time and prime cost of a control system. Also, we need additional computing power for the extrapolation. If we try to simplify of the model, it results in a decrease of the control effectiveness and accuracy. Also, the large work (for updating a model) needs again after changing or upgrade of a system.

If during the delay time some noticeable influence of the control system exists (not inert systems), then the method becomes even more complex and expensive. In this case, we need the complex iterative schemes, these iterations can do not converge, the long calculation time is necessary, this time can result in an additional time delay. We can get loss of control in the system.

Application of Azbelev's methods [2] for the stability of functional differential equations can help to open new page in the robot control (see the work [11])). In the work [2], based on this theory, the stability analysis and methods for finding solutions of systems of differential equations with time delay were developed. The following advantages of this method can be described:

1. We can decrease costs and time for the development of a control system.
2. If a system has been changed, we can easily to update the system control.
3. The technique is universal for many types of systems.
4. Due to fact that mathematical methods have high accuracy, a system would have precise and efficient control.
5. For complex cases of not inert controlled system, where the control effect is noticeable during the delay time, we do not get no additional delays or control problems.

In this paper, we give the description of autopilot for stabilization of robot autonomous path (polygonal chain). We use vision-based navigation for the finding robot path parameters [12–21]. We used the new patented technology for ground robot navigation using airborne vision-based control [22–27]. The software package was developed that includes approaches for visual navigation control of a ground robot (see Fig. 1) from top position (from a tower, a tethered drone, a balloon, an antenna). Two physical prototypes (where camera is on a top position) controlled by this software were also developed. The system is an example of control systems for robots and can be used for the coordination of the ground robots (automated agricultural machines, automated transport, aerodrome and municipal vehicles, garden lawnmowers and so on).

We develop robot visual navigation using the cameras located on tethered aerial apparatus or towers, tracking the robots on the operation area and observing their environment including artificial and natural landmarks. Two prototypes of these navigation systems were created in the Laboratory of Applied Mathematics of Ariel University in teamwork with TRANSIST VIDEO LLC (Skolkovo, Moscow) [22, 23].

The main insight is that "eyes" of a robot are not positioned on the robot but can be separate autonomous system. Hence, the "eyes" can come up and observe the robot position from above. We describe in the paper algorithms that can be used for the physical prototypes of the system. The system includes a camera in the upper position connected to computer, the computer that can control the robot. Computer software can analyze image obtained from the camera, looks for difference between

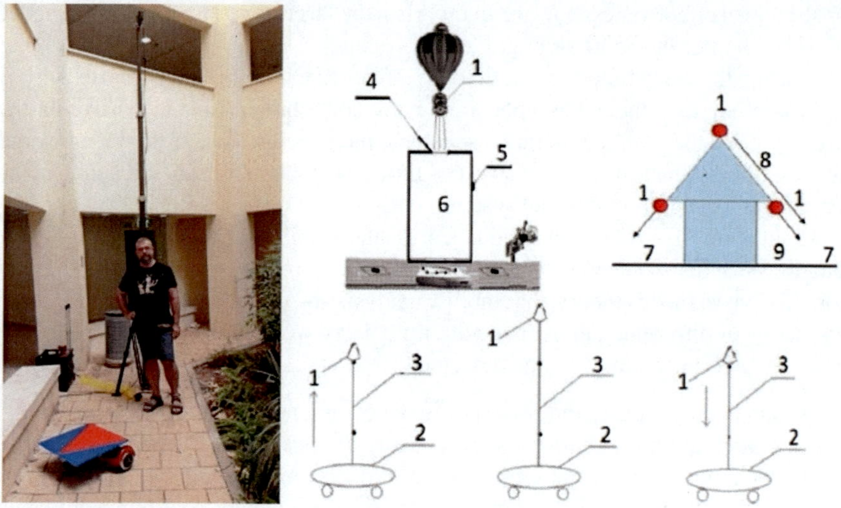

Fig. 1 Airborne terrestrial robot control, some possible camera dispositions; 1—camera, 2—lawnmower; 3—moving up and down long antenna; 4—open door; 5—press button; 6—ground-based energy charged device (ECD); 7—lawn; 8—line of sight; 9—blind spot

actual ground robot position and desirable position, and send Wi-Fi command to reduce the difference.

We present here how to create a relevant controlling signal for autopilot if the information time delay from navigation system exists. We will use the described autopilot for the robot control of path parameters found from the vision-based navigation.

In this paper, description of motion equations and parameters of robot path controlled by airborne autopilot is based on the results of the presentation [28].

The structure of the paper is the following. The first section is Introduction. We described here the theme of the paper—state of the art with references in the field of stability theory methods and methods of description for robot path controlled by autopilot with time delay. The second section gives detailed preliminary results for mathematical stability theory methods, which are used for stabilization of the ground robot path controlled by airborne autopilot. The equation for the ground robot path is described. In the third section, we use the mathematical stability theory for finding parameters controlled by airborne autopilot, which are necessary for the robot path stability and finding upper boundary for the time delay. The fourth section is the conclusion.

2 Preliminary Results of the Investigation for Ground Robot's Flight Stability

2.1 Mathematical Preliminary Results: Stability of Systems with Time Delays

Sign L_∞ denotes the space of essentially bounded measurable functions: $[0, \infty) \rightarrow R$. In the paper, sign "$e$" is the Euler number. Let us investigate the non-homogenous system of differential equations

$$x_i^{'}(t) = \sum_{j=1}^{n} p_{ij}(t)x_j(t - \theta_{ij}(t)), t \in (0, +\infty)$$

$$x_i(\xi) = x_i(0), \xi < 0, i = 1, \ldots, n. \quad (2.1)$$

Here, the components $x_i : [0, +\infty) \rightarrow R$ of the vector $x = col\{x_1, \ldots, x_n\}$ are assumed to be absolutely continuous and their derivatives $x_i^{'} \in L_\infty$.

$P_k(t) = p_{ij}(t)_{i,j=1,\ldots,n}$ are $n \times n$ matrices with entries $p_{ij}(t) \in L_\infty, \theta_{ij}(t) \in L_\infty$ for $k = 1, \ldots, m$ and $i, j = 1, \ldots, n$.

A vector function x is a solution of (2.1) if it satisfies system (2.1) for almost all $t \in [0, +\infty)$.

Let us define

$\theta_i^* = esssup_{t \geq 0}\{\theta_{ii}(t)\}$

It was demonstrated in Proposition 2.3 in [1]:

Assume that the following conditions are correct:

1.1 The matrix P is Hurwitz, i.e., all its eigenvalues have negative real parts, for $t \geq 0$. The matrix P is Metzler, i.e., all its off-diagonal elements are nonnegative for $t \geq 0$: $p_{ij}(t) \geq 0$ for every $i \neq j$, here $i, j = 1, \ldots, n$.

1.2 For every $i = 1, \ldots, n$ the conditions is correct: $|p_{ii}(t)|\theta_i^* \leq \frac{1}{e}$

then system (2.1) is exponentially stable.

Let us investigate the system of second-order differential equations

$$x_i^{''}(t) = q_i(t)x_i^{'}(t - \tau_i(t)) + \sum_{j=1}^{n} p_{ij}(t)x_j(t - \theta_{ij}(t)) = 0, t \in [0, +\infty)$$

$$x_i(\xi) = x_i(0); x_i^{'}(\xi) = x_i^{'}(0), \xi < 0, i = 1, \ldots, n, \quad (2.2)$$

Here, the components $x_i : [0, +\infty) \to R$ of the vector $x = col\{x_1, \ldots, x_n\}$ are assumed to be absolutely continuous and their derivatives $x_i' \in L_\infty$.

$P(t) = \{p_{ij}(t)\}_{i,j=1,\ldots,n}$ are $n \times n$ matrices with entries $p_{ij}(t) \in L_\infty, q_i(t) \in L_\infty$, $\theta_{ij}(t) \in L_\infty, \tau_i(t) \in L_\infty, i, j = 1, \ldots, n$.

A vector function x is a solution of (2.2) if it satisfies system (2.2) for almost all $t \in [0, +\infty)$.

Let us denote

$$\tau_i^* = esssup_{t\geq 0}\{\tau_i(t)\}.$$

It was shown in Theorem 1.1 in [6]: (some misprinting from this paper is corrected here):

Assume that the following conditions are correct:

1.1 The matrix P is Hurwitz, i.e., all its eigenvalues have negative real parts, for $t \geq 0$

$p_{ii}(t) << -\epsilon < 0, q_i(t) << -\epsilon < 0, 4|p_{ii}(t)| < q_i^2(t)$.

The matrix P is Metzler, i.e., all its off-diagonal elements are nonnegative for $t \geq 0$: $p_{ij}(t) \geq 0$ for every $i \neq j$, where $i, j = 1, \ldots, n$.

1.2 For every $i = 1, \ldots, n$ the following conditions are correct:

$|q_i(t)|\tau_i^* \leq \frac{1}{e}, \theta_{ii}(t) \leq \tau_i(t) \leq \tau_i^* < \infty$

then system (2.2) is exponentially stable.

2.2 Engineering Preliminary Results: Motion Equations of Ground Robot

2.2.1 Nonlinear Equations for Robot Motion

We can introduce the following parameters and variables, which are used in ground robot equations of motion (see Fig. 2) [28]:

1. for variables describing motion:

   x and y—coordinates of ground robot
   α—angular of rotation of the robot on the plane
   v—translation velocity of the robot
   ω—angular velocity of the robot

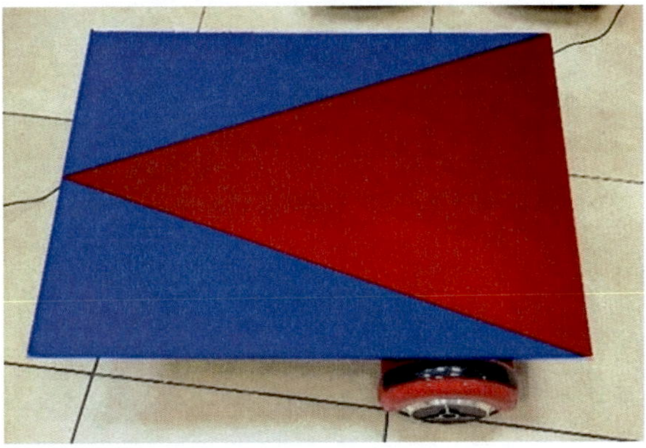

Fig. 2 Ground robot

2. Ground robot parameters:

   R—wheel radius,
   l—distance between wheels.

3. Controlling signals:

   ω_R and ω_L—angle velocities of rotation of the right and left wheels
   From [28] rotation and forward movement are described by the following system of equations:

$\dot{x} = v\cos\alpha$

$\dot{y} = v\sin\alpha$

$\dot{\alpha} = \omega$

$v = \frac{R(\omega_R+\omega_L)}{2}$ (2.2)

$\omega = \frac{2R(\omega_R-\omega_L)}{l}$.

2.2.2 Stationary Desirable Trajectory

2.2.2.1 Solution of Nonlinear Equation

Path curve can be described by formulas (Fig. 3):

$x = f_x(s)$; $y = f_y(s)$

where s is some parameter ($0 \le s \le S_{max}$), f_x, f_y are some functions.

Let us define some variables describing stationary path:

$\overrightarrow{V}(s) = (\frac{dx}{ds}, \frac{dy}{ds})$ is s-vector velocity

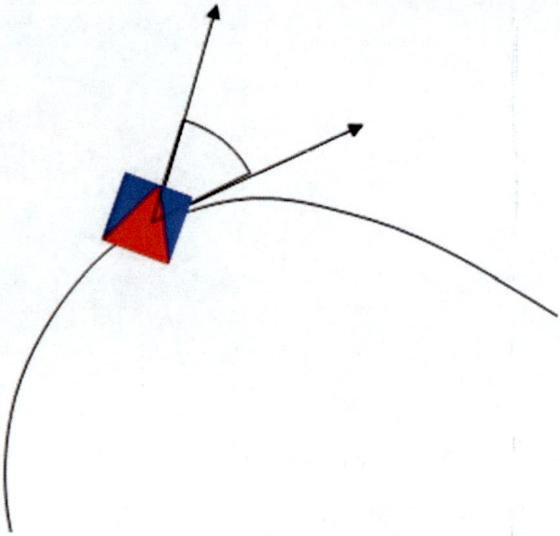

Fig. 3 Ground robot path

$V(s) = \sqrt{(\frac{dx}{ds})^2 + (\frac{dy}{ds})^2}$ is s-velocity
$\vec{\tau}(s) = \frac{\vec{V}(s)}{V(s)}$ is tangent to trajectory
$\sin(\alpha(s)) = (\vec{e}_y \cdot \vec{\tau}(s))$
$\cos(\alpha(s)) = (\vec{e}_x \cdot \vec{\tau}(s))$
where $\vec{e}_x, \vec{e}_y$ are orts of axes OX and OY
$\vec{a}(s) = (\frac{d^2x}{ds^2}, \frac{d^2y}{ds^2})$ is s-vector acceleration
$a(s) = \sqrt{(\frac{d^2x}{ds^2})^2 + (\frac{d^2y}{ds^2})^2}$ is s-acceleration
$\vec{a}_n(s) = \vec{a}(s) - (\vec{a}(s) \cdot \vec{\tau}(s))\vec{\tau}(s)$ is s-normal component of acceleration
$\vec{n}(s) = \frac{\vec{a}_n(s)}{|\vec{a}_n(s)|}$ is normal to trajectory
$\omega(s) = \frac{|\vec{a}_n(s)|}{V(s)}$ is s-angular velocity
$S_W = \int_0^{Smax} V(s)ds$ is full length of path.

2.2.2.2 Time Dependence of Stationary Trajectory

We use variable s as time, but it is not time. So, we need to define real dependence of motion parameters on time t.

Let us choose some mean velocity of ground robot

$$V_{min} \leq V_{mean} \leq V_{max}$$

where V_{min}, V_{max} are minimal and maximal velocity of robot.

Maximal time of the motion can be found from formula:

$T_{max} = \frac{S_W}{V_{mean}}$, so $0 \leq t \leq T_{\max}$.

If we suppose motion with constant velocity $V(t) = V_{mean}$, as a result we can find dependence of time t on s:

$t(s) = \frac{\int_0^s V(s)ds}{V_{mean}}$.

However, some maximal angular velocity $\omega_{\max}$ exists and robot can exceed this maximal angular velocity in some points with high curvature if robot moves with constant velocity V_{mean}. So, robots need to decrease its constant velocity (V_{mean}) to prevent from exceeding.

Angular velocity as function of time can be found from formula:

$\omega(t) = \frac{\omega(s)}{\frac{dt(s)}{ds}}$.

Namely, suppose that $\omega_m = \max_{0 \leq t \leq T_{max}} \omega(t)$ and $\omega_m > \omega_{\max}$. Then $K = \frac{\omega_m}{\omega_{max}}$ and new value of maximal time is $T_{max_{new}} = K \cdot T_{max}$ and $V_{mean_{new}} = \frac{S_W}{T_{max_{new}}}$.

If $V_{mean_{new}} \leq V_{min}$ robot needs to stop in points with high curvature and change only angle in these points. We recommend estimating our trajectory by polygonal chain path. This path consists of linear motion along with line segments with constant translational velocity and zero angular velocity, and rotations in vertices with constant angular velocity and zero translational velocity.

2.2.3 Linear Equations for Perturbations with Respect to Stationary Solution

2.2.3.1 Concluding Linear Equation for the Perturbations

As a result, that the system (1) is nonlinear, the analysis of stability is too hard for these equations. So, we need to make linearization of the equations. Let us suppose that the parameters $x(t)$, $y(t)$, $\alpha(t)$, $v(t)$, $\omega(t)$ corresponds to steady flight and have also some small increments $\delta x(t)$, $\delta y(t)$, $\delta\alpha(t)$, $\delta v(t-\tau)$, $\delta\omega(t-\tau)$. These small increments are the results of perturbation forces changing the path.

We can define the following deviations of the stationary path:

$$\begin{aligned}&\dot{x}(t) + \delta\dot{x}(t) = (v(t) + \delta v(t-\tau))\cos(\alpha(t) + \delta\alpha(t))\\ &= (v(t) + \delta v(t-\tau))(\cos(\alpha(t))\cos(\delta\alpha(t)) - sin(\alpha(t))\sin(\delta\alpha(t)))\end{aligned}$$

$$\begin{aligned}&\dot{y}(t) + \delta\dot{y}(t) = (v(t) + \delta v(t-\tau))\sin(\alpha(t) + \delta\alpha(t))\\ &= (v(t) + \delta v(t-\tau))(\cos(\alpha(t))\sin(\delta\alpha(t)) + sin(\alpha(t))\cos(\delta\alpha(t)))\end{aligned}$$

$$\dot{\alpha}(t) + \delta\dot{\alpha}(t) = \omega(t) + \delta\omega(t-\tau).$$

The resulting nonlinear equations for the perturbations:

$$\delta\dot{x}(t) = \delta v(t-\tau)(\cos(\alpha(t))\cos(\delta\alpha(t)) - sin(\alpha(t))\sin(\delta\alpha(t)))$$
$$+v(t)(\cos(\alpha(t))(\cos(\delta\alpha(t)) - 1) - sin(\alpha(t))\sin(\delta\alpha(t)))$$

$$\delta\dot{y}(t) = \delta v(t-\tau)(\cos(\alpha(t))\sin(\delta\alpha(t)) + sin(\alpha(t))\cos(\delta\alpha(t)))$$
$$+ v(t)(\cos(\alpha(t))\sin(\delta\alpha(t)) + sin(\alpha(t))(\cos(\delta\alpha(t))) - 1)$$

$$\delta\dot{\alpha}(t) = \delta\omega(t-\tau).$$

After linearization of nonlinear equations for perturbation, we get the following linear equations:

$$\begin{aligned}\delta\dot{x}(t) &= \delta v(t-\tau)cos(\alpha(t)) - v(t)\sin(\alpha(t))\delta\alpha(t)\\ \delta\dot{y}(t) &= \delta v(t-\tau)\sin(\alpha(t)) + v(t)\cos(\alpha(t))\delta\alpha(t)\\ \delta\dot{\alpha}(t) &= \delta\omega(t-\tau).\end{aligned} \tag{2.3}$$

The control parameters $\delta v(t-\tau)$, $\delta\omega(t-\tau)$are defined by equations:

$$\begin{aligned}\delta v(t-\tau) &= a_x(t)\delta x(t-\tau) + a_y(t)\delta y(t-\tau) + a_\alpha(t)\delta\alpha(t-\tau)\\ \delta\omega(t-\tau) &= b_x(t)\delta x(t-\tau) + b_y(t)\delta y(t-\tau) + b_\alpha(t)\delta\alpha(t-\tau)\end{aligned} \tag{2.4}$$

where parameters $a_x(t), a_y(t), a_\alpha(t), b_x(t), b_y(t), b_\alpha(t)$ can be chosen freely by autopilot to provide the stationary path to be stable. We will find in the next section an algorithm finding these parameters for polygonal chain path.

2.2.3.2 Description of Controlled Parameters in Linear Equations Defined by Autopilot

If stationary parameters cannot guarantee themselves stability of the desirable stationary trajectory, it is necessary to use autopilots (Fig. 4). The autopilot makes so that the controlling parameters $\delta v(t-\tau)$, $\delta\omega(t-\tau)$ will be some functions of the output-controlled parameters $(\delta x(t); \delta y(t); \delta\alpha(t))$, which are perturbations according to the desirable stationary trajectory. The output parameters values can be obtained by the autopilot from navigation measurements, for example, i.e., from vision-based navigation, satellite navigation, inertial navigation and so on. Using the navigation measurements, the autopilot creates controlling signals to reduce undesirable perturbation. However, time delay exists in getting output-controlled parameters by autopilot for any navigation measurements, which has noticeable value for the visual navigation. So, some problem exists because of the lack of the information for the control. In the paper, we present that even when the time delay exists, we can create control signals giving a stable path.

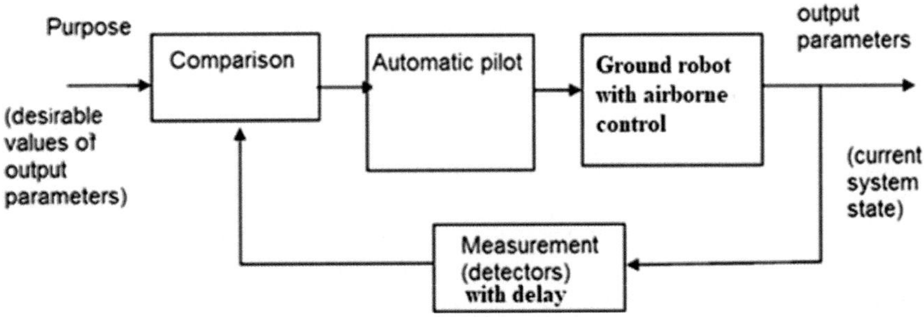

Fig. 4 Automatic control

3 Analysis of Robot Path Stability

3.1 Adjusting the System of Linear Differential Equation for Perturbations to the Form Appropriate for Using Preliminary Mathematical Theory

We recommend estimating our trajectory by polygonal chain path. This path consists of linear motion along with line segments with constant translational velocity and zero angular velocity (Rotation), and rotations in vertices with constant angular velocity and zero translational velocity (Linear motion).

3.1.1 Rotation

Stationary solution

$$\alpha(t) = \omega t + \phi$$

$$v(t) = 0.$$

Let us choose the initial time t=0 in such a way that $\phi = 0$ So, we get following equations for perturbations

$$\delta \dot{x}(t) = \delta v(t-\tau) cos(\omega t)$$

$$\delta \dot{y}(t) = \delta v(t-\tau) \sin(\omega t)$$

$$\delta \dot{\alpha}(t) = \delta \omega(t-\tau)$$

$$\delta v(t-\tau) = a_x(t)\delta x(t-\tau) + a_y(t)\delta y(t-\tau) + a_\alpha(t)\delta\alpha(t-\tau)$$

$$\delta\omega(t-\tau) = b_x(t)\delta x(t-\tau) + b_y(t)\delta y(t-\tau) + b_\alpha(t)\delta\alpha(t-\tau).$$

Let us pass to the rotating system of coordinate:

$$x_r(t) = x(t)\cos(\omega t) + y(t)\sin(\omega t)$$

$$y_r(t) = -x(t)\sin(\omega t) + y(t)\cos(\omega t).$$

Let us find differential equations for $x_r(t)$ and $y_r(t)$:

$$\delta\dot{x}_r(t) = \delta\dot{x}(t)\cos(\omega t) + \delta\dot{y}(t)\sin(\omega t) - \omega\delta x(t)\sin(\omega t) + \omega\delta y(t)\cos(\omega t) =$$
$$=\delta v(t-\tau) + \omega\delta y_r(t)$$

$$\delta\dot{y}_r(t) = -\delta\dot{x}(t)\sin(\omega t) + \delta\dot{y}(t)\cos(\omega t) - \omega\delta x(t)\cos(\omega t) - \omega\delta y(t)\sin(\omega t) =$$

$$=-\omega\delta x_r(t).$$

Finally,

$$\delta\dot{x}_r(t) = \delta v(t-\tau) + \omega\delta y_r(t)$$

$$\delta\dot{y}_r(t) = -\omega\delta x_r(t)$$

$$\delta\dot{\alpha}(t) = \delta\omega(t-\tau)$$

We see that this system can be divided into two independent system of equations

$$\delta\dot{x}_r(t) = \delta v(t-\tau) + \omega\delta y_r(t)$$

$$\delta\dot{y}_r(t) = -\omega\delta x_r(t)$$

and

$$\delta\dot{\alpha}(t) = \delta\omega(t-\tau).$$

3.1.1.1 Differential equations for $\delta x_r(t)$, $\delta y_r(t)$

Let us suppose

$$a_x(t) = -2a_r cos(\omega(t-\tau))$$

$$a_y(t) = -2a_r sin(\omega(t-\tau))$$

$$a_\alpha(t) = 0$$

So

$$\delta v(t-\tau) = a_x(t)\delta x(t-\tau) + a_y(t)\delta y(t-\tau) =$$
$$-2a_r cos(\omega(t-\tau))\delta x(t-\tau) - 2a_r sin(\omega(t-\tau))\delta y(t-\tau) = -2ar\delta x_r(t-\tau).$$

Finally, differential equations for $x_r(t)$ and $y_r(t)$:

$$\delta \dot{x}_r(t) = -2a_r\delta x_r(t-\tau) + \omega\delta y_r(t)$$
$$\delta \dot{y}_r(t) = -\omega\delta x_r(t).$$

We can find second-order differential equation:

$$\delta \ddot{x}_r(t) = -2a_r\delta \dot{x}_r(t-\tau) + \omega\delta\dot{y}_r(t) = -2a_r\delta\dot{x}_r(t-\tau) - \omega^2\delta x_r(t)$$
$$\delta y_r(t) = \frac{\delta\dot{x}_r(t) + 2a_r\delta x_r(t-\tau)}{\omega}. \tag{3.1}$$

3.1.1.2 Differential Equations for $\delta\alpha(t)$

For angle of rotation, we get

$$\delta\dot{\alpha}(t) = \delta\omega(t-\tau)$$

Control parameter for coordinates velocity is following

$$b_x(t) = 0$$

$$b_y(t) = 0$$

$$b_\alpha(t) = b_\alpha$$

$$\delta\omega(t-\tau) = b_\alpha\delta\alpha(t-\tau).$$

Differential equation for $\delta\alpha(t)$

$$\delta\dot{\alpha}(t) = b_\alpha\delta\alpha(t-\tau). \tag{3.2}$$

3.1.2 Linear Motion

Stationary solution

$$\alpha(t) = \alpha = const$$

$$v(t) = v = const.$$

So, we get the following equations for perturbations

$$\delta \dot{x}(t) = \delta v(t-\tau) cos(\alpha) - v \sin(\alpha) \delta\alpha(t)$$

$$\delta \dot{y}(t) = \delta v(t-\tau) \sin(\alpha) + v \cos(\alpha) \delta\alpha(t)$$

$$\delta \dot{\alpha}(t) = \delta\omega(t-\tau)$$

$$\delta v(t-\tau) = a_x(t)\delta x(t-\tau) + a_y(t)\delta y(t-\tau) + a_\alpha(t)\delta\alpha(t-\tau)$$

$$\delta\omega(t-\tau) = b_x(t)\delta x(t-\tau) + b_y(t)\delta y(t-\tau) + b_\alpha(t)\delta\alpha(t-\tau).$$

Let us pass to rotated system of coordinate:

$$x_l(t) = x(t)\cos(\alpha) + y(t)\sin(\alpha)$$

$$y_l(t) = -x(t)\sin(\alpha) + y(t)\cos(\alpha).$$

Let us find differential equations for $x_l(t)$, $y_l(t)$ and $\alpha(t)$:

$$\delta \dot{x}_l(t) = \delta v(t-\tau)$$

$$\delta \dot{y}_l(t) = v\delta\alpha(t)$$

$$\delta \dot{\alpha}(t) = \delta\omega(t-\tau).$$

We see that this system can be divided to two independent system of equation

$$\delta \dot{x}_l(t) = \delta v(t-\tau)$$

and

$$\delta \dot{y}_l(t) = v\delta\alpha(t)$$

$$\delta \dot{\alpha}(t) = \delta\omega(t-\tau).$$

3.1.2.1 Differential Equations for $\delta x_l(t)$

Let us suppose

$$a_\alpha(t) = 0$$

$$a_x(t) = -a_l cos(\alpha)$$

$a_y(t) = -a_l sin(\alpha).$

So, control signal for coordinates velocity is following:

$\delta v(t-\tau) = -a_l \delta x_l(t-\tau).$

Differential equation for $\delta x_l(t)$:

$$\delta \dot{x}_l(t) = -a_l \delta x_l(t-\tau). \quad (3.3)$$

3.1.2.2 Differential Equations for $\delta y_l(t)$, $\delta\alpha(t)$

Let us suppose

$b_x = 0$

$b_y = -a$

$b_\alpha = -2b.$

So, control signal for angular velocity is following:

$\delta\omega(t-\tau) = -a\delta y_l(t-\tau) + -2b\delta\alpha(t-\tau).$

Differential equations for $\delta y_l(t)$ and $\delta\alpha(t)$
$\delta\dot{y}_l(t) = v\delta\alpha(t)$

$\delta\dot{\alpha}(t) = -a\delta y_l(t-\tau) - 2b\delta\alpha(t-\tau).$

We can find second-order differential equation:

$$\delta\ddot{\alpha}(t) = -2b\delta\dot{\alpha}(t-\tau) - a\delta\dot{y}_l(t-\tau) = -2b\delta\dot{\alpha}(t-\tau) - av\delta\alpha(t-\tau)$$
$$\delta y_l(t-\tau) = -\frac{2b\delta\alpha(t-\tau) + \delta\dot{\alpha}(t)}{a}. \quad (3.4)$$

3.2 Applying the Mathematical Theory for Stabilization of Ground Robot Real Trajectory with Respect to the Chosen Desirable Stationary Trajectory

3.2.1 Constrains for the Differential Equation Constants

3.2.1.1 Rotation

3.2.1.1.1 Differential Equations for $\delta x_r(t)$, $\delta y_r(t)$

Apply the condition 1.1 of Theorem 1.1 in [6], presented in Sect. 2 of the paper, to the system (3.1):

$$\delta \ddot{x}_r(t) = -2a_r \delta \dot{x}_r(t-\tau) - \omega^2 \delta x_r(t).$$

Condition 1.1 of Theorem 1.1 is fulfilled if:

$$-2a_r < 0$$

$$-\omega^2 < 0$$

$$(-2a_r)^2 > 4\omega^2.$$

As a result, we get

$$\omega \neq 0$$
$$a_r > |\omega|. \tag{3.5}$$

Because $y_r(t)$ is function of $x_r(t)$ and its derivative

$$\delta y_r(t) = \frac{\delta \dot{x}_r(t) + 2a_r \delta x_r(t-\tau)}{\omega}$$

if $x_r(t)$ is exponentially stable then $y_r(t)$ is also exponentially stable.

3.2.1.1.2 Differential Equations for $\delta\alpha(t)$

Apply the condition 1.1 of Proposition 2.3 in [1], presented in Sect. 2 of this paper, to the system (3.2)

$$\delta \dot{\alpha}(t) = b_\alpha \delta \alpha(t-\tau)$$

Condition 1.1 of Proposition 2.3 is fulfilled if:

$$b_\alpha < 0 \tag{3.6}$$

3.2.1.2 Linear motion

3.2.1.2.1 Differential Equations for $\delta x_l(t)$

Apply the condition 1.1 of Proposition 2.3 in [1], presented in Sect. 2 of this paper, to the system (3.3)

$$\delta \dot{x}_l(t) = -a_l \delta x_l(t - \tau)$$

Condition 1.1 of Proposition 2.3 is fulfilled if:

$$-a_l < 0.$$

Finally

$$a_l > 0 \tag{3.7}$$

3.2.1.2.2 Differential Equations for $y_l(t), \delta\alpha(t)$

Apply the condition 1.1 of in Theorem 1.1 in [6], presented in Sect. 2 of this paper, to the system (3.4)

$$\delta\ddot{\alpha}(t) = -2b\delta\dot{\alpha}(t - \tau) - av\delta\alpha(t - \tau).$$

Condition 1.1 of in Theorem 1.1 is fulfilled if:

$$-2b < 0$$

$$-av < 0$$

$$(-2b)^2 > 4av$$

Finally

$$\begin{aligned} b &> 0 \\ b^2 &> av > 0. \end{aligned} \tag{3.8}$$

Because $y_l(t)$ is function of $\delta\alpha(t)$ and its derivative

$$\delta y_l(t - \tau) = -\frac{2b\delta\alpha(t-\tau) + \delta\dot{\alpha}(t)}{a}$$

if $\delta\alpha(t)$ is exponentially stable then $y_l(t)$ is also exponentially stable.

3.2.2 Constrains for Time Delay

3.2.2.1 Rotation

Differential Equations for $\delta x_r(t)$, $\delta y_r(t)$

Apply the condition 1.2 of in Theorem 1.1 in [6], presented in Sect. 2 of this paper, to the system

$$\tau < \infty$$

$$\tau \leq \frac{1}{e|2a_r|} = \frac{1}{2e|a_r|}. \tag{3.9}$$

Differential Equations for $\delta\alpha(t)$

Apply the condition 1.2 of Proposition 2.3 in [1], presented in Sect. 2 of this paper, to the system

$|b_\alpha|\tau \leq \frac{1}{e}$.

Finally

$$\tau \leq \frac{1}{e|b_\alpha|}. \tag{3.10}$$

Final

From (3.9), (3.10)

$$\tau \leq \frac{1}{e|b_\alpha|} \text{ and } \tau \leq \frac{1}{2e|a_r|}. \tag{3.11}$$

3.2.2.2 Linear Motion

Differential Equations for $\delta x_l(t)$

Apply the condition 1.2 of Proposition 2.3 in [1], presented in Sect. 2 of this paper, to the system

$|a_l|\tau \leq \frac{1}{e}$.

Finally

$$\tau \leq \frac{1}{e|a_l|}. \tag{3.12}$$

Differential Equations for $y_l(t)$, $\delta\alpha(t)$

Apply the condition 1.2 of in Theorem 1.1 in [6], presented in Sect. 2 of this paper, to the system

$\tau < \infty$

$2|b|\tau \leq \frac{1}{e}$.

Finally

$$\tau \leq \frac{1}{2e|b|} \tag{3.13}$$

Final
From (3.12), (3.13)

$$\tau \leq \frac{1}{2e|b|} \text{ and } \tau \leq \frac{1}{e|a_l|} \tag{3.14}$$

.

3.3 *Final Solution*

3.3.1 For Rotation

Stationary solution

$\alpha(t) = \omega t + \phi$

$v(t) = 0$

$\phi = 0$.

Control parameters: From (3.5), (3.6)

$\delta v(t-\tau) = -2a_r cos(\omega(t-\tau))\delta x(t-\tau) - 2a_r sin(\omega(t-\tau))\delta y(t-\tau)$

$\omega \neq 0$

$a_r > |\omega|$

$\delta\omega(t-\tau) = b_\alpha \delta\alpha(t-\tau)$

$b_\alpha < 0$.

3.3.2 For Linear Motion

Stationary solution

$$\alpha(t) = \alpha$$

$$v(t) = v.$$

Control parameters:
From (3.7), (3.8)

$$\delta v(t-\tau) = -a_l cos(\alpha)\delta x(t-\tau) - a_l sin(\alpha)\delta y(t-\tau)$$

$$a_l > 0$$

$$\delta\omega(t-\tau) = a\sin(\alpha)\delta x(t-\tau) - a\cos(\alpha)\delta y(t-\tau) - 2b\delta\alpha(t-\tau)$$

$$av > 0$$

$$b > \sqrt{av}.$$

3.3.3 Delay Time

From (3.11), (3.14)

$$\tau \leq \frac{1}{2e|b|}, \text{ and } \tau \leq \frac{1}{e|a_l|}, \text{ and } \tau \leq \frac{1}{e|b_\alpha|}, \text{ and } \tau \leq \frac{1}{2e|a_r|}.$$

4 Conclusion

We demonstrated the possibility to get stable ground robot path using airborne automatic control when there exists the delay of time in transfer of motion parameter information from navigation measurement system to autopilot. We can find the controlled parameters for some types of path (polygonal chain) and estimated upper boundary of time delay for such system.

It should be mentioned that all these results were obtained in linear assumption for perturbations. If noise is large and as a result, the correspondent perturbations are too large to use linear assumption, then our results are incorrect.

Conflict of interest statement

On behalf of all authors, the corresponding author states that there is no conflict of interest.

References

1. Domoshnitsky, A., Fridman, E.: A positivity-based approach to delay-dependent stability of systems with large time-varying delays. Elsevier J. Syst. Control Lett. **97**, 139–148 (2016). https://www.sciencedirect.com/science/article/pii/S0167691116301384?via
2. Agarwal, R.P., Berezansky, L., Braverman, E., Domoshnitsky, A.: Nonoscillation Theory of Functional Differential Equations with Applications. Springer (2011)
3. Agarwal, R.P., Domoshnitsky, A., Maghakyan, A.: On exponential stability of second order delay differential equations. Czechoslovak Math. J. **65**(140), 1047–1068 (2015). https://link.springer.com/article/10.1007
4. Domoshnitsky, A.: Nonoscillation, maximum principles, and exponential stability of second order delay differential equations without damping term. J. Inequalities Appl. 361 (2014)
5. Domoshnitsky, A., Maghakyan, A., Berezansky, L.: W-transform for exponential stability of second order delay differential equations without damping terms. J. Inequal. Appl. 20 (2017)
6. Domoshnitsky, A., Kupervasser, O., Kutomanov, H.: A positivity-based approach to delay-dependent stability of systems of second order equations. In: AIP Conference Proceedings, Vol. 2159(1), p. 030009 (2019)
7. Avasker S., Domoshnitsky A., Kogan M., Kupervasser O., Kutomanov H., Rofsov Y., Volinsky I., Yavich R.: A method for stabilization of drone flight controlled by autopilot with time delay. Springer Nat. Appl. Sci. **2**(225) (2020). https://link.springer.com/article/10.1007/s42452-020-1962-6?shared-article-renderer
8. Samoilov, L.K.: Classical Method of the Account of Influence Time Delays of Signals in Devices of Control Systems. Izvestiya SFedU. Eng. Sci. **4**(177), 40–49 (2016). http://old.izv-tn.tti.sfedu.ru/?p=22284&lang=en
9. Krushel, E.G., Stepanchenko, I.V.: Informacionnoe zapazdyvanie v cifrovyh sistemah upravlenija, pp. 1–124. Monograph, VolgGTU, Volgograd (2004)
10. Edited by Heinz Unbehauen—Araki, M.: Control systems, robotics and automation, vol. II. Eolss Publishers/UNESCO, PID Control, UK (2009)
11. Azbelev, N.V., Simonov, P.M.: Stability of Differential Equations with Aftereffect. Stability and Control: Theory, Methods and Applications. vol. 20. Taylor & Francis, London (2003)
12. Kupervasser, O., Rubinstein, A.: Correction of inertial navigation on system's errors by the help of video-based navigator based on digital terrain map. Positioning **4**, 89–108. http://file.scirp.org/pdf/POS_2013022811491738.pdf
13. Kupervasser, O., Lerner, R., Rivlin, E., Rotstein, H.: Error analysis for a navigation algorithm based on optical-flow and a digital terrain map. In: The Proceedings of the 2008 IEEE/ION Position, Location and Navigation Symposium, pp. 1203–1212 (2008)
14. Kupervasser, O., Voronov, V.: A navigation filter for fusing DTM/correspondence updates. In: Proceedings of the IEEE International Conference on Robotics and Biomimetics (ROBIO), pp. 1591–1596 (2011)
15. Lerner, R., Kupervasser, O., Rivlin, E.: Pose and motion from omnidirectional optical flow and a digital terrain map. In: The Proceedings of the 2006 IEEE/RSJ International Conference on Intelligent Robots and Systems, pp. 2251–2256 (2006)
16. Lerner, R., Kupervasser, O., Rivlin, E.: Pose and motion from omnidirectional optical flow and a digital terrain map. In: The Proceedings of the 2006 IEEE/RSJ International Conference on Intelligent Robots and Systems, pp. 2251–2256 (2012)
17. Lerner, R., Rivlin, E., Rotstein, H.P.: Pose and motion recovery from correspondence and a digital terrain map. IEEE Trans. Pattern Anal. Mach. Intell. **28**(9), 1404–1417 (2006)

18. Kupervasser, O., Sarychev, V., Rubinstein, A., Yavich, R.: Robust positioning of drones for land use monitoring in strong terrain relief using vision-based navigation. Int. J. Geomate **14**(45), 10–15 (2018). https://www.geomatejournal.com/sites/default/files/articles/10-15-7322-Kupervasser-May-2018-g1.pdf
19. Liu, Y., Rodrigues, M.A.: Statistical image analysis for pose estimation without point correspondences. Pattern Recogn. Lett. **22**, 1191–1206 (2001)
20. David, P., DeMenthon, D., Duraiswami, R., Samet, H.: SoftPOSIT: Simultaneous pose and correspondence determination. ECCV 2002, LNCS 2352, pp. 698–714 (2002). https://link.springer.com/chapter/10.1007/3-540-47977-5_46
21. Shin, D., Park, S.G., Song, B.S., Kim, E.S., Kupervasser, O., Pivovartchuk, D., Gartseev, I., Antipov, O., Kruchenkov, E., Milovanov, A., Kochetov, A., Sazonov, I., Nogtev, I., Hyun, S.W.: Precision improvement of MEMS gyros for indoor mobile robots with horizontal motion inspired by methods of TRIZ. In: Proceedings of 9th IEEE International Conference on Nano/Micro Engineered and Molecular Systems (IEEE-NEMS 2014) 13–1 April 2014 ,Hawaii, USA, pp. 102–107 (2014)
22. Ehrenfeld I., Kogan M., Kupervasser O., Sarychev V., Volinsky I., Yavich R., Zangbi B., Visual navigation for airborne control of ground robots from tethered platform: creation of the first prototype, Proceeding of the IEEE International Conference on New Trends in Engineering and Technology, GRTIET, Tirupathi, Chennai, Tamil Nadu, India (2018). https://www.ieeemadras.org/wp-content/uploads/2018/09/May2018_NL.pdf
23. Ehrenfeld, I., Kupervasser, O., Kutomanov, H., Sarychev, V., Yavich, R.: Algorithms developed for two prototypes of airborne vision-based control of ground robots. In: Proceeding of the 9 Conference on GEOMATE 2019. Tokyo, Japan (2019). http://vixra.org/pdf/1906.0416v1.pdf
24. Kupervasser, O.Y., Kupervasser, Y.I., Rubinstein, A.A.: Russian Utility model: apparatus for coordinating automated devices, Patent 131276, Russia (2012). 12 Nov 2012
25. Kupervasser, O.Y., Kupervasser, Y.I., Rubinstein, A.A.: German Utility model: vorrichtng fur Koordinierung automatisierter Vorrichtungen. Patent Nr. 21 2013 000 225, Germany (2015). 10 July 2015
26. Kupervasser, O.Y.: Franch utility model application: coordination system for ground movable automated devices. Utility model, publication number 3037157, France (2016). 4 June 2016. http://bases-brevets.inpi.fr/en/document-en/FR3037157.html?s=1482862654519&p=5&cHash=1f6f99c78cfb2124b5accb92be338388
27. Kupervasser, O.Y.: Russian invention: method for coordinating terrestrial mobile automated devices, Patent RU 2691788 C2, Russia (2015). 5 June 2015. https://findpatent.ru/patent/269/2691788.html, https://patentimages.storage.googleapis.com/62/98/2f/74b4635a8dde8b/RU2691788C2.pdf
28. Prof. Longoria, R.G.: Turning Kinematically. Spring (2015). https://docplayer.net/31417021-Turning-kinematically.html

Some Properties of the Solution of the Nonlinear Equation of Oscillations in Modeling the Magnetic Separation

Yaroslav Petrivskyi and Volodymyr Petrivskyi

Abstract A qualitative analysis of the equation simulating the process of dry enrichment of raw materials with weak magnetic properties on a drum magnetic separator is carried out. The parametric nature of the role of the free term of the equation, which is the bifurcation point for the model, is established. The study of the properties of the singular point made it possible to allow to build a function to characterize a periodic partial solution and an algorithm for calculating the separation angle of the particle from the surface of the drum during enrichment by the dry separation method, which is convenient for practical use. From a physical point of view, in the process of magnetic separation, when there is friction proportional to the square of the angular velocity in the system, with the force acting on the particles of constant magnetic force, the work expended on overcoming the friction forces increases with the square of the angular velocity, while the operation of the external forces remains unchanged.

Keywords Bifurcation point · Phase portrait · Period function

1 Main Part

A differential equation is described that describes the motion of a particle with weak magnetic properties on the surface of a drum of a magnetic separator during dry magnetic enrichment on a separator with an overhead feed. According to Murrariu [1], such an equation has the following form:

Y. Petrivskyi (✉)
Rivne state humanitarian university, Rivne, Ukraine
e-mail: prorectorsgu@ukr.net

V. Petrivskyi
Taras Shevchenko National University of Kyiv, Kyiv, Ukraine

A. Domoshnitsky et al. (eds.), *Functional Differential Equations and Applications*, Springer Proceedings in Mathematics & Statistics 379,
https://doi.org/10.1007/978-981-16-6297-3_5

$$\frac{d^2\theta}{dt^2} = \frac{g}{R+b}\sin\theta \pm \mu_{d,s}\frac{g}{R+b}\cos\theta - \left(\frac{d\theta}{dt}\right)^2 + \frac{F_m}{\rho_\rho V_\rho (R+b)}, \quad (1)$$

where θ—the angle of rotation of the drum, R—the radius of the drum, b—the particle size, the "+" sign and the coefficient of dynamic friction μ_d characterize the first phase, when the particle hits the drum, the "−" sign and the coefficient of static friction μ_s characterize the third phase, namely the process of separation of the particle from the drum, F_m—the magnetic force acting on the particle, V_ρ—the volume of the particle and g—the acceleration of gravity.

At the stage of changing the nature of particle friction from dynamic to static, the second phase, the particle moves with a constant angular velocity equal to the speed of rotation of the drum. The corresponding equation has the following form:

$$\frac{d^2\theta}{dt^2} = 0 \; or \; \frac{d\theta}{dt} = \omega_\rho. \quad (2)$$

To conduct a qualitative study and create an algorithm that is convenient for applied research and simulate the process of particle motion on a drum, we transform the particle motion equation (1) defined relative to the angle θ into the equivalent equations of motion of a particle relative to the angular velocity of this particle. For this purpose, we denote

$$\frac{g}{R+b} = a, \; \mu_{d,s}\frac{g}{R+b} = b.$$

Then

$$a\sin\theta \pm b\cos\theta = A\sin(\theta \pm \phi),$$

where $A = \sqrt{a_1^2 + a_2^2}$, angle ϕ denoting like

$$\cos\phi = \frac{a_1}{\sqrt{a_1^2 + a_2^2}}, \; \sin\phi = \frac{a_2}{\sqrt{a_1^2 + a_2^2}}.$$

Given these transformations, Eq. (1) is simplified:

$$\frac{d^2\theta}{dt^2} = A\sin\left(\theta \pm \phi_{d,s}\right) - \left(\frac{d\theta}{dt}\right)^2 + B, \; B = \frac{F_m}{\rho_\rho V_\rho (R+b)}, \; A > 0, \; B > 0. \quad (3)$$

In general, Eq. (1), when the free term is equal to zero or when the value of the magnetic force F_m is similar to the equation simulating the damped oscillations of a pendulum immersed in a medium, which, when the pendulum moves, creates a force proportional to the square of its speed and directed opposite to this speed. Quite detailed qualitative studies of this equation are known, for example, given in [1, 2], where the case of dynamical systems, reduced under certain simplifying assumptions to the mathematical model of the oscillation of a pendulum with "linear friction" under the influence of constant torque, is also studied. It also shows cases where physical analogies can be reduced to the indicated type of model with "linear friction", for example, the problem of the synchronous motor, the problem of parallel operation of generators, etc.

For the convenience of studying Eq. (1), we introduce new $x = \theta \pm \phi_{d,s}$, $y = \frac{dx}{dt}$ variables. We obtain a non-conservative system of two first-order equations:

$$\begin{cases} \frac{dx}{dt} = y \\ \frac{dx}{dt} = A \sin x - y^2 + B \end{cases}. \tag{4}$$

Equivalent to system (4) is the equation of integral curves on the cylinder, which, like system (4) and initial Eq. (3), cannot be directly integrated.

$$\frac{dy}{dx} = \frac{A \sin x + y^2 + B}{y}. \tag{5}$$

Note that isoclines $\frac{dy}{dx} = 0$ are biased quadratic sinusoids whose equations are of the form

$$y^2 = A \sin x + B. \tag{6}$$

Equivalent to system (4) is the equation of integral curves on the cylinder, which, like system (4) and initial Eq. (3), cannot be directly integrated.

The graph of curve (7) suppresses the OX axis, respectively, the axis θ, only at values $\frac{B}{A} < 1$. When $\frac{B}{A} > 1$, the isocline does not cross the OX axis.

The coordinates of the singular points, the equilibrium state of the system, can be found by solving the system:

$$\begin{cases} y = 0 \\ A \sin x - y^2 + B = 0 \end{cases} \tag{7}$$

Obviously, for $\frac{B}{A} > 1$, singular points do not exist. At $\frac{B}{A} < 1$, there are two distinct points: (1) $y_1 = 0, x_1 = \arcsin \frac{B}{A}$; (2) $y_2 = 0, x_2 = \pi - \arcsin \frac{B}{A}$, $0 < \arcsin \frac{B}{A} < \frac{\pi}{2}$. When $x_{1,2} = \frac{pi}{2}$, equilibrium positions merge. Thus, the $\frac{B}{A}$ parameter value is bifurcation for Eq. (1). To clarify the nature of the equilibrium states of system (4), we linearize the right-hand side of the second equation of system (4) in a neighborhood of singular points, making the change $x = x_i + \xi, i = 1, 2$. Expanding $\sin x$ in a series of powers and limiting ourselves to linear terms, we obtain

$$\begin{cases} \frac{dx}{dy} = y \\ \frac{dy}{dt} = A \cos x_i \xi, i = 1, 2 \end{cases}. \tag{8}$$

The corresponding characteristic equation for system (9) has the following form [3]:

$$\lambda^2 - A\cos x_i = 0, \tag{9}$$

whereas $\cos x_i > 0$ then we have two real roots of different signs, which means that the first singular point with coordinates $y_1 = 0, x_1 = \arcsin\frac{B}{A}$ is a singular point of the saddle type, and the characteristic lines on the phase plane are a family of hyperbolas. For the second case $\cos x_2 < 0$, the characteristic equation has purely conjugated roots and, accordingly, a singular point with coordinates $y_2 = 0, x_2 = \pi - arcsin\frac{B}{A}$ is a center point.

The results obtained make it possible to construct a phase portrait of the system (4) (Fig. 1).

Denoting right parts of Eq. (4) like $F(x, y)$ and $G(x, y)$, we will get the corresponding value of Bendixson's criterion parameter:

$$F_x^{'} + G_y^{'} = -2y.$$

Thus, for the trajectories on the phase cylinder in the region where $y > 0$, dynamic system (4) does not have closed paths on the phase cylinder that do not cover the cylinder and can have the largest one limit cycle covering the cylinder. When it exists, such a cycle is necessarily stable, according to the obtained characteristic indicator.

Further, the following calculation algorithm is proposed for finding the necessary parameters of the dry magnetic separation process on a drum separator. To do this, we carry out further transformations of Eq. (3). According to [4], we set

$$\nu(\theta) = (\theta')^2, \left(\theta^2 = \frac{d\theta}{dt}\right). \tag{10}$$

Then differentiating the right and left side of (8),

$$\nu'\theta' = 2\theta'\theta''. \tag{11}$$

Substituting the substitution (10) and (11) in Eq. (3), we obtain

$$\nu'\theta + 2\nu(\theta) = 2A\sin(\theta \pm \phi_{d,s}) + 2B. \tag{12}$$

Equation (12) is a linear, inhomogeneous first-order differential equation with constant coefficients, defined with respect to the function $\nu = \nu(\theta)$. In accordance with notation (10), $\omega = \sqrt{\nu(\theta)}$ is the particle angular velocity. The solution of Eq. (12) accordingly has the following form:

$$\nu(\theta) = C\exp^{-2\theta} + \frac{4}{5}A\sin(\theta \pm \phi_{d,s}) - \frac{2}{5}A\cos(\theta \pm \phi_{d,s}) + B, \tag{13}$$

where C—const., "+" sign and parameter ϕ_d characterize the first phase when the particle hits the drum and the "–" sign and the parameter ϕ_s characterize the third phase.

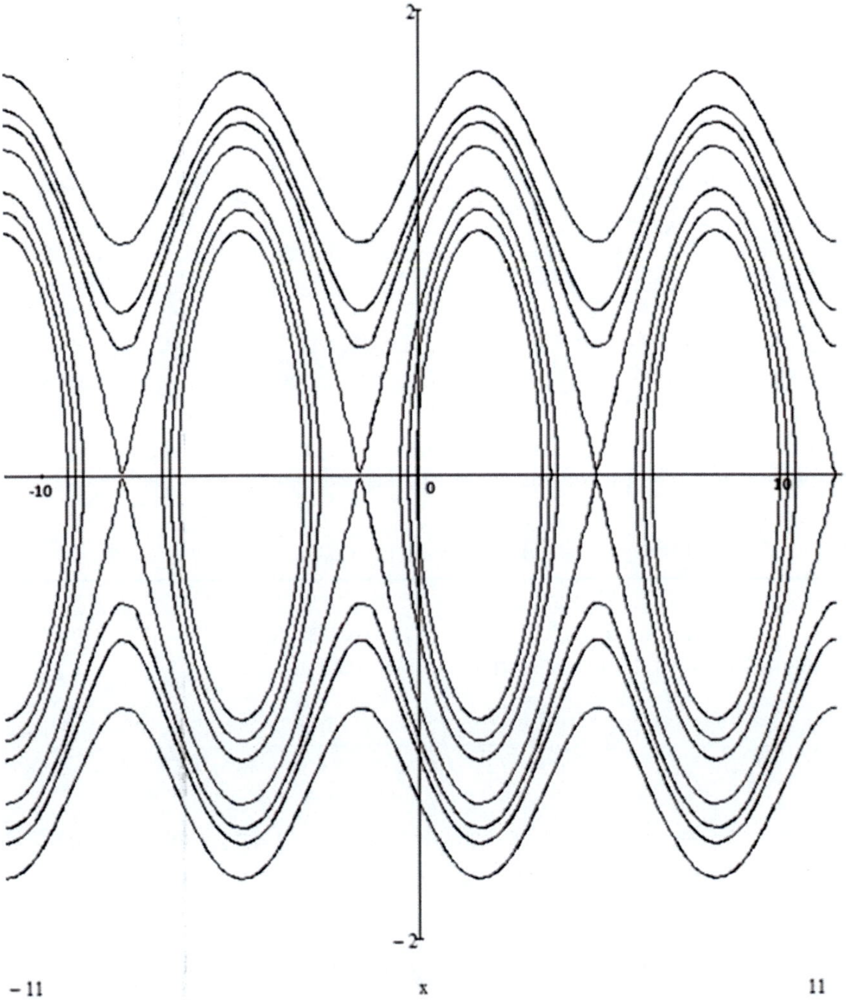

Fig. 1 Phase portrait of a non-conservative system (4)

The result allows us to conduct research on the existence of a function of a periodic solution of Eq. (1). Considering the substitution (10), for the case when the periodic solution curve (periodic orbit) intersects the axis Θ at a point $\theta = \theta_0$, there is a curve on the phase plane

$$\frac{d\theta}{dt} = \sqrt{Ce^{-2\theta} + \frac{4}{5}A\sin(\theta \pm \phi_d, s) - \frac{2}{5}A\cos(\theta \pm \phi_d, s) + B}.$$

Given the symmetry of the periodic orbit, for the period function, we obtain the expression

$$T(\theta_0) = \int_0^{\theta_0} \frac{d\theta}{\sqrt{Ce^{-2\theta} + \frac{4}{5}A\sin(\theta \pm \phi_{d,s}) - \frac{2}{5}A\cos(\theta \pm \phi_{d,s}) + B}}.$$

We carry out the transformation of the radical expression for the case of a particular solution to Eq. (1). Marking $A_1 = \frac{4}{5}A$, $A_2 = \frac{2}{5}A$, $D = \sqrt{A_1^2 + A_2^2} = \frac{2}{\sqrt{5}}A$, $\cos\alpha = \frac{A_1}{D}$ and $\sin\alpha = \frac{A_2}{D}$, we get $\frac{4}{5}A\sin(\theta \pm \phi_{d,s}) - \frac{2}{5}A\cos(\theta \pm \phi_{d,s}) + B = D\sin(\theta \pm \phi_{d,s} - \alpha) + B = \frac{2}{\sqrt{5}}$ $A\sin(\theta \pm \phi_{d,s} - \alpha) + \frac{\sqrt{5}}{2}\frac{B}{A}$.

Given that special points $y_1 = 0$, $x_1 = \arcsin\frac{B}{A}$ and $y_2 = 0$, $x_2 = \pi - \arcsin\frac{B}{A}$ defined for $0 < \arcsin\frac{B}{A} < \frac{\pi}{2}$, where $\sin(x) = \frac{B}{A}$ and $\sin(\theta \pm \phi_{d,s} - \alpha) = \cos(\frac{\pi}{2} - \theta \pm \phi_{d,s} + \alpha)$, denoting $\frac{\sqrt{5}}{2}\frac{B}{A} = \frac{\sqrt{5}}{2}\sin(x) = \sin(x) = -\cos(\frac{\pi}{2} + x)$, we get $\frac{2}{\sqrt{5}}$ $A\cos(\frac{\pi}{2} - \theta \pm \phi_{d,s} + \alpha) + \sin\theta_0$. Labeling for convenience $\frac{\pi}{2} - \theta \pm \phi_{d,s} + \alpha = \theta$ and $\frac{\pi}{2} + x = \theta_0$, finally, the root expression takes the form

$$\frac{4}{5}A\sin(\theta \pm \phi_{d,s}) - \frac{2}{5}A\cos(\theta \pm \phi_{d,s}) + B = \frac{2A}{\sqrt{5}}(\cos\theta - \cos\theta_0).$$

Next, applying the formula for each of the cosines $\cos\gamma = 1 - 2\sin^2\frac{\gamma}{2}$ and by changing the integration variable in the transformed integral,

$$T(\theta_0) = \frac{\sqrt[4]{5}}{\sqrt{2A}} \int_0^{\theta_0} \frac{d\theta}{\sqrt{\cos\theta - \cos\theta_0}}.$$

Using $\sin\phi = \frac{\sin\frac{\theta}{2}}{\sin\frac{\theta_0}{2}}$ [5], we finally obtain the value of the period function, which is expressed through the full elliptic integral of the first kind.

$$T(\theta_0) = \frac{\sqrt[4]{5}}{\sqrt{2A}} \int_0^{\frac{\pi}{2}} \frac{d\phi}{\sqrt{1 - k^2\sin^2(\phi)}} = \frac{\sqrt[4]{5}}{\sqrt{2A}} K(k),\ k = \sin\left(\frac{\theta_0}{2}\right).$$

The boundary value of this integral is known [6]:

$$\lim_{k\to 0} \int_0^{\frac{\pi}{2}} \frac{d\phi}{\sqrt{1 - k^2\sin^2(\phi)}} = \int_0^{\frac{\pi}{2}} d\phi = \frac{\pi}{2}.$$

The results obtained allow us to assert the validity of the following theorem.

Theorem 1 *For non-negative parameter values $\frac{B}{A} < 1$ for a partial periodic solution of Eq. (1), when periodic orbits tend to a singular point of the center type, there exists a boundary value of the period equal to $\frac{\sqrt[4]{5}\pi}{\sqrt{8A}}$, and with the unlimited approximation of periodic orbits to the separatrix, the period function increases unlimitedly.*

Consider the practical application of the results for a specific example of the separation process.

Thus, taking into account the initial conditions, the process of particle motion on the drum is modeled by the following problems with initial conditions for linear equations of the first order. I phase.

$$\nu'(\theta) + 2\nu(\theta) = 2A\sin(\theta + \phi_d) + B \tag{14}$$

$$\nu|_{t=0} = 0.$$

At the initial time $t = 0$, when the particle hits the drum, its speed is zero. The end of the first phase is a point in time t_1 when the particle velocity becomes equal to the speed of the drum. Then the dynamic friction coefficient μ_d or accordingly parameter ϕ_d in Eq. (13) changes to the coefficient of static friction μ_s or parameter ϕ_s.

II phase.

$$\omega_r = \sqrt{\nu(\theta)}, t > t_1, \tag{15}$$

where ω_r—drum angular velocity.

III phase.

$$\nu'(\theta) + 2\nu(\theta) = 2A\sin(\theta - \phi_s) + 2B \tag{16}$$

$$\nu|_{t=t_2} = \omega_r^2.$$

The condition for the separation of the particle from the drum (the end of the third phase) is the condition that the value of the centrifugal force exceeds the value of the magnetic force with a decrease in the radial component of gravity. Figure 2 schematically shows the phases of particle motion on the drum corresponding to problems (14)–(16).

Consider an example for the model values of the physico-mechanical parameters of the separation process [7]: $R = 0.25\,\text{m}, b = 0.002\,\text{m}, \mu_s = 0.75, \mu_d = 0.15, \rho = 4700\,\frac{\text{kg}}{\text{m}^3}, F = 3.2 * 10\_4H$.

An illustration of the solution to problem (14) simulating the first phase of particle motion is shown in Fig. 3. The point of intersection of the particle's velocity curve with the straight line, the speed of rotation of the drum, characterizes the value of the angle of rotation at which the first phase of the process ends, namely, the angular velocity of the particle is equal to the angular velocity of the drum. Based on the general solution (13), the initial condition of problem (16), we find the solution to the problem for this case.

$$\nu(\theta) = 53.6\exp^{-2\theta} + \frac{4}{5}A\sin(\theta + \phi_d) - \frac{2}{5}A\cos(\theta + \phi_d) + B. \tag{17}$$

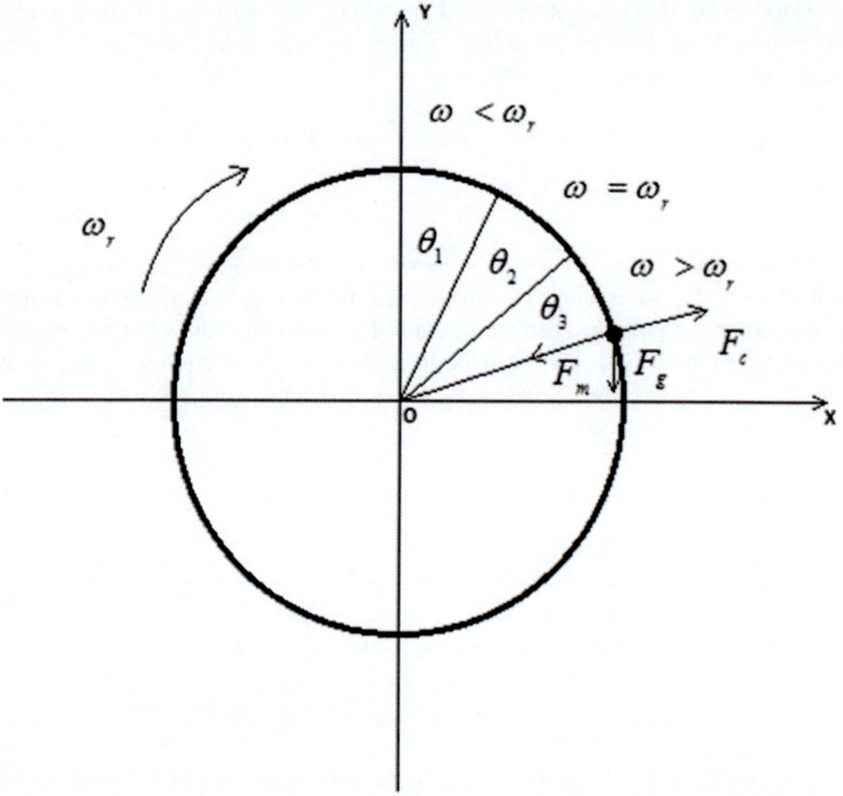

Fig. 2 The phase state of the particle on the drum magnetic separator: 1—the angular velocity of the drum is greater than the angular velocity ω_r of the particle ω; 2—the angular velocity of the particle is equal to the angular velocity of the drum; 3—the angular velocity of the particle is greater than the angular velocity of the drum (separation of particles from the drum)

From the condition $\frac{2\pi 70}{60} = \sqrt{53.6 \exp_{-2\theta} + \frac{4}{5} A \sin(\theta + \phi_d) - \frac{2}{5} A \cos(\theta + \phi_d) + B}$, we can find value $\theta_1 = 0.53$ rad.

After the transitional second phase, solving problem (14), under the initial condition determined by the angular velocity of the drum, we obtain the solution to the problem in the following form:

$$\nu(\theta) = 37.3 \exp^{-2\theta} + \frac{4}{5} A \sin(\theta + \phi_s) - \frac{2}{5} A \cos(\theta + \phi_s) + B. \tag{18}$$

In Fig. 4, the curve characterizing the motion of the particle during phase III is located above the straight line—the angular velocity of the drum.

When the value of the centrifugal force acting on the particle exceeds the value of the magnetic force, the value of the radial component of the gravity decreases, the particle detaches from the drum, namely

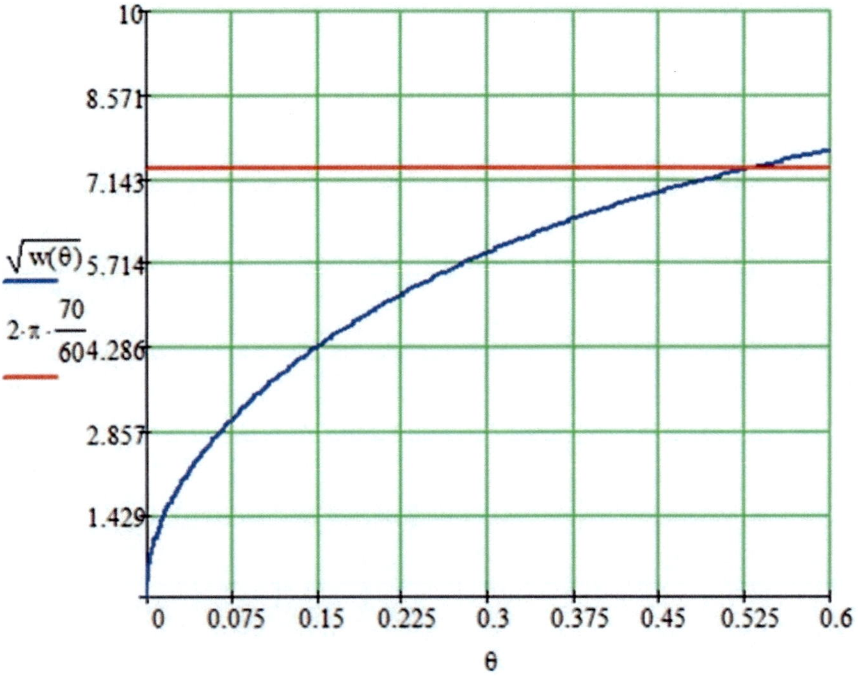

Fig. 3 Particle and drum velocity plots

$$F_c = m\omega^2(R + b) > F_m + F_{gr}. \tag{19}$$

Based on formula (13) and condition (16) for our case, we find the angle θ_3 when there is a separation of particles from the drum, namely $\theta_3 = 1.29$ rad.

In Fig. 5, graphs are constructed that illustrate the increase in the value of the centrifugal force acting on the particle and the decrease in the value of the radial component of gravity with a constant magnetic force. The point of intersection of the graphs is the value of the separation angle of the particle from the drum.

2 Conclusion

It is complemented by the existing universality of the process of mathematical modeling, which consists in the fact that the ordinary equation of oscillation covers a wide range of processes—from the operation of a system of synchronous motors to the enrichment of minerals. An expression is found for the period function for a particular solution of oscillation equation (1). The practical significance of the results of the research is as follows. From a physical point of view, in the process of magnetic separation, when there is friction proportional to the square of the angular veloc-

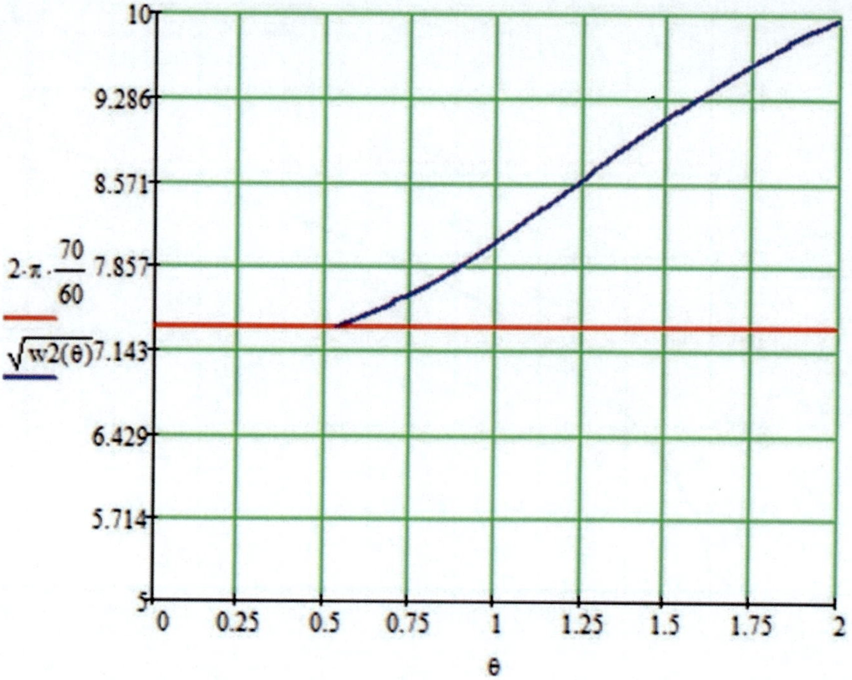

Fig. 4 Particle and drum velocity plots

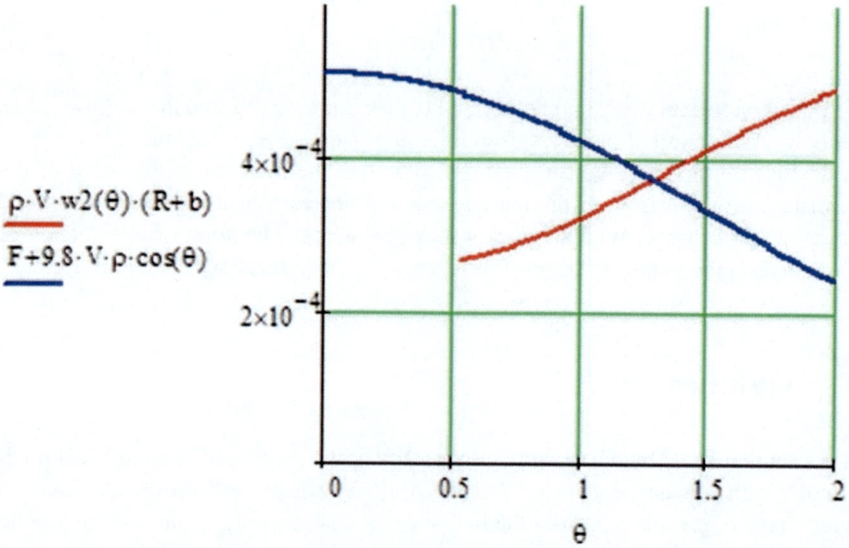

Fig. 5 Values of the centrifugal force acting on the particle and the radial component of gravity with a constant magnetic force

ity in the system, with the force acting on the particles of constant magnetic force, the work expended on overcoming the friction forces increases with the square of the angular velocity, while the operation of the external forces remains unchanged. Therefore, if the value of the parameter in Eq. (3) $\frac{B}{A} > 1$, the permanent component of the magnetic force is so large that it exceeds the maximum value of the centrifugal force acting on the magnetic particle, the separation of the magnetic particle will not occur—the separation process is absent under any initial conditions. This condition will be observed until a balance is established between the friction scattering forces and the magnetic force. The value of the parameter $\frac{B}{A} = 1$ is bifurcation point. When the value of the parameter is $\frac{B}{A} < 1$, the moment characterizing the value of the radial component of gravity exceeds the corresponding moment of magnetic force.

References

1. Jan, S.: Magnetic Techniques for the Treatment of Materials; Svoboda, J.: Kluwer Academic Publishers, Johannesburg (2004), 576p
2. Stocker, J.: Nonlinear Oscillations in Mechanical and Electrical Systems; Stocker, J.: - M.: Inostrannaya literatura (1952), 264p
3. Andronov, A., Witt, A., Haikin, S.: Oscillations Theory. Nauka (1981), 918p
4. Kamke, E.: Handbook of ordinary differential equations. Nauka (1971), 576p
5. Chicone, C.: Ordinary Differential Equations with Applications. Springer, 545p (1999)
6. Fichtenholz, G.: Differential and Integral Calculus. Nauka (1966), 800p
7. Petrivskyi, Y.: Algorithm for drawing a trajectory of weakly magnetic particles in case of illiteration by the method of dry magnetic separation. Petrivskyi, Y., Zubarev, A.: Newsletter of the National University of Water and Environmental Engineering, pp. 374–380 (2016)

Diffusion-Kinetic Model for Curing of Epoxy Polymer

S. V. Rusakov, V. G. Gilev, and A. Yu. Rakhmanov

Abstract A diffusion-kinetic model is presented in the form of a system of partial differential equations of the parabolic type, which allows one to estimate the ablation of the components of the epoxy polymer in the liquid phase under conditions of imbalance in the stoichiometric equilibrium simulating the effect of vacuum. An analysis of the proposed model showed that the boundaries of the mass fraction of the hardener, at which a transition to the gel fraction is possible, are between 10 and 60%, which corresponds to the results of a full-scale experiment. Additionally, using the constructed mathematical model, the effective values of the kinetic parameters were determined at which the estimated time of yield loss is in good agreement with the experimental one.

Keywords Reokinetic · Viscosity · Epoxy resin · Hardening · Numerical modeling

1 Introduction

Epoxy-based composites are widely used due to their high durability properties, good adhesion to different materials, resistance to external factors, and low shrinkage rate. They are also of low weight, which makes them promising for creating large deployable structures in Earth orbit and in outer space. In outer space, such objects are exposed to high vacuum, the flow of charged particles and atomic oxygen, rapid temperature changes. All these factors can significantly affect the polymerization result of the target structure [1, 2]. The cost of conducting full-scale experiments directly in outer space is too expensive, and sometimes an experiment is not possible at all. In this case, the mathematical modeling is almost the only way to research the process.

S. V. Rusakov (✉) · V. G. Gilev · A. Yu. Rakhmanov
Perm State National Research University, 15 Bukirev st., 614990 Perm, Russia
e-mail: rusakov@psu.ru

A. Domoshnitsky et al. (eds.), *Functional Differential Equations and Applications*, Springer Proceedings in Mathematics & Statistics 379,
https://doi.org/10.1007/978-981-16-6297-3_6

To estimate the kinetics of curing of epoxy oligomers, various models are used, a review of which is given in [3]. Reference [4] presents the mathematical model of curing kinetics and viscosity changes during curing of a binder, based on the dimensions theory. As another example, [5–7] describes molecular dynamics methods for exploring the curing process of thermosetting polymers which allow predicting the evolution of the degree of crosslinking depending on the curing time, and the gelation time depending on the molecular structure of the copolymer and curing conditions. Reference [8] a describes mathematical model of the curing process, based on statistical methods of analysis. The model allows predicting the evolution of the degree of crosslinking depending on the curing time and the gelation time depending on the molecular structure of the copolymer.

However, all these models are not applicable for considering the polymerization processes occurring in open space. Reference [9]. In particular, high vacuum leads to the entrainment of part of the components of the mixture, i.e., violation of the stoichiometric balance of the components of the mixture. In this regard, the problem of the contribution of each of the components of the binder in the polymerization process, as well as the research of the dependence of the gelation time and other physical properties of the material on the ratio of the proportions of epoxy resin and hardener is still relevant. The problem of entrainment of the substance during the curing of epoxy in open space was posted in [10]. The mathematical model based on a special choice of initial and boundary conditions for describing the process of ablation of hardener molecules was described in [11]. Reference [12] has presented the investigation of the effect of hardener concentration on the viscosity of the epoxy binder in the initial portion of the polymerization process. It has also shown that the calculation results are in good agreement with experimental data.

This paper focuses on the new model in the form of a system of partial differential equations of the parabolic type, which allows one to estimate the ablation of the components of the epoxy polymer in the liquid phase under conditions of imbalance in the stoichiometric equilibrium simulating the effect of vacuum. Estimates of the model parameters obtained as a result of processing full-scale experiments are presented.

2 Problem Formulation

In this paper, we'll consider an endless layer of epoxy polymer in the liquid phase. In this case, we can restrict ourselves to a one-dimensional formulation of the problem with a spatial coordinate that is directed across the layer (Fig. 1).

As a mathematical model, we consider an initial-boundary value problem of the form

- system of equations

$$\frac{\partial C^{ep}(t,x)}{\partial t} = \frac{\partial}{\partial x}\left(D\left(C^{ep},C^{am}\right)\frac{\partial C^{ep}(t,x)}{\partial x}\right) - K^{ep}\left(C^{ep},C^{am}\right)C^{ep}(t,x)C^{am}(t,x),$$

Fig. 1 Geometric view

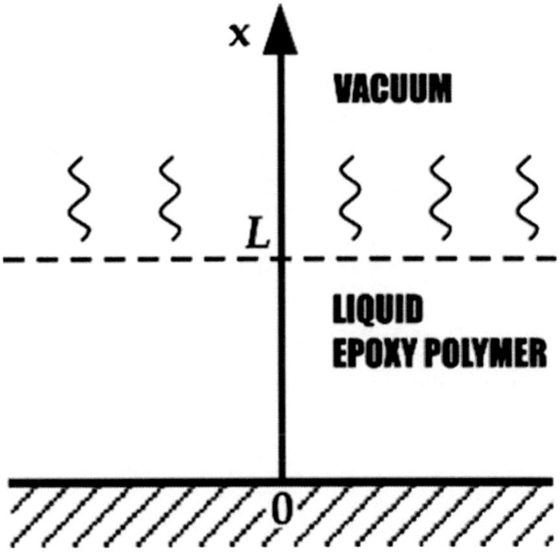

$$\frac{\partial C^{am}(t,x)}{\partial t} = \frac{\partial}{\partial x}\left(D\left(C^{ep},C^{am}\right)\frac{\partial C^{am}(t,x)}{\partial x}\right) - K^{am}\left(C^{ep},C^{am}\right)C^{ep}(t,x)C^{am}(t,x), \quad (1)$$

$t > 0, \quad x \in (0, L(t))$

• nitial conditions

$$C^{ep}(0,x) = C_0^{ep} = const, \quad C^{am}(0,x) = C_0^{am} = const,$$
$$C_0^{ep} + C_0^{am} = 1, \quad x \in [0, L(0)], \quad (2)$$

• boundary conditions

$$D\Big(C^{ep},C^{am}\Big)\frac{\partial C^{ep}(t,0)}{\partial x} = D\Big(C^{ep},C^{am}\Big)\frac{\partial C^{am}(t,0)}{\partial x} = 0,$$
$$D\Big(C^{ep},C^{am}\Big)\frac{\partial C^{ep}(t,L(t))}{\partial x} = -\alpha_0(t)\alpha_{ep}\Big(C^{ep},C^{am}\Big)C^{ep}(t,L(t)),$$
$$D\Big(C^{ep},C^{am}\Big)\frac{\partial C^{am}(t,L(t))}{\partial x} = -\alpha_0(t)\alpha_{am}\Big(C^{ep},C^{am}\Big)C^{am}(t,L(t)) \quad (3)$$

Introduce the following notations:

$C^{ep}(t,x)$, $C^{am}(t,x)$—mass fraction of unreacted (in the liquid phase) epoxy resin and hardener molecules;

$K^{ep}(C^{ep}, C^{am})$, $K^{am}(C^{ep}, C^{am})$—kinetic parameters that determine the speed of the polymerization reaction;

$D(C^{ep}, C^{am})$—diffusion coefficient;

$L(t)$—epoxy polymer layer thickness;

$\alpha_{ep}\left(C^{ep}, C^{am}\right)$, $\alpha_{am}\left(C^{ep}, C^{am}\right)$—ablation factor coefficient;

$\alpha_0(t) = 1 - \exp\left(-\left(\frac{t}{\tau}\right)^2\right)$—(empirical) coefficient determining the "start" of the ablation mechanism, τ—parameter of adjustment.

To determine the diffusion coefficient, we will use the well-known Einstein formula:

$$D\left(C^{ep}, C^{am}\right) = \frac{D_0 T}{\eta\left(C^{ep}, C^{am}\right)}, \quad D_0 = \frac{C_B}{6\pi r} \tag{4}$$

where $C_B = 1.38 \cdot 10^{-23}$ J/K—Boltzmann const, r—effective molecular radius, $\eta(C^{ep}, C^{am})$—dynamic viscosity of epoxy polymer.

We'll determine the ablation coefficients, including work to overcome the energy barrier [11]

$$\alpha_{ep}\left(C^{ep}, C^{am}\right) = \frac{1}{2} \cdot erfc\left(\sqrt{\frac{A_{ep}\eta(C^{ep}, C^{am})}{T}}\right), \quad A_{ep} = \frac{m_{ep}\gamma_{ep}}{C_B},$$

$$\alpha_{am}\left(C^{ep}, C^{am}\right) = \frac{1}{2} \cdot erfc\left(\sqrt{\frac{A_{am}\eta(C^{ep}, C^{am})}{T}}\right), \quad A_{am} = \frac{m_{am}\gamma_{am}}{C_B}, \tag{5}$$

where m_{ep}, m_{am}—mass of molecules, γ_{ep}, γ_{am}—empirical constants.

An epoxide composite is a high molecular compound. Accordingly, values r, m_{ep}, and m_{am} in expressions (4)–(5) could only be considered as some effective values. In this case, full-scale experiments are required to estimate the parameters D_0, A_{ep} and A_{am}. In this model, there are three time-based characteristics that determine the nature of the curing process and

$\left[D/L^2\right]^{-1}$, $\left[\alpha_{ep}/L,\ \alpha_{am}/L\right]^{-1}$, $\left[K_{ep},\ K_{am}\right]^{-1}$. Since all these functions are solution dependent, the ratios of the rates of the processes (diffusion, ablation, and kinetic) can change at different points in time of the polymerization process.

The curing process of the epoxy polymer is characterized by a monotonic increase in viscosity until there is a loss of fluidity: as $t \to t^*$ viscosity $\eta(t) \to \infty$ (t^* is the solidification time). From formula (4.1), it follows that the diffusion coefficient tends to zero as $t \to t^*$, and the system of equations (1) degenerates into a system of distributed dynamic equations.

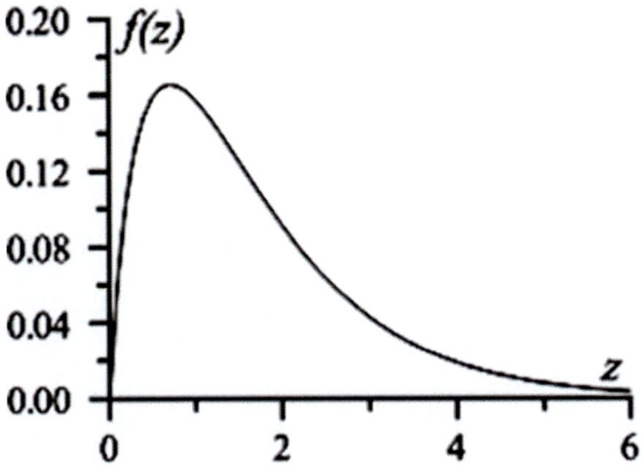

Fig. 2 Dependence of function $f(z)$ on parameter

Viscosity is included in both the diffusion coefficient and the entrainment coefficient. Therefore, we propose to consider the asymptotic behavior of the function $f(z) = z \cdot erfc\left(\sqrt{z}\right) \sim \alpha/D,$, where $z \sim \eta$. It's easy to show that as $z \to \infty$ $f(z) \to 0$ in the corresponding boundary conditions (3), the singularity does not occur. The graphical representation of this function is in Fig. 2. It shows that the ratio of diffusion time to ablation time is not monotonous and depends on the viscosity.

3 Kinetic Model

Let's consider the kinetic component of the system of equations in more detail (1). It is a result of the use of the simplest two component dynamic model of the form

$$\frac{d\varphi^{ep}(t)}{dt} = -K^{ep}\varphi^{ep}(t)\varphi^{am}(t), \quad \varphi^{ep}(0) = 1,$$

$$\frac{d\varphi^{am}(t)}{dt} = -K^{am}\varphi^{ep}(t)\varphi^{am}(t), \quad \varphi^{am}(0) = 1, \tag{6}$$

where $\varphi^{ep}(t),\ \varphi^{am}(t)$—the proportion of unreacted molecules (oligomers) of epoxy resin and hardener, which are associated with the desired functions of equations (1) by simple relations $C^{ep}(t) = C_0^{ep}\varphi^{ep}(t), \quad C^{am}(t) = C_0^{am}\varphi^{am}(t)$.

In the zero approximation, we can assume that $K^{ep},\ K^{am} = const$, then the problem (2.1) has an analytical solution. So, if $K^{am} > K^{ep}$ then

$$\varphi^{am}(t) = \frac{C}{(1+C)\exp(K^{ep}Ct) - 1}, \quad C = \frac{K^{am}}{K^{ep}} - 1,$$

$$\varphi^{ep}(t) = \frac{K^{ep}}{K^{am}}\varphi^{am}(t) + \left(1 - \frac{K^{ep}}{K^{am}}\right); \quad (7)$$

if $K^{ep} > K^{am}$ then

$$\varphi^{ep}(t) = \frac{C}{(1+C)\exp(K^{am}Ct) - 1}, \quad C = \frac{K^{ep}}{K^{am}} - 1, \quad (8)$$

$$\varphi^{am}(t) = \frac{K^{am}}{K^{ep}}\varphi^{ep}(t) + \left(1 - \frac{K^{am}}{K^{ep}}\right).$$

In the case of stoichiometric balance $K^{ep} = K^{am} = K$ we obtain: $\varphi^{ep}(t) = \varphi^{am}(t) = \varphi(t)$, then we get the Cauchy problem for one equation, the solution of which is

$$\varphi(t) = (1 + Kt)^{-1}. \quad (9)$$

4 The Research Subject

The experiments used an epoxy composition of "cold" curing: oligomer-epoxy resin L, hardener—EPH 161 certified for use in structural composite materials for aviation purposes. The mixture that used in experiments was prepared in the weight ratio of epoxy resin to hardener 4:1, recommended by the manufacturer. Measurements of the mass of solutions and their components required to calculate the concentration of the mixture were performed using analytical scales LV-210 having 2nd accuracy class, the absolute measurement error of which is ±0.4 mg. After preparation, the mixture was mixed for 1–2 min with an Electromechanical mixer and an additional 1–2 min in an ultrasonic bath Digital Ultrasonic Cleaner CD 4820 at a frequency of 40 kHz. The last operation also contributes to the degassing of the mixture.

5 Experimental Settings and Procedure

The control of the polymerization process of the binder was carried out by rheological method on a rotary rheometer Physica MCR 501. Used geometry is $\ll cone - plate \gg$. The cone has a diameter of $25mm$ and angle of 1^0. Such geometry ensures uniformity of the shear rate gradient in the measuring gap. A special temperature device $H - PTD200$ based on the Peltier effect was used to maintain and change the temperature regime. During the experiment, $0.07ml$ of solution was placed on the working surface of the rheometer plate. The thickness of the solution layer along its outer radius was $0.04mm$ for rapid establishment of the working temperature in the sample even under the conditions of heat generation due to the polymerization reaction. All experiments were carried out under the conditions of shear deformation of the mixture according to the harmonic law with a frequency of $1\,Hz$ in the deformation control mode, which allows to evaluate not only viscous, but also viscoelastic characteristics of the samples. In order to minimize heat generation in the sample due to shear flow, all measurements were carried out in a discrete mode of temperature change. The measurement time of each experimental point was no more than 15 s. The interval between measurements (standby mode) was in the range from 5 to 30 min. The process continued until the reaction mixture began to enforce a strong resistance to shear deformation.

6 Result and Discussion

The results of changes in the complex viscosity of the polymerized adhesive composition at a temperature of 40 °C in the case of stoichiometric balance of the components is in Fig. 3. It can be seen that over a period of time the viscosity changes slightly. Then it grows intensive by several orders of magnitude.

In [13], in order to approximate the dependence of the viscosity of the polymerizing composition, it was proposed to use the empirical Chong formula [14]

$$\eta(t) = \eta_0 \left(1 + \frac{a\psi(t)}{1 - \psi(t)/\psi_m}\right)^2, \tag{10}$$

where $\psi(t)$—the mass fraction of epoxy resin and hardener molecules reacting which in the absence of vacuum ablation creating spatial concentration inhomogeneity. $\psi(t)$ is determined by a simple relation

$$\psi(t) = 1 - C_0^{ep}\varphi^{ep}(t) - C_0^{am}\varphi^{ep}(t), \tag{11}$$

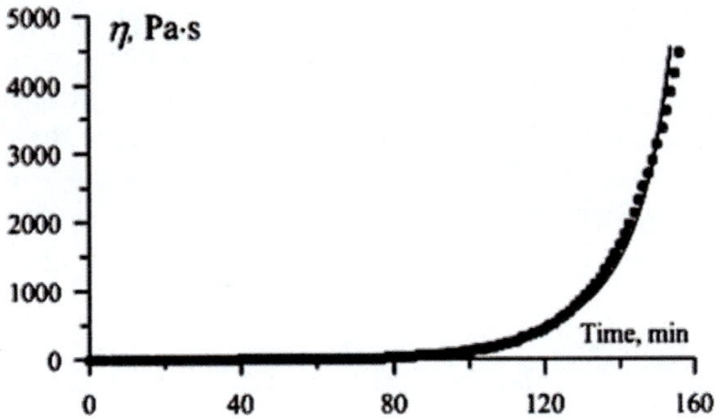

Fig. 3 Changes in the viscosity of the adhesive composition in the polymerization process

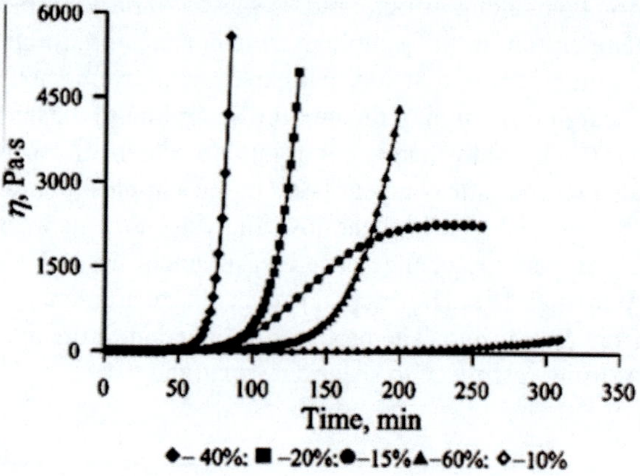

Fig. 4 Changes in the viscosity of the binder during polymerization depending on the mass concentration of the hardener. $T = 45\,°\mathrm{C}$

where η_0—viscosity value at the initial time, ψ_m—the part of oligomers (epoxy resin and hardener molecules) that have entered into a polymerization reaction, in which the transition from the liquid phase to the gel-fraction phase. The parameter a is empirical and is used to approximate the experimental data. In Fig. 3, the solid curve shows the dependence (10), where $a = 16$ and $\psi_m = 0.51$. The analytical solution of Eq. (9) with $K = 0.0059$ was used to determine the function $\psi(t)$ according to (11).

During vacuum ablation near the surface of the polymer layer, the stoichiometric balance may be disturbed, which will affect the value of the kinetic parameters K^{ep} and K^{am}.

To evaluate these changes, a series of experiments were performed with various mass fraction of hardener. Reference [4] shows an example of changing the viscosity of the polymerized adhesive composition depending on the concentration of the hardener. Stoichiometric equilibrium corresponds to a hardener concentration of 20%. Experiments have shown that at hardener (Cam) concentrations of less than 10% and more than 60%, the mixture does not solidify (Fig. 4).

The change in viscosity in the curing process of epoxy resins is complex, but in most cases, the purely practical interest is not the entire range of viscosity changes, but only the critical point of loss of fluidity: solidification time t^*. The used geometry of the rheometer does not allow to register the change of rheological properties from the beginning of the reaction to its almost complete end. A well-known method for determining the hardening time is to determine the maximum achievable value of the viscosity of the material and to construct the inverse viscosity dependence $1/\eta$ at the final stages of curing [15], which, as a rule, is well approximated by a straight line (see, for example, Fig. 5). The intersection of this line with the abscissa axis determines the moment of reaching infinite viscosity. The results of this process are presented in Fig. 6.

Based on the chemical composition of the epoxy composition it is possible to show the validity of the ratio

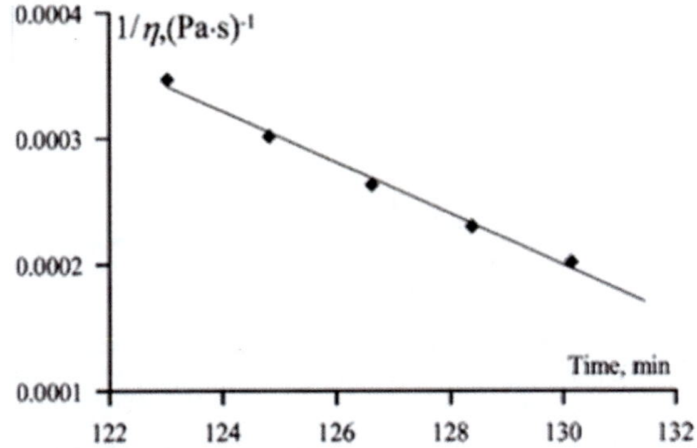

Fig. 5 Determination of solidification time of samples. $T = 45\,^{\circ}\mathrm{C}$. The hardener concentration is 20%

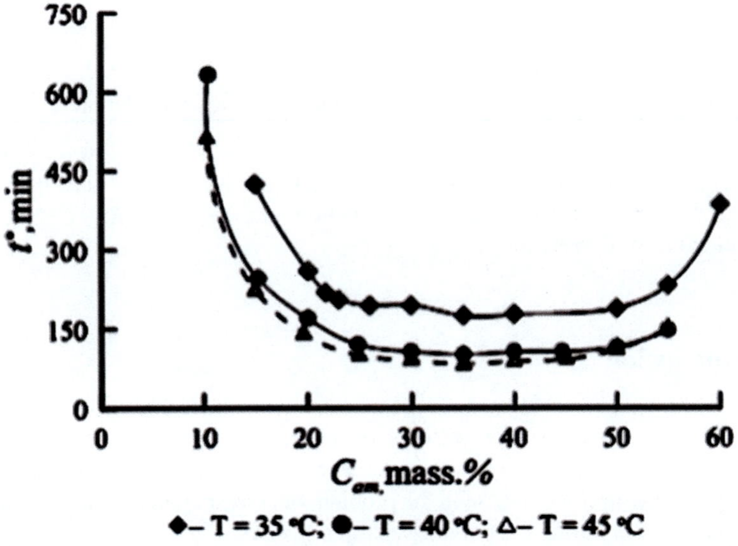

Fig. 6 Change of curing time t^* depending on mass concentration of hardener and temperature

$$\frac{K^{ep}}{K^{am}} = \frac{4C_0^{am}}{C_0^{ep}} \tag{12}$$

Analysis of the proposed model (6), given the fact that stoichiometric balance is determined by the ratio of 4:1 showed that the boundaries of the mass fraction of curing agent in which the transition in gel fraction is possible are in range of 10% and 60%, which corresponds to the results of the experiment (see Fig. 6). In addition, with using the constructed mathematical model, it is possible to determine the effective value of the kinetic parameters, at which the estimated time of flow loss is in good

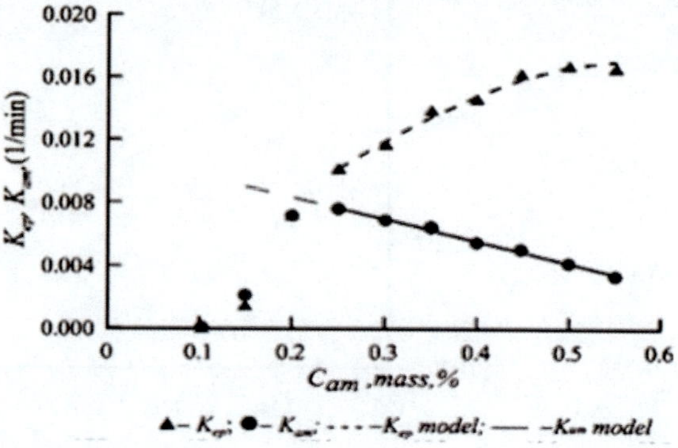

Fig. 7 Values of kinetic parameters of the model (6) depending on the mass concentration of the hardener at temperature $T = 40\,^{\circ}\mathrm{C}$

agreement with the experimental one. In this case, the reasoning was based on the following:

- analytical solutions (7)–(9)
- the fact that at time t the value of the function defined by the relation (11) is equal to $\psi(t) \approx 0.5$

The results of these calculations are shown in Fig. 7. It can be seen that the dependence $K^{am}(C^{am})$ is a piecewise linear function and the values $K^{ep}(C^{am})$ can be determined from the relation (12).

7 Conclusion

The diffusion-kinetic model presented in the paper and the parameters included in it can be identified accurately enough using laboratory experiments, which allows to hope for the possibility of its use in the problem of assessing the vacuum ablation of an epoxy polymer during the polymerization process.

Acknowledgements The study was carried out with the financial support of the Government of the Perm region in the framework of the research project No. C-26/793 and the RFBR grant No. 17-41-590649.

References

1. Kondiurin, A.V., Nechitailo, G.S.: Composite material for inflatable structures, photocurable in conditions of orbital space flight. Osmonautics Rocket Eng. **3**(56), 182–190 (2009)
2. Pestrenin, V.M., Pestrenina, I.V., Rusakov, S.V., Kondyurin, A.V.: Curing of large prepreg shell in solar synchronous low earth orbit: precession flight regimes. Acta Astronautica **151**, 242–247 (2018)

3. Francis, B.: Cure kinetics of epoxy thermoplastic blends. Handbook of Epoxy Blends, pp. 649–674 (2017). 23 June 2017
4. Studentsov, V.N.: Kinetika otverzhdeniya termoreaktivnyh smol i izmenenie ih vyazkosti v processe otverzhdeniya. Plasticheskie massy, pp. 24–26 (2019)
5. Unger, R., Braun, U., Fankhanel, J., Daum, B., Arash, B., Rolfes, R.: Molecular modelling of epoxy resin crosslinking experimentally validated by near-infrared spectroscopy. Comput. Mater. Sci. **161**, 223–235 (2019)
6. Li, C., Strachan, A.: Molecular scale simulations on thermoset polymers: a review. J. Poly. Sci. Part B: Poly. Phys. **53**(2), 103–122 (2015)
7. Okabe, T., Oya, Y., Tanabe, K., Kikugawa, G., Yoshioka, K.: Molecular dynamics simulation of crosslinked epoxy resins: curing and mechanical properties. Eur. Poly. J. **80**, 78–88 (2016)
8. Chen, N., Lee, N., Bortolato, S.A., Martino, D.M.: Experimental studies and mathematical modeling of the curing reaction of bioinspired copolymers. Green Chem. Lett. Rev. **11**(4), 387–398 (2018)
9. Kondyurin, A., Lauke, B., Richter, E.: Polymerization processes of epoxy matrix composites under simulated free space conditions. High Perform. Poly. **16**, 163–175 (2004)
10. Rusakov, S.V.: The effect of ablation of material on the process of cure of epoxy resin in the conditions of open space. In: Proc. Nozzl. Jets Trans. X Int. Conf. on Nonequilibr. (NPNJ' 2014), pp. 567–569. MAI, Alushta (2014)
11. Svistkov, A.L., Komar, L.A., Kondyurin, A.V., Mal'tsev, M.S., Terpugov, V.N.: Isparenie molekul otverditelia v reaktsii polimerizatsii epoksidnoi smoly// Materialy XI Mezhdunarodnoj konferencii po neravnovesnym processam v soplah i struyah (NPNJ–2016), pp. 385–387, MAI (2016)
12. Gilev, V.G., Kondyurin, A.V., Rusakov, S.V.: Investigation of epoxy matrix viscosity in the initial stage of its formation. Mater. Sci. Eng. **208**, 012014 (2017). https://doi.org/10.1088/1757-899X/208/1/012014
13. Kondyurin, A.V., Komar, L.A., Svistkov, A.L.: Simulation of the curing kinetics of a composite material based on epoxy binder. Mekhanika kompozitsionnykh materialov i konstruktsii, vol. 16, No. 4, pp. 597–611 (2010), in Russian
14. Chong, J.S., Christiansen, E.B., Baer, A.D.: Rheology of concentrated suspension. J. Appl. Polym. Sci. **15**, 2007–2021 (1971). https://doi.org/10.1002/app.1971.070150818
15. Malkin, A.J., Kulichikhin, S.G.: Rheokinetics of curing. Adv. Polym. Sci. **101**, 217–257 (1991)

Part II
Dynamics of Models in Biology, Medicine and Ecology

Modeling of Control of the Immune Response in the Acute Form of an Infectious Disease Under Conditions of Uncertainty

M. V. Chirkov and S. V. Rusakov

Abstract The paper considers a numerical solution of the problem of discrete control of the immune response in an infectious disease under conditions of uncertainty. The immune processes are described by ordinary differential equations with a retarded argument. Conditions of uncertainty imply that values of the parameters of the model are unknown and their estimates are adjusted by new clinical and laboratory data. We use an algorithm that allows one, within the framework of the mathematical model of infectious disease, to construct the control function and at the same time to identify parameters. We deal with the control of the immune response in the acute form of an infectious disease. Immunotherapy consisting in the injection of donor antibodies is chosen as a controlling factor. In doing so, it is shown that immunotherapy is an effective treatment for the acute form of an infectious disease. The presented results are based on simulation of experimental data.

Keywords Mathematical model of immune response · Discrete control · Immunotherapy

1 Introduction

Mathematical models of the immune response in infectious diseases are applied to study the dynamics of the immune defense of the organism against viral and bacterial infections. As a rule, these models are represented by nonlinear systems of ordinary differential equations [1, 5, 7–10, 16] that contain a large number of parameters. The values of the parameters of models characterize the properties of the immune system and antigens. Using estimates of parameters calculated on the basis of laboratory data,

M. V. Chirkov · S. V. Rusakov (✉)
Perm State National Research University, 15 Bukirev st., 614990 Perm, Russia
e-mail: rusakov@psu.ru

M. V. Chirkov
e-mail: chirkov@psu.ru

A. Domoshnitsky et al. (eds.), *Functional Differential Equations and Applications*, Springer Proceedings in Mathematics & Statistics 379,
https://doi.org/10.1007/978-981-16-6297-3_7

one can analyze the immune response in a patient and develop the most appropriate treatment regimen.

The solving of these problems is complicated by the fact that the conventional approach allows one to obtain estimates of parameters only at the end of the disease when the prognosis and recommendations on the choice of treatment lose their relevance. Therefore, it is of interest to develop methods that enable one to construct a control under conditions of uncertainty where values of parameters are unknown but the range they belong to is known and their estimates are adjusted by new clinical and laboratory data.

2 Basic Mathematical Model of Infectious Disease

The main defense mechanism that eliminates antigens from the organism is an immune reaction targeted only against a specific antigen causing a given disease. After penetrating into an organism, antigens begin to proliferate in the cells of a target organ, which leads to the damage of an organ. An immune response involves the formation of antibodies that can neutralize antigens. The generation of antibodies is preceded by stimulation of the immune system, which consists of the formation of plasma cells that produce antibodies. An antibody binds an antigen and the outcome of the disease depends on the struggle between them.

The immune response described above is reflected in the basic mathematical model of an infectious disease proposed by Marchuk [7]. Using the model, one can predict the course of the disease and its outcome and the introduction of control functions allows one to give recommendations on the choice of treatment.

The model describes the dynamics of the following variables: $\nu(t)$, $s(t)$, and $f(t)$ are the relative concentrations of antigens, plasma cells, and antibodies, respectively, and $m(t)$ is the proportion of target organ cells destroyed by antigens.

Including the control [3, 4], the basic model of an infectious disease in the nondimensional form can be represented by a system of ordinary differential equations

$$\begin{aligned}
\frac{d\nu}{dt} &= a_1\nu - a_2 f\nu, \\
\frac{ds}{dt} &= \xi(m)a_3 f(t-\tau)\nu(t-\tau) - a_5(s-1), \\
\frac{df}{dt} &= a_4(s-f) - a_8 f\nu + u, \\
\frac{dm}{dt} &= a_6\nu - a_7 m
\end{aligned} \tag{1}$$

with initial conditions for $t \in [-\tau, 0]$

$$\nu(t) = \nu_0\theta(t), \quad s(t) = 1, \quad f(t) = 1, \quad m(t) = 0, \tag{2}$$

where $\theta(t)$ is the Heaviside function defined by the formula

$$\theta(t) = \begin{cases} 1 & for\ t \geq 0, \\ 0 & for\ t < 0. \end{cases} \tag{3}$$

The control function $u = u(t) \in U$ describes the injection of donor antibodies from the external environment (immunotherapy) and $\xi(m)$ characterizes the malfunction of the immune system due to considerable damage of an organ

$$\xi(m) = \begin{cases} 1, & 0 \leq m < m^*, \\ \dfrac{m-1}{m^*-1}, & m^* \leq m \leq 1, \end{cases} \tag{4}$$

where m^* is the maximum proportion of cells destroyed by antigens when the normal functioning of the immune system is still possible.

Values of the parameters of the model are unknown but there is a range of admissible values

$$a_i \in [a_i^-, a_i^+], \quad i = \overline{1, L}, \tag{5}$$

where L is the number of parameters ($L = 8$ in the basic model of an infectious disease).

Model (1) describes general regularities common to all infectious diseases. The infectious disease is assumed to be a conflict between pathogenic multiplying antigens and the immune system of an organism. The study of the basic model of infectious disease resulted in obtaining qualitatively different types of solution, which were interpreted as forms of the disease: subclinical, acute, chronic, and lethal.

The form of the solution is uniquely determined by initial conditions and parameter values, which are referred to as immunological status of an organism. In this regard, it is necessary to solve the problem of control under conditions of uncertainty where estimates of parameters are adjusted during the construction of a control.

3 Algorithm of Control Under Conditions of Uncertainty

We assume that the laboratory data can be obtained at certain time moments corresponding to the grid nodes

$$\prod = \{t_i : t_i = i\Delta t, \quad i = \overline{1, N}, \quad \Delta t = T/N\}. \tag{6}$$

Thus, the input data are discrete. The control function $u = u(t)$ that characterizes the rate of injection of donor antibodies is chosen from the set

$$U = \{u(t) : u(t) = u_{i-1} \in [0, b],\ t \in [t_{i-1}, t_i),\ i = \overline{1, N},\ u(T) = u_{N-1}\}, \quad (7)$$

where the parameter $b > 0$ denotes the physiologically admissible doses of medicaments.

To construct a control function with uncertainty, we use the algorithm proposed in [13–15]. The algorithm for the construction of a program of treatment is as follows:

On the set of admissible values of parameters, the K sets of parameters are randomly generated

$$\alpha^{(k)} \in \Theta = \left\{\alpha : a_{ij} = a_i^- + jh_i,\ j = \overline{0, M_i},\ h_i = \frac{a_i^+ - a_i^-}{M_i},\ i = \overline{1, L}\right\},\ k = \overline{1, K}. \quad (8)$$

For $t \in \prod$, we define the admissible parameter sets that satisfy the following condition:

$$\left|\nu(t_i, \alpha^{(k)}) - \nu^{\exp}(t_i)\right| < \varepsilon,\ \ i = \overline{0, N},\ \ k = \overline{1, K_i}, \quad (9)$$

where $\nu^{\exp}(t_i)$ is the laboratory data obtained on the basis of medical analyses; ε is the value of acceptable deviation of the calculated values of the antigen concentration from laboratory data, K_i is the number of parameter sets at time t_i. The estimate of parameters is defined by the formula

$$\overline{a}_j^{(i)} = \frac{\sum_{k=1}^{J_i} a_j^{(k)}}{J_i},\quad j = \overline{1, L},\quad i = \overline{0, N}, \quad (10)$$

where J_i represents the number of admissible parameter sets at time t_i; and $J_i \le K_i$, $i = \overline{0, N}$, $K_i = J_{i-1}, i = \overline{1, N}$, $K_0 = K$, where K is the initial number of parameter sets; $J_i = K_i - H_i, i = \overline{0, N}$, where H_i is the number of inadmissible parameter sets at time t_i.

To construct a control function, we apply the algorithm proposed in [11, 12]. The idea of the algorithm is as follows: the dynamics of antigens must be put into a necessary level corresponding to a certain solution of the basic mathematical model of an infectious disease. This solution is considered to be the support solution. The antigen concentration values obtained from the support solution are given on the grid (6)

$$\nu^*(t_i),\quad i = \overline{1, N}. \quad (11)$$

If the predicted level of the antigen concentration does not coincide with the support value, then as a control u_i, we choose the value that leads the solution curve of the antigen concentration to a necessary level.

4 Results of Computational Experiments

The support solution is determined by solving the system

$$\begin{aligned}\frac{d\nu}{dt} &= a_1\nu - a_2 f\nu,\\ \frac{ds}{dt} &= \xi(m)a_3 f(t-\tau)\nu(t-\tau) - a_5(s-1),\\ \frac{df}{dt} &= a_4(s-f) - a_8 f\nu - cu,\\ \frac{dm}{dt} &= a_6\nu - a_7 m,\\ \frac{du}{dt} &= f - 1 - ku\end{aligned} \tag{12}$$

with initial conditions for $t \in [-\tau, 0]$

$$\nu(t) = \nu_0\Theta(t), \quad s(t) = 1, \quad f(t) = 1, \quad m(t) = 0, \quad u(t) = 0 \tag{13}$$

with $c = 1, k = 4$. This variation of the basic model is proposed in [2, 6]. If $a_1 < a_2,\ c > 0,\ k > 0$ then system (12) is exponentially stable. In this case, the control provides stability of the state of a healthy organism. The status of a healthy body is determined by the following stationary solution of the basic model:

$$\nu = 0, \ s = 1, \ f = 1, \ m = 0.$$

To construct a control function under conditions of uncertainty, we utilize an imitation of clinical and laboratory indicators. The experimental values in criterion (9) are determined by the solving of model (12)–(13) for a specific set of parameter values. Further, considering that the parameter values are unknown, we construct a control and simultaneously correct values of the parameters for problem (1)–(5) using algorithm (6)–(11). Calculations are carried out with parameters corresponding to the acute form of a disease.

The various solution curves correspond to different values of parameters of the model. The model parameters are identified in terms of values of concentration of antigens. It follows that the solution curves of the chosen characteristic must lie in a certain neighborhood of experimental values. Thus, we obtain a range where the solution curves of concentration of antigens must be. In Fig. 1, the bounds of this range are depicted by the dashed curves. If the solution curve for some set of parameters goes beyond the range at a point of an interval of integration, further computations with this set of parameters are not carried out. Figure 1 shows the possible solution curves going beyond the allowable ranges.

Figure 2a–d illustrate the dynamics of the immune response during the natural course of the disease (solid curves) and in the case of the obtained program of

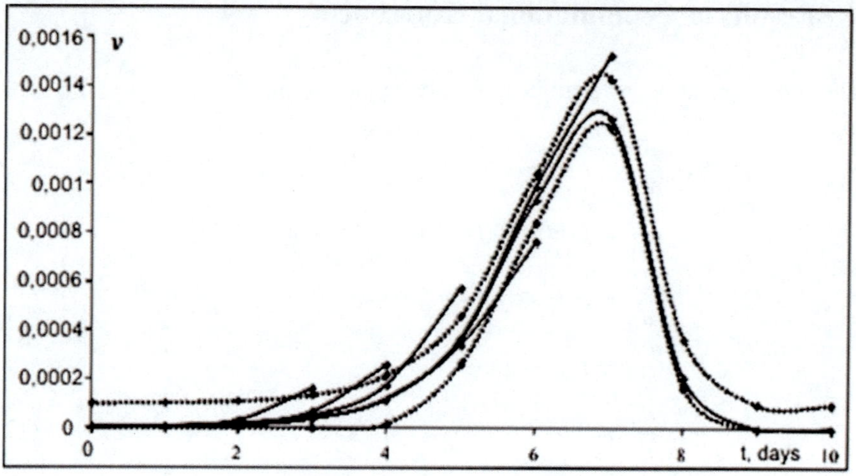

Fig. 1 Solution curves of antigen concentration

treatment (dashed curves). The following values were used: $m^* = 0,1$; $b = 5$; $\tau = 0,5$; $\nu_0 = 10^{-6}$; $\varepsilon = 2,5 \cdot 10^{-4}$. The acute form of the disease is characterized by a rapid increase of antigen concentration in an organism, by a strong and effective immune response, and by a rapid decrease of antigen quantity down to the values tending to zero. This situation is understood as recovery. The control function is shown in Fig. 2e. The treatment program consists of the increasing injection of donor antibodies.

The obtained estimates of parameters and the exact values are presented in Table 1. The bounds of the range of admissible values and the step of grid Θ are determined for each parameter. The average error of the estimates of parameters is 1.93%.

Thus, the algorithm proposed makes it possible to construct the treatment program and to estimate the immunological status of the organism, i.e., values of the model parameters.

5 Conclusion

Results of numerical experiments show that immunotherapy is an effective treatment in the acute form of a disease. Using the considered algorithm, one can construct the control of the immune response under conditions of uncertainty correcting the estimates of parameters by using the new clinical and laboratory data.

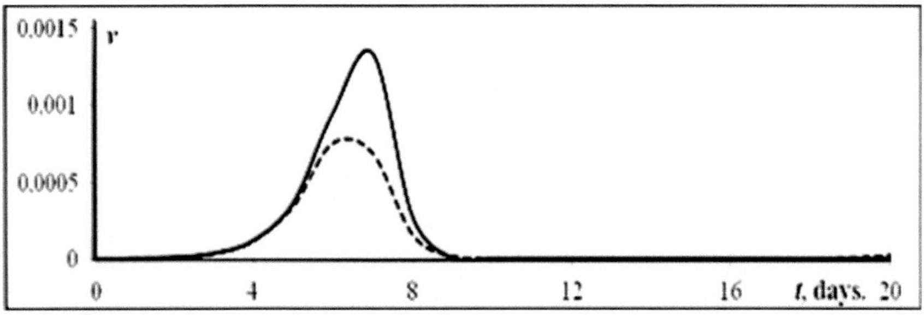

(a)

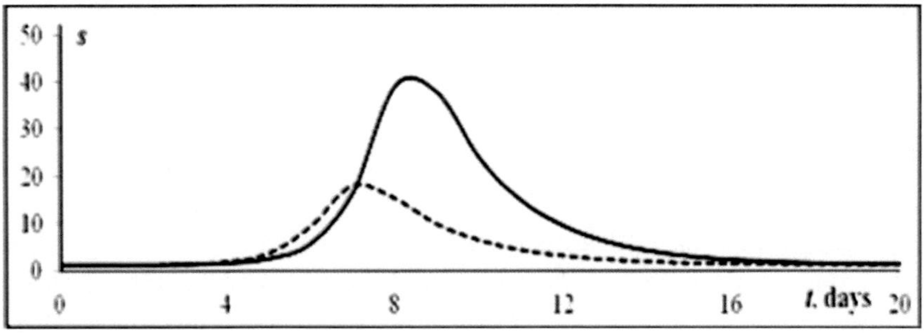

(b)

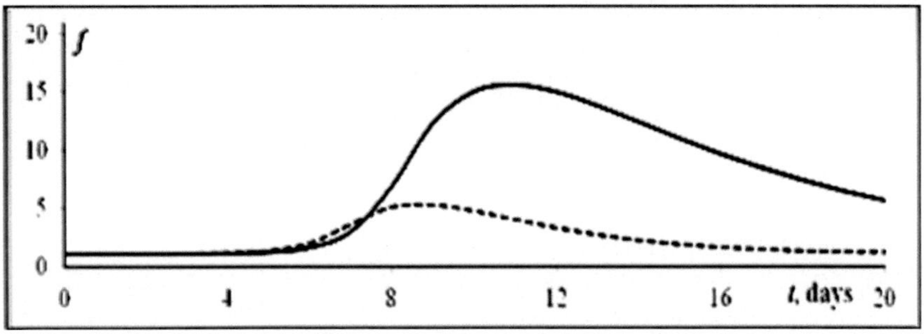

(c)

Fig. 2 Dynamics of the immune response: **a** antigen, **b** plasma cells, **c** antibodies, **d** proportion of destroyed cells, **e** control function

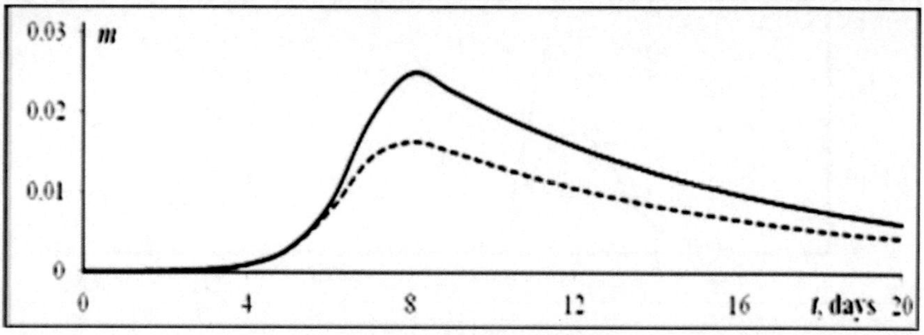

(c)

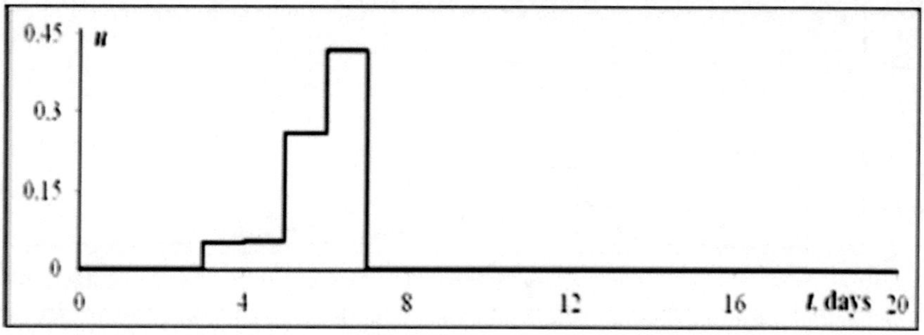

(d)

Fig. 2 (continued)

Table 1 Parameters of the model

Parameter	a_i^-	a_i^+	h_i	Estimate	Exact value	Error,%
a_1	1,75	2,25	0,1	2,024	2,000	1,20
a_2	0,55	1,05	0,1	0,799	0,800	0,13
a_3	9550	10550	100	10033	10000	0,33
a_4	0,145	0,195	0,01	0,168	0,170	0,20
a_5	0,25	0,75	0,1	0,513	0,500	2,60
a_6	7,5	12,5	1	10,214	10,000	2,14
a_7	0,095	0,145	0,01	0,124	0,120	3,33
a_8	5,5	10,5	1	8,443	8,000	5,54

References

1. Belyh, L.N.: Analiz matematicheskikh modeley v immunologii. Nauka, Moscow (1988).(in Russian)
2. Bershadsky, M., Chirkov, M., Domoshnitsky, A., Rusakov, S., Volinsky, I.: Distributed control and the Lyapunov exponents in the model of infectious diseases. Complexity (2019). https://doi.org/10.1155/2019/5234854
3. Bolodurina, I.P., Lugovskova, Y.P.: Mathematical model of management immune system. Rev. Appl. Ind. Math. **15**(6), 1043–1044 (2008) (in Russian)
4. Bolodurina, I.P., Lugovskova, Y.P.: Optimum control of human immune reactions. Problemy upravleniya. **5**, 44–52 (2009). (in Russian)
5. Dasgupta, D.: Artificial Immune Systems and Their Applications. Springer, Berlin (1999)
6. Domoshnitsky, A., Volinsky, I., Bershadsky, M.: Around the model of infectious disease: the Cauchy matrix and its properties. Symmetry (2019). https://doi.org/10.3390/sym11081016
7. Marchuk, G.I.: Matematicheskiye modeli v immunologii. Nauka, Moscow (1980).(in Russian)
8. Marchuk, G.I.: Matematicheskiye modeli v immunologii. Vychislitelnyye metody i eksperimenty, Nauka, Moscow (1991).(in Russian)
9. Marchuk, G.I.: Mathematical Modeling of Immune Response in Infectious Diseases. Springer Science & Business Media, Dordrecht (2013)
10. Pogozhev, I.B.: Primenenie matematicheskikh modeley zabolevaniy v klinicheskoy praktike (Using of the mathematical models of deseases in clinical practice). Nauka, Moscow (1988).(in Russian)
11. Rusakov, S.V., Chirkov, M.V.: Discrete control in the simplest mathematical model of infectious disease. Vestnik Permskogo universiteta. Matematika. Mekhanika. Informatika. **4**(8), 59–63 (2011) (in Russian)
12. Rusakov, S.V., Chirkov, M.V.: Mathematical model of immunotherapy on the dynamics of immune response. Problemy upravleniya. **6**, 45–50 (2012). (in Russian)
13. Rusakov, S.V., Chirkov, M.V.: Parameter identification and control in mathematical models of immune response. Russ. J. Biomech. **18**(2), 226–235 (2014)
14. Rusakov, S.V., Chirkov, M.V.: Parameter identification in the mathematical model of immunotherapy effect on the dynamics of the immune response. Vestnik Permskogo universiteta. Matematika. Mekhanika. Informatika. **1**(24), 46–51 (2014) (in Russian)
15. Rusakov, S.V., Chirkov, M.V.: Discrete control of a dynamical system with delay under conditions of uncertainty. J. Math. Sci. **230**(5), 762–765 (2018)
16. Zuev, S.M.: Statisticheskoe otsenivanie parametrov matematicheskikh modeley zabolevaniy [Statistical rejection of the parameters of the mathematical models of deseases]. Nauka, Moscow (1988).(in Russian)

Some Problems of Mathematical Modeling of Radiophysical Sounding Signals

Alexey Kolchev and Ivan Egoshin

Abstract The greatest attention in radiophysical sounding is paid to determining parameters of known phenomena. Therefore, the main task in sounding signal processing is to detect known signals and evaluate their parameters. However, an important scientific problem is the problem of the discovery of still unknown phenomena. In this paper, we propose a mathematical model of the received signal in the form of a mixture of two probability distributions to detect unknown signals. The random process that describes the signal is assumed to be substantially unsteady (different sections of the process have different probability distributions). Parameters in the system of Kolmogorov differential equations (process intensity) randomly depend on time under these conditions. It is shown that the intensity in the proposed model of the mixture does not depend on the fraction of samples of the selected component. Machine learning methods to detect sounding signals and methods for detecting outliers of random processes to select signal samples are proposed. The developed methods are used in an ionosphere sounding equipment with a chirp signal. The equipment kits were provided to various scientific institutions of Russia and also placed at geophysical stations along the Northern Sea Route, Russia. These methods can be used to extract arbitrary signals with their similar statistical characteristics.

Keywords Modeling · Signal processing · Hazard function

1 Introduction

A standard task in radiophysical research is a task of evaluating signal parameters when sounding known processes or phenomena. In this case, the shape of received signals is known, and their parameters are unknown. The research task of radio-

A. Kolchev (✉)
Kazan Federal University, Kazan, Russia
e-mail: kolchevaa@mail.com

I. Egoshin
Mari State University, Yoshkar-Ola, Russia
e-mail: jungl91@mail.com

A. Domoshnitsky et al. (eds.), *Functional Differential Equations and Applications*, Springer Proceedings in Mathematics & Statistics 379,
https://doi.org/10.1007/978-981-16-6297-3_8

physical sounding consists in detecting signals of an unknown form from unknown processes (since the phenomenon has not yet been discovered, no one has already observed similar signals) or from complex and non-stationary processes (for example, round-the-world HF signals of ionospheric propagation, signals of angular ionospheric scattering [1, 2]).

The purpose of this work is to develop methods for detecting and extracting unknown signals during radiophysical sounding.

2 Chirp Sounding of the Ionosphere

In the paper, signals of the oblique sounding of the ionosphere are considered as radiophysical signals. Figure 1 shows a diagram of such sounding. A transmitter continuously emits a chirp sounding signal. Typically, the frequency tuning speed is 100–500 kHz/s, and the sounding range is from 3 to 30 MHz [3, 4]. The transmitted signal is reflected from the ionosphere and enters the receiving antenna. At the same time, signals of all radio equipments operating in the range of 3–30 MHz also get to the receiving antenna.

Figure 2 shows a spectrogram of a signal at the input of the receiver. The time is plotted on the vertical axis and the sounding frequency on the horizontal axis. The chirp sounding signal has the form of an oblique line, and the signals of extraneous radio equipment have the form of vertical lines.

Figure 2 shows that the sounding signal crosses signals from other radio equipment. These can be signals of various types: speech, music and discrete signals. The signals are completely different and unknown to the receiving side.

The signal is processed by compression in the frequency domain in the receiver. This method consists in the fact that the received signal is multiplied by a local oscillator signal, and then an element-wise spectral analysis of the difference frequency signal is performed [5].

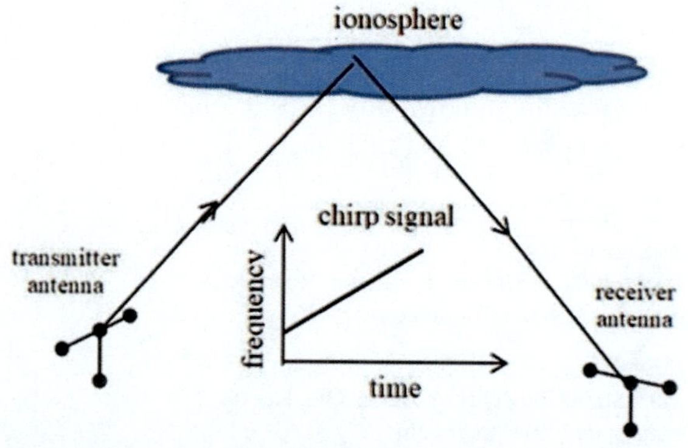

Fig. 1 Diagram of ionosphere chirp sounding

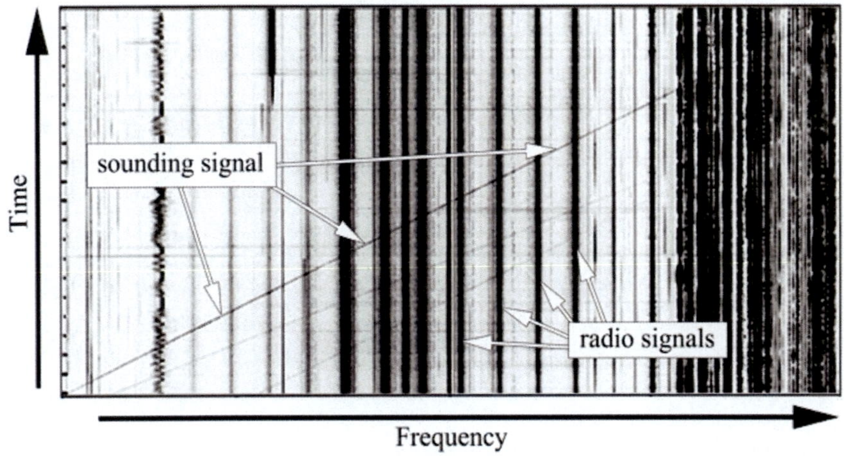

Fig. 2 Signal at the input of the receiver

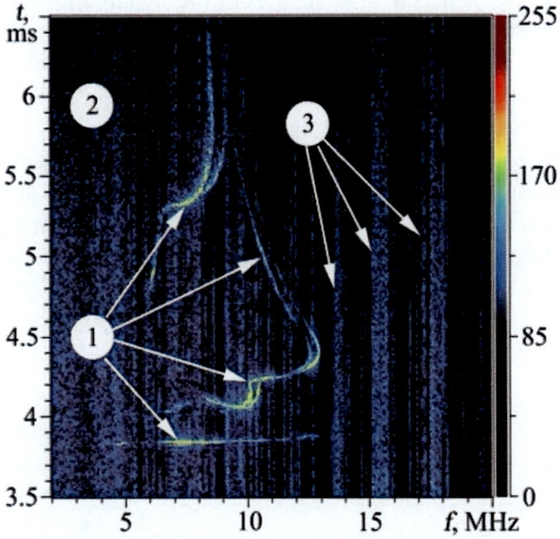

Fig. 3 The ionogram obtained on Gorkovskaya (60.27°N, 29.38°E)—Yoshkar-Ola (56.62°N, 47.87°E) path, September 13, 2013, 08:48:00 UTC. The main elements of the ionogram image: 1—tracks of signal propagation modes; 2—background noise; 3—spectrally lumped (station) interference

The result of the ionosphere sounding is an ionogram (see Fig. 3), which characterizes the dependence of the amplitude of the radio signal A from the sounding frequency f and the group delay time τ. The amplitude of the signal corresponds to a gradation of brightness in the image.

Spectrally lumped noise is located in separate columns of the ionogram. The power of fluctuation noise varies with frequency.

Figure 4a shows an example of a 4-second signal fragment at the output of the ionosonde receiver with a chirp signal, which is divided into four parts corresponding to the individual signal processing elements. Figure 4b shows corresponding

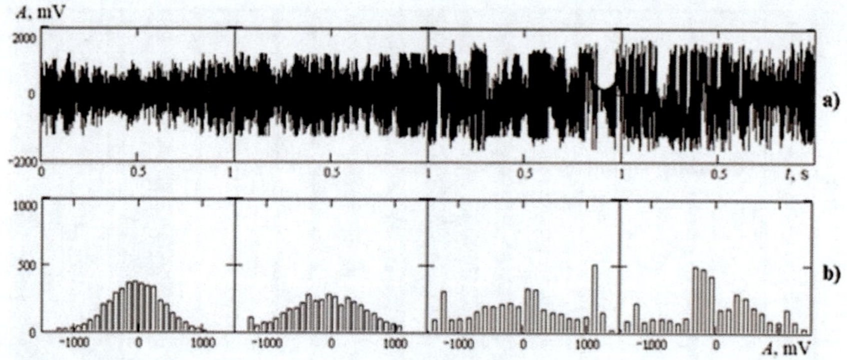

Fig. 4 Elements of the signal (**a**) at the output of the chirp receiver of the ionosonde and their corresponding histograms (**b**)

histograms of the samples distribution of this signal. Figure 4 shows that not only do the distribution parameters of the signal samples vary from element to element, but the distribution laws themselves change. Thus, the signal is processed under conditions of a priori nonparametric uncertainty [6].

During the build of the ionogram, the processing of the received signal is done element-wise and each signal element has its own central frequency f_c and occupies its own frequency band, different from other elements (the results of processing the signal element correspond to a separate column on the ionogram). Therefore, the spectrally lumped noise (label 3 in Fig. 3), the characteristics of the received signals (label 1 in Fig. 3), and the fluctuation noise power (label 2 in Fig. 3) will change from one element to another in the frequency band of the processed signal. This causes a change in the statistical characteristics of the signal at the output of the ionosonde receiver.

3 Statistical Signal Model

Processing in the receiver by some transformations (filtering, decomposition over a certain basis, etc.) leads to the fact that the detected signals have significantly different values of a certain physical parameter.

Since, in modern devices, signals at the final stage of processing are always presented in digital form, the problem of signal detection at the output of a device can be formulated as follows.

Consider a set of n numbers $x_1, x_2, ..., x_n$ that are results of some observations (we assume that digital samples are uncorrelated and independent). If the signal of propagation modes is absent, then all samples $x_1, x_2, ..., x_n$ are considered as a realization of independent and identically distributed random variables $X_1, X_2, ..., X_n$ with cumulative distribution function $F(x)$ (or as n realizations of a random variable

X). If there are samples in the set that are associated with the signal, then we will suppose that the random variables $X_1, X_2, \ldots, X_n$ are also independent, the samples $X_1, X_2, \ldots, X_m$ $(m < n)$ have the distribution $F_1(x)$, while the samples $X_{m+1}, \ldots, X_n$ have a distribution $F_2(x)$ "shifted to the right" relative to $F_1(x)$ (or there are $(n - m)$ realizations of a random variable Y).

In this formulation, the problem of detecting the presence of samples of the signal in the set becomes the problem of determining whether the set is obtained from samples of the random variable X or from samples of the random variable Z which is a two-component mixture of the random variables X and Y.

Moreover, if $f_1(x)$ is a probability density function of the random variable X, and $f_2(x)$ is a probability density function of the random variable Y, then a probability density function of the random variable Z can be written as $f_Z(x) = (1 - h) \cdot f_1(x) + h \cdot f_2(x)$, where $h = (n - m)/n$ is a fraction of samples of the random variable Y in the mixture.

The subsequent simplification of the model is linked to the characteristic of the data structure: we assume that the samples are non-negative and the values of the selected samples are not less than those of the rest of the samples.

Under these assumptions, the distribution density function $f_z(x)$ and the distribution function $F_z(x)$ of the resulting random variable Z can be written as follows:

$$f_Z(x) = \begin{cases} (1-h) f_1(x), 0 \le x < x_a, \\ h \cdot f_2(x), x_b \le x \le x_c, \\ 0, x \notin [0, x_a) \cup [x_b, x_c], \end{cases} \tag{1}$$

$$F_Z(x) = \begin{cases} 0, x < 0, \\ (1-h) F_1(x), 0 \le x < x_a, \\ 1 - h \cdot (1 - F_2(x)), x_b \le x \le x_c, \\ 1, x > x_c, \end{cases} \tag{2}$$

where $[0, x_a)$ is an interval of possible values of the random variable X, and $[x_b, x_c]$ is an interval of possible values of the random variable $Y (x_b \le x_c)$. $F_1(x)$ is a distribution function of the random variable X, and $F_2(x)$ is a distribution function of the random variable Y.

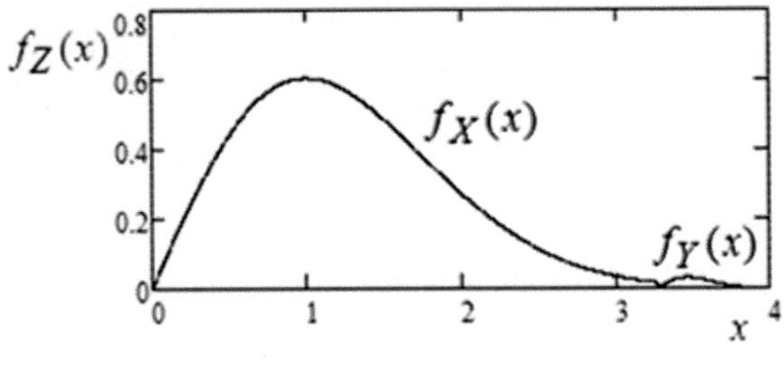

Fig. 5 The distribution density function of the component Z

It follows from (1) and (2) that if $h \ll 1$, then the effect of the second component on $f_Z(x)$ and $F_Z(x)$ is small (see Fig. 5).

4 Signal Detection Method

Thus, the task is reduced to finding a functional characterization of the mixture components for which the contribution of the component Y to the functional characterization of the whole mixture would not depend on h. That is, for $x \geq x_b$, the dependence was only on the properties of the random variable Y and it should be given relative to the fraction of samples. Therefore, it should be a local characteristic in the vicinity of some $x = x_0$, depending on the proportion of samples exceeding x_0. The main local characteristic of a continuous random variable Z at a point x_0 is its distribution density function $f_z(x_0)$. Find the probability that $Z > x_0$:

$$P(Z > x_0) = \int_{x_0}^{+\infty} f_Z(x)dx = 1 - P(Z \leq x_0) = 1 - F_Z(x_0).$$

Next, build their ratio: $\frac{f_Z(x_0)}{1-F_Z(x_0)}$.

A similar construction is found in reliability theory, where the uptime T is a random variable [7].

There is a correspondence with the failure rate; $\lambda(t)$ is the ratio of the number of failed objects per unit time $\Delta n(\Delta t)/\Delta t$ to the average number of objects that continue to work properly in a given time interval $N_{\Delta t}(t)$: $\lambda(t) = \frac{\Delta n(\Delta t)}{N_{\Delta t}(t)\Delta t}$, where $\Delta n(\Delta t)$ is a number of object failures over a period of time from $t - \Delta t/2$ to $t + \Delta t/2$; $N_{\Delta t}(t) = \frac{N(t-\frac{\Delta t}{2})+N(t+\frac{\Delta t}{2})}{2}$ and $N(t)$ is a number of objects that work properly on the interval $[0; t]$ from the original N_0 objects.

When $\Delta t \to 0$, the value $\frac{\Delta n(\Delta t)}{N_0 \Delta t}$ tends to the density of the distribution of failures: $\lim\limits_{\Delta t \to 0} \frac{\Delta n(\Delta t)}{N_0 \Delta t} = f_n(t)$, and $\lim\limits_{\Delta t \to 0} \frac{N_{\Delta t}(t)}{N_0} = P(t)$ determines the probability of uptime over time $T < t$. The probability of failure during this time: $Q(t) = 1 - P(t)$.

When $\Delta t \to 0$, $\lambda(t) = \frac{f_n(t)}{P(t)} = \frac{f_n(t)}{1-Q(t)} = \frac{f_n(t)}{1-\int_{-\infty}^{t} f_n(t)dt} = \frac{f_n(t)}{1-F_n(t)}$, $F_n(t)$ is a distribution function of failures. The failure rate $\lambda(t)$ is often called a hazard function.

For a random variable Z, the hazard function is written as follows:

$$\lambda_Z(x) = \frac{f_Z(x)}{1 - F_Z(x)} = \begin{cases} 0, x < 0, \\ \frac{(1-h)\cdot f_1(x)}{1-(1-h)\cdot F_1(x)}, 0 \leq x < x_a, \\ \lambda_2(x), x_b \leq x \leq x_c, \\ 0, x > x_c, \end{cases} \tag{3}$$

where $\lambda_2(x) = \frac{f_2(x)}{1-F_2(x)}$ is the hazard function of the random variable Y.

Equation (3) shows that the hazard function of the random variables mixture expressed by (1) for $x_b \leq x \leq x_c$ does not depend on the fraction of the component

Y in the mixture. The hazard function $\lambda(x)$ can, therefore, be used to detect samples of the second component in the mixture for small values of h [8].

The practical application of the hazard function in detecting ionosphere sounding signals can be made on the basis of training sets. Let there be a set of experimental data samples containing only a component X (there is a set of realizations of a random

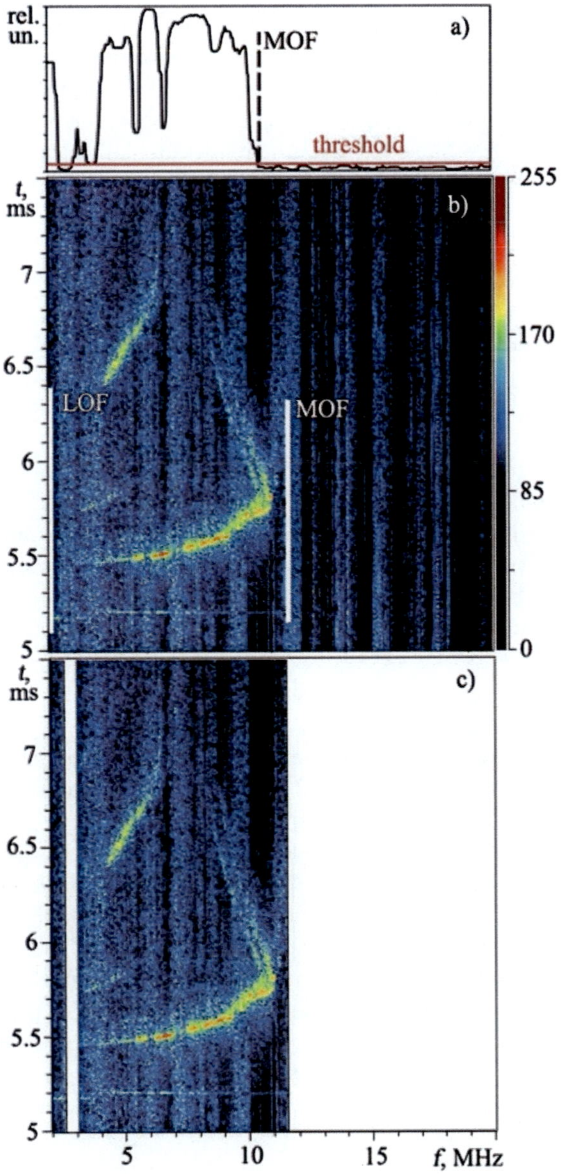

Fig. 6 Result of the signal detection method

process). Construct an averaged normalized sample hazard function $\lambda_{mean}(x)$ using this set. Normalization is carried out so that $0 \leq \lambda_{mean}(x) \leq 1$. A normalized hazard function $\lambda_N(x), 0 \leq \lambda_N(x) \leq 1$, is determined for a set of N samples of a signal that does not belong to the set of the training set.

The difference $\lambda_N(x)$ from $\lambda_{mean}(x)$ can be estimated by a distance d between these functions by defining it, for example: $d = \int_0^1 |\lambda_N(x) - \lambda_{mean}(x)|dx$.

If the value d does not exceed a certain threshold value $\varepsilon (d \leq \varepsilon)$, then it is considered that the set contains only one component X. If $d > \varepsilon$, then it is considered that the set contains two components X and Y. The threshold value ε can be determined by the training set. An example of the operation of this method is presented in Fig. 6.

Figure 6b shows a source ionogram. An operator marked the extreme observed frequencies (LOF and MOF) on it. Figure 6a shows a plot of the d versus frequency and threshold level. Figure 6c shows the result of the algorithm.

5 Signal Extraction Method

In the framework of the proposed statistical model in the form of a mixture of distributions, the sign of a sample corresponding to the sample of the extracted signal is that it exceeds the interval of possible values of a random variable X. From the point of view of a random process, such a sample can be considered an outlier of the random process or interpreted as an anomalous sample. In this work, the threshold was chosen according to the method of [9], which was constructed for a wide class of distribution laws and takes into account the excess of distribution:

if for the sample x, an estimate

$$x - \overline{x} > \sigma \cdot (1.55 + 0.8\sqrt{\varepsilon - 1} \cdot \lg(n/10)) \quad (4)$$

is performed, then this sample is considered abnormal (i.e., the sample is considered to be the sounding signal) where n is a number of samples; $\overline{x}$ is a sample mean; σ is a standard deviation and ε is an excess.

So that the calculation of the excess, the sample mean and the standard deviation is not affected by abnormal samples, the calculation is first performed on the initial part of the variation series of the size in the M samples ($M > 50\%$ of the total number of samples N). Removing from the original set abnormal samples gives an estimate $\widetilde{h}_1$ for the exact value of the quantity h. The procedure for extracting abnormal samples can be repeated with a new value $M = 1 - \widetilde{h}_1$ and a new estimate $\widetilde{h}_2$ of the exact value of the value h can be obtained. That is, the method can be considered iterative. Since each subsequent iteration reduces the error of false selection, but increases the error of skipping the "useful" signal, the number of iterations is established experimentally.

6 The Experimental Results

After applying the method for extracting abnormal samples, the image remains "pepper" noise—single noise emissions (random process) with an intensity comparable to the useful signal. To remove such noise, often use spatial filters with different shapes and sizes of apertures. These filters are also considered by the authors, but the conducted experiments have shown that their use entails a loss (deletion) of "weak" signal. It will if the time dispersion range of weak signal corresponds to 1–2 elements (pixels). The use of different morphological transformations or smoothing filters (such as averaging, Gaussian filter) distorts tracks of propagation signal modes.

This reason has been considered an alternative method of digital image processing—The Progressive Probabilistic Hough Transform.

The Progressive Probabilistic Hough Transform [10] allows to find any set straight lines and curves on the image. The classical Hough transform is based on the representation of a required object in the form of the parametrical equation:

$$x \cdot cos\Theta + y \cdot sin\Theta = R, \tag{5}$$

where x, y are the point coordinates on the image, R is the normal distance to the line from the origin and Θ is the angle between the normal and the x-axis.

The straight line may be interpreted as a parametric transformation between an image space (x, y) and a parameter space or a Hough space (R, Θ). Any point in the image space can be transformed to a sinusoidal curve in the Hough space given by this equation. Conversely, a point (R, Θ) in the Hough space uniquely describes a straight line in the image plane. Thus, collinear points in the image will be transformed to sinusoidal curves in the Hough space which intersect at a common point. The parameters of a line in the image can be found by finding this intersection.

The result of the classical Hough transform is endless straight lines, which makes it impossible to extract signal propagation mode on the ionogram. Thus, additional analysis must be carried out consisting in comparison of found endless lines with the original image and approval of their lengths with existing points of the original image. However, such analysis requires additional computational costs. In this connection, an essential step to optimize the algorithm is that will not take all the ionogram image points for the Hough transform, and only a part p, where $0\% \leq p \leq 100\%$ [10]. That is, at first "control" points from the image are extracted, and the Hough transform is carried out for it.

Figure 7 shows the operation of the stages of the algorithm. Figure 7a shows the source ionogram, Fig. 7b is a stage of signal detection (use of the hazard function), Fig. 7c is a stage of signal extraction (using the method of extraction of abnormal samples), and Fig. 7d is the result of using the Hough transform to extract signals.

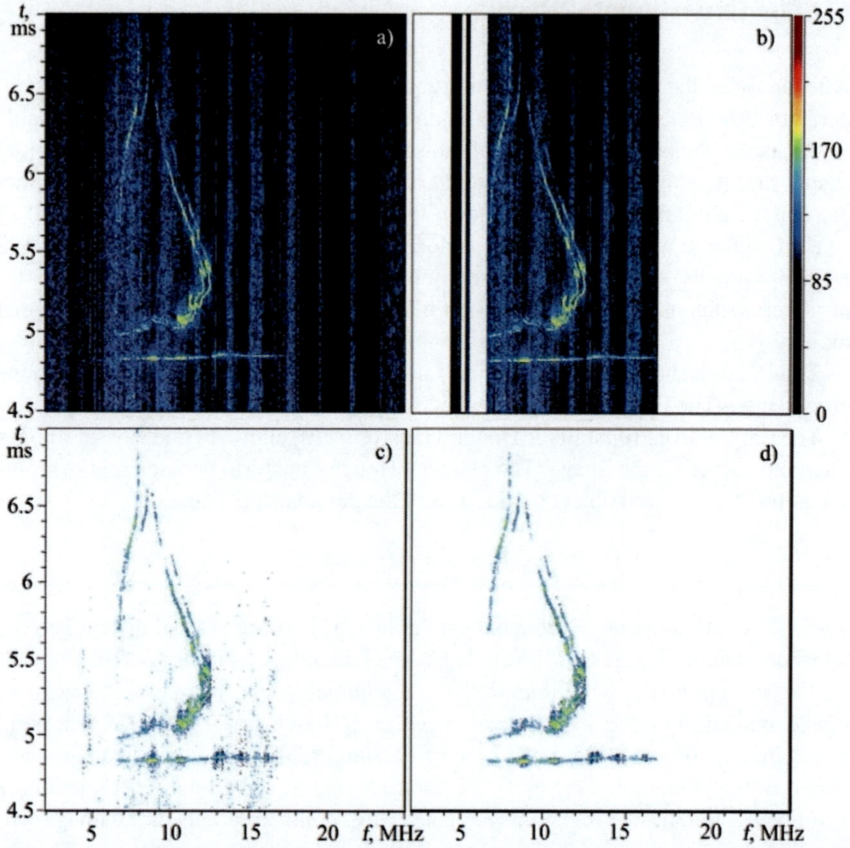

Fig. 7 The effectiveness of the method for signal extracting on ionograms. The bottom figures show that the signal is extracted quite efficiently

7 Conclusions

The hazard function can be used to detect signals within the framework of the model of a mixture of probability distributions. The algorithm for detecting sounding signals using the hazard function is proposed to be implemented on the basis of training sets. Using the same model, the method for extracting the tracks of signal propagation modes on the ionogram without losing (deleting) the "weak" signal has been developed. The frequency range or the time scattering range of the weak signal corresponds to 1–2 spectral elements of the ionogram. The method can also be used for other image signals with similar statistical properties, for example, in the reliability theory, in the analysis of economic processes.

References

1. Chen, K., Zhu, Z., Ning, B., Lan, J.: Fenglou Sun Developing a new mode for observation of ionospheric disturbances by digital ionosonde in ionospheric vertical sounding. Radio Sci. **47**(3), 1–10 (2012)
2. Uryadov, V.P., Vybornov, F.I., Kolchev, A.A., Vertogradov, G.G., Sklyarevsky, M.S., Egoshin, I.A., Shumaev, V.V., Chernov, A.G.: Impact of heliogeophysical disturbances on ionospheric HF channels. Adv. Space Res. **61**(7), 1837–1849 (2018)
3. Ivanov, V.A., Kurkin, V.I., Nosov, V.E., Uryadov, V.P., Shumaev, V.V.: Chirp Ionosonde and Its Application in the Ionospheric Research. Radiophys. Quantum Electron. **46**(11), 821–851 (2003)
4. Ponyatov, A.A., Vertogradov, G.G., Uryadov, V.P., Vertogradova, E.G., Shumaev, V.V., Chernov, A.G., Chaika, E.G.: Features of Superlong-Distance and Round-the-World Propagation of HF Waves. Radiophys. Quantum Electron. **57**(6), 417–434 (2014)
5. Afanasyev, N.T., Grozov, V.P., Nosov, V.E., Tinin, M.V.: The structure of a chirp signal in the randomly inhomogeneous earth-ionosphere channel. Radiophys. Quantum Electron. **43**(11), 847–857 (2000)
6. Kolchev, A.A., Nedopekin, A.E., Khober, D.V.: Application of techniques for separating anomalous samples during the processing of SW LFM Signal. Radioelectron. Commun. Syst. **55**(9), 418–425 (2012)
7. Miller, R.G., Jr., Gong, G., Munoz, A.: Survival Analysis. Wiley, New York (1981)
8. Kolchev, A.A., Egoshin, I.A.: Use of hazard function for signal detection on ionograms. IEEE Geosci. Remote Sens. Lett. **15**(6), 803–807 (2018)
9. Kolchev, A.A., Tsiry, A.O.: Rejection of spectrally lumped noise during chirp sounding of the ionosphere. Radiophys. Quantum Electron. **49**(9), 675–682 (2006)
10. Kiryat, N., Eldar, Y., Bruckstein, A.M.: A Probabilistic Hough Transform. Pattern Recognit. **24**(4), 303–316 (1991)

PDE Modeling of Bladder Cancer Treatment Using BCG Immunotherapy

T. Lazebnik, S. Yanetz, and S. Bunimovich-Mendrazitsky

Abstract Immunotherapy with Bacillus Calmette-Guérin (BCG)—an attenuated strain of Mycobacterium bovis (M. bovis) used for anti-tuberculosis immunization—is a clinically established procedure for the treatment of superficial bladder cancer. Bunimovich-Mendrazitsky et al. [16] studied the role of BCG immunotherapy in bladder cancer dynamics in a system of nonlinear ODEs. The purpose of this paper is to develop a first mathematical model that uses PDEs to describe tumor-immune interactions in the bladder as a result of BCG therapy considering the geometrical configuration of the human bladder. A mathematical analysis of the BCG as a PDE model identifies multiple equilibrium points, and their stability properties are identified so that treatment that has the potential to result in a tumor-free equilibrium can be determined. Estimating parameters and validating the model using published data are taken from in vitro, mouse, and human studies. The model makes clear that the intensity of immunotherapy must be kept within limited bounds. We use numerical analysis methods to find the solution of the PDE describing the tumor-immune interaction; in particular, analysis of the solution's stability for given parameters is presented using Computer Vision methodologies.

Keywords Numerical analysis · PDE's solution stability · PDE's parameters' sensitivity analysis · 34A34 · 35A25 · 35A30 · 65M60 · 68W25

T. Lazebnik (✉)
Department of Cancer Biology, University College London, London, UK
e-mail: t.lazebnik@ucl.ac.uk

S. Yanetz
Department of Mathematics, Bar-Ilan University, Ramat Gan, Israel

S. Bunimovich-Mendrazitsky
Department of Mathematics, Ariel University, Ariel, Israel

A. Domoshnitsky et al. (eds.), *Functional Differential Equations and Applications*,
Springer Proceedings in Mathematics & Statistics 379,
https://doi.org/10.1007/978-981-16-6297-3_9

1 Introduction and Related Work

Bladder Cancer (BC) is the seventh most common cancer worldwide. It is estimated that around 400,000 new cases are diagnosed annually and 150,000 people die directly from BC every year [1]. Bacillus Calmette–Guérin (BCG) has been used to treat non-invasive BC for more than 40 years [2]. It is one of the most successful biotherapies for cancer in use. Despite long clinical experience with BCG, the mechanism of its therapeutic effect is still under investigation. BCG immunotherapy has proven to reduce both recurrence and progression of BC and, therefore, represents an important tool in the treatment of BC. BCG treatment protocols differ mainly by the amount of the injected dosage, the injection rate, and the schedule of the treatment [3].

Mathematical modeling of biological processes in general and medical processes, in particular, is an active field of study. The benefit gained from describing a system using mathematical modeling is the ability to analyze and understand it better by using only theoretical analysis, which decreases the need for clinical experiments to further understand the system in question [4]. Several mathematical models that describe the interactions of the immune system with tumor cells based on ODE are [5–11]. Study of the bladder cancer using mathematical modeling has been researched in the past from different angles [12–15].

One of the models was invented by Bunimovich-Mendrazitsky et al. [16]. Their model assumed continuous BCG instillation and allowed both exponential and logistic growth for tumor cells inside the bladder. They studied the equilibria when the stability and analysis of the system's bifurcation was the main focus. It was found that bistability excises so that a treatment may result in the tumor-free equilibrium or high-tumor state, depending on the initial tumor size reflected by the cancer cell count. The equations describe a balance between a high dosage which caused the patient to suffer from side effects and too little dosage caused inefficient treatment.

The mathematical model proposed by Bunimovich-Mendrazitsky et al. [16] is as follows:

$$\frac{dB(t)}{dt} = -p_1 E(t)B(t) - p_2 B(t)T_u(t) - \mu_1 B(t) + b \tag{1}$$

$$\frac{dE(t)}{dt} = -\mu_2 E(t) + \alpha T_i(t) + p_4 E(t)B(t) - p_5 E(t)T_i(t) \tag{2}$$

$$\frac{dT_i(t)}{dt} = p_2 B(t)T_u(t) - p_3 T_i(t)E(t) \tag{3}$$

$$\frac{dT_u(t)}{dt} = \lambda(t)T_u(t) - p_2 B(t)T_u(t). \tag{4}$$

The state variables $B(t)$, $E(t)$, $T_i(t)$, and $T_u(t)$ represent the concentration of BCG in the bladder, effector cell population, tumor cell population that has been infected with BCG, and tumor cell population that is uninfected with BCG, respectively. The parameter p_1 is the rate of BCG killed by effector cells; p_2 is the infection rate of uninfected tumor cells by BCG; p_3 is the rate of destruction of tumor cell infected by BCG by effector cells; p_4 is the immune response activation rate; p_5 is the rate of effector cells deactivation after binding with infected tumor cells. α is the growth rate of effector cell population; λ is the tumor population growing rate; b is the strength of BCG instillation.

Several attempts of modeling the problem have taken under consideration only the population's size of different cells in the system over time, based on the biological dynamic of the system using Ordinary Differential Equations (ODEs) [6, 7]. One approach to improve the model is taking under consideration an approximation of the geometry configuration of the bladder in the mathematical modeling yielding in Partial Differential Equations (PDEs). The PDEs Model's parameters sensitivity and solution's stability for given parameters is the main focus of this paper. We combine numerical calculations with computer vision algorithms to find the PDE's model solution's stability for a non-Lyapunov PDE system.

2 Mathematical Modeling and Numerical Analysis

The mathematical model differs from the Bunimovich-Mendrazitsky et al. [16] model by taking under consideration the geometrical configuration of the human bladder. The new model can be described by the following system of PDEs:

$$\begin{aligned}\frac{\partial B(r,t)}{\partial t} &= -p_1 E(r,t)B(r,t) - p_2 B(t)T_u(r,t) \\ &\quad -\mu_1 B(r,t) + b + D_1 \frac{1}{r^2}\frac{\partial}{\partial r}(r^2 \frac{\partial B(r,t)}{\partial r})\end{aligned} \tag{5}$$

$$\begin{aligned}\frac{\partial E(r,t)}{\partial t} &= -\mu_2 E(r,t) + \alpha T_i(r,t) + p_4 E(r,t)B(r,t) \\ &\quad -p_5 E(r,t)T_i(r,t) + D_2 \frac{1}{r^2}\frac{\partial}{\partial r}(r^2 \frac{\partial E(r,t)}{\partial r})\end{aligned} \tag{6}$$

$$\frac{\partial T_i(r,t)}{\partial t} = p_2 B(r,t)T_u(r,t) - p_3 T_i(r,t)E(r,t) + D_3 \frac{1}{r^2}\frac{\partial}{\partial r}(r^2 \frac{\partial T_i(r,t)}{\partial r}) \tag{7}$$

$$\frac{\partial T_u(r,t)}{\partial t} = \lambda T_u(r,t) - p_2 B(r,t)T_u(r,t) + D_4 \frac{1}{r^2}\frac{\partial}{\partial r}(r^2 \frac{\partial T_u(r,t)}{\partial r}). \tag{8}$$

All the variables with the same notation and meaning as described in Eqs. (1)–(4). D_1, D_2, D_3, D_4 are the diffusion rate in the system for $B(r,t)$, $E(r,t)$, $T_i(r,t)$, and $T_u(r,t)$ respectively. The variable t stands for the time of the system and r stands for the euclidean distance in $\mathbb{R}^3$ from the point (0, 0, 0) in polar coordinates. The center of the system's geometry is defined to be (0, 0, 0).

In the scope of this paper, it will be assumed that the bladder has a form of a perfect sphere ring satisfying the following condition:

$$r_0^2 \leq x^2 + y^2 + z^2 \leq R^2. \tag{9}$$

The variables x, y, z are the *Cartesian* coordinates system, r_0 and R are the radius of the internal and external spheres of the geometrical configuration, respectively. We ignore the three tunnels connected to the approximately ellipsoidal shape of the bladder's epithelium and approximate the ellipsoidal shape with a sphere shape.

The PDE system differs from the ODE system in two ways: 1) the PDE model adds another dimension (r); 2) the PDE model takes under consideration the geometry of the problem, and the diffusion factor added to each population, respectively.

The inner sphere boundary condition is given to be:

$$\frac{\partial B(r,t)}{\partial r} = b, \quad \frac{\partial E(r,t)}{\partial r} = 0, \quad \frac{\partial T_i(r,t)}{\partial r} = 0, \quad \frac{\partial T_u(r,t)}{\partial r} = 0. \tag{10}$$

The initial condition is assumed to be

$$B(r, t_0) = 0, \; E(r, t_0) = 0, \; T_u(r, t_0) = \frac{cr}{R - r_0}, \; T_i(r, t_0) = 0, \tag{11}$$

where $c > 0$ is the tumor cells distribution factor.

2.1 *Biological Border and Start Conditions*

The boundary condition of the external sphere is unknown. It is assumed that naturally the cell population spread over time satisfies diffusion equations. Therefore, one can find the boundary condition of the external sphere by reverse engineering the values that best satisfy the known start conditions and internal boundary sphere conditions. Algorithm 1 addresses this problem (Fig. 1).

2.2 *Numerical Analysis*

The set of equations can be classified as a set of nonlinear, second order, partial differential equations from $\mathbb{R}^2$ to $\mathbb{R}^4$, where $\mathbb{R}^2$ is the space of time (marked by t) and radial distance from the center of the bladder's geometry configuration (marked by r) and $\mathbb{R}^4$ is the populations' counts of all four populations (marked by E, B, T_i, T_u). In such case, it is possible to use Galerkin-Petrov's method [17] taking the form

Algorithm 1 Find external sphere boundary conditions

procedure EXTERNALBOUNDARY($startConditions, internalBoundaryCondition$) ▷ The external sphere boundary conditions
  sample uniformly points from the inner and outer sphere and mark as P
  $i \leftarrow 1$
  while start condition not satisfied **do**
    $t_{start} \leftarrow t_0 - i$
    run diffusion equations with system's start conditions and internal boundary condition at t_{start} and the points P
    $i \leftarrow i + 1$
  return P ▷ The external boundary conditions

$$C(r, t, u, \frac{\partial u}{\partial r})\frac{\partial u}{\partial t} = r^{-2}\frac{\partial}{\partial r}(r^2 f(r, t, u, \frac{\partial u}{\partial r})) + s(r, t, u, \frac{\partial u}{\partial r}). \qquad (12)$$

This method is a numerical process allowing to retrieve the populations' size of all four cell populations given the start condition, boundary condition, and Eqs. (5)–(8).

The calculation has been performed on a software by *Matlab* (version 2012b) using the *pdepe* method [17] while replacing the build in matrix exponent function with Lazebnik and Yantz's function [18]. Few tests have been conducted to examine the results and differences between the ODE model and the PDE model.

In Figure (1), the x-axis represents the time that has been passed from the beginning of the treatment in weeks and the y-axis is the size of the cell population. This graph

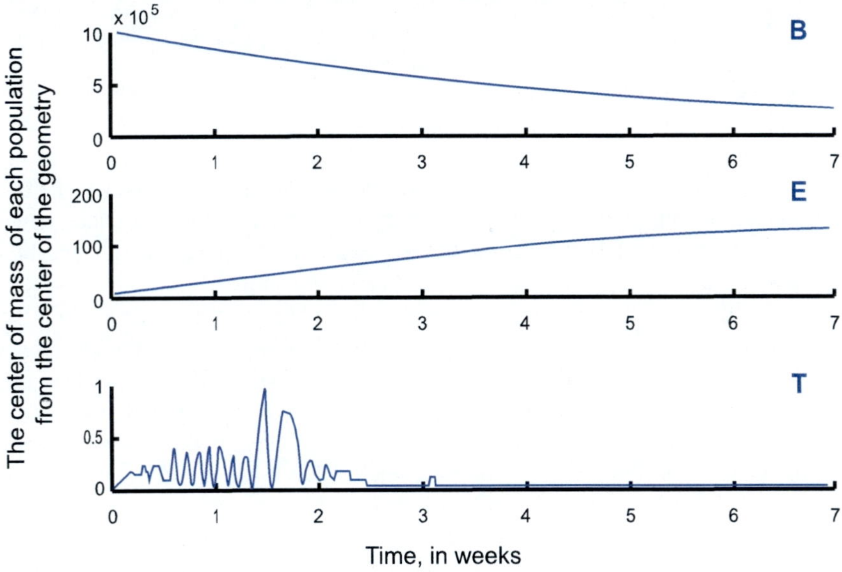

Fig. 1 Cell population over time

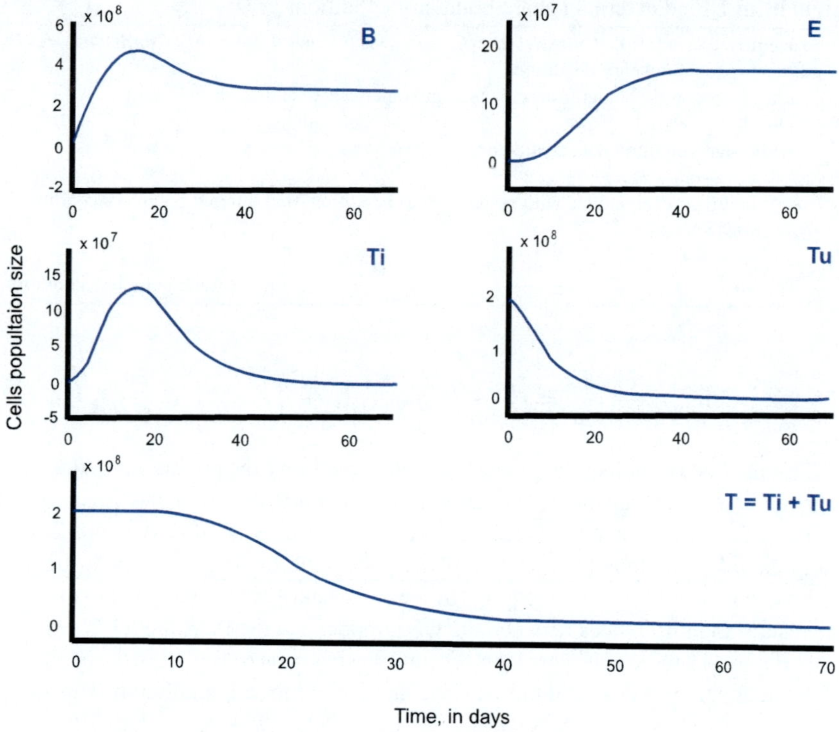

Fig. 2 Delta in cell population over time between the ODE and PDE models

averages a thousand iteration results in order to reduce the error which inherently takes place in numerical calculation. One can notice a reduction in the cancer cell population decrements over time reflecting the effect of the treatment. Furthermore, the decrements of the BCG-infected cell in the first graph can be explained by the fact that the BCG is injected into the system in the same place, but the immune system increases its effort to fight the disease as described in the second graph (E), which in turn leads to a decrease in the BCG-infected cell population.

The PDE model provides further understanding of the system as its predicts the population size to be two orders of magnitude bigger than the original ODE model prediction. A Pearson correlation between each individual population size between the ODE and the PDE models provides poor results showing that there is no linear correlation between the models and they provide different predictions for the system. On the other hand, the difference between the models converges to a constant for all the cell population after the fifth week, basically indicating a correlation which converges to one between the ODE and PDE models in long enough treatments.

Figure (2) shows the deltas in the different populations between the two models when the x-axis is the time passed from the beginning of the treatment in days and the y-axis is the difference between the sizes of the cells populations (Fig. 2).

3 PDE Model Parameters Sensitivity Analysis

The numerical calculation of the PDE allows to analyze the system's sensitivity to different parameters. The first parameter is the influence of the insert rate of BCG into the bladder (b). From clinical experiments [16], it is known that $b \in [10^5, 10^7]$. The *least squares* [20] analysis method has been used to calculate the effect on the system's output. Note that $[t_i]_0^{70} \in [0, 70]$ such that $\forall i : \Delta(t_{i+1} - t_i) = c$. The function family used to approximate the real function is

$$\alpha, \beta, \gamma, \delta \in \mathbb{R} : f(\alpha, \beta, \gamma, \delta) = \alpha e^{\beta b} + \gamma e^{\delta b} \tag{13}$$

The algorithm to calculate function f which minimizes the sum of the square of the errors between the function value and model's value is

Algorithm 2 Find best fitting function to parameter's behavior

procedure PDELEASTSQUARES(PDE model, boundaries, f, h)
▷ f is the approximation function and h is the sample step size
$T \leftarrow$ empty list
$i \leftarrow 0$
$b \leftarrow boundaries[0]$
while b < boundaries[1] **do**
$t_{start} \leftarrow t + 0 - i$
$T[i] \leftarrow$solve(PDEmodel)
$R^2[i], T[i] \leftarrow$LeastSquaresFit(t, $T[i]$, normal distribution))
$b \leftarrow b + h$
$i \leftarrow i + 1$
R^2, bestModel $\leftarrow$ LeastSquaresFit([boundaries[0], boundaries[1], h], T, f)
return $bestModel$

Running the algorithm given a sampling step in the size of $\frac{10^7 - 10^5}{10^4} = 990$ provides the following results:

$$R^2 = 0.993, T(t, b) = (23.104 e^{1.6985*10^{-9} b} - 357.4288 * 10^3 e^{-373.2229*10^{-9} b}) * e^{-(\frac{(t - (2.919 e^{-565.818*10^{-9} b} + 4.152 e^{-33.018*10^{-1}}) b}{(26.537 e^{-1.374*10^{-9} b} + 13.15 e^{-74.532*10^{-1}}) b})^2}, \tag{14}$$

$$R^2 = 0.976, T_u(t, b) = (27.107 e^{926.29*10^{12} b} - 3.09 * 10^6 e^{-1.295*10^{-6} b}) * e^{t*(269.118*10^{-3} e^{53.978*10^{-9} b} - 245.631*10^{-3} e^{-341.431*10^{-9} b})}, \tag{15}$$

$$R^2 = 0.983, T_i(t, b) = (1.783 * 10^7 e^{1.434*10^{-8} b} - 9.264 * 10^6 e^{-7.944*10^{-7} b}) * e^{-\left(\frac{(t-(15.502 e^{-1.06*10^{-6} b} + 10.623 e^{-6.538*10^{-9}}) b}{(12.788 e^{-1.183*10^{-6} b} + 7.616 e^{-6.754*10^{-8}}) b}\right)^2}. \quad (16)$$

A smaller b produces a smaller BCG-infected population (B) and also a smaller effector cell population (E). The decrease in the tumor cell population (T) over time has a lower rate for smaller b. Not enough injected BCG can even produce the unwanted result that tumor cell population decrements will not reset at the end of the treatment. On the other hand, bigger b produces a higher peak around the end of the first week of the treatment risking the penitent immune system.

To approximate the influence function of parameter b, the *least square* analysis method can be used again with function (13). From clinical experiments [16], it is known that this treatment is reasonable when $T_u(t_0) \in [2 * 10^5, 3 * 10^9]$. Running algorithm (2) provides the following results:

$$R^2 = 0.995, T_u(r, t_0) = (1.93 * 10^1 1 e^{8.85*10^{-6} T_u(r,t_0)} - 1.64 * 10^{11} e^{2.11*10^{-4} T_u(r,t_o)}) * e^{-\left(\frac{(t-(2.73 e^{3.02*10^{-6} T_u(r,t_0)} + 3.45 e^{-2.18*10^{-4}}) T_u(r,t_0)}{(10.72 e^{8.22*10^{-6} T_u(r,t_0)} - 8.77 e^{1.88*10^{-4}}) T_u(r,t_0)}\right)^2}. \quad (17)$$

4 PDE Model Solution's Stability Analysis

The PDE does not satisfy the conditions needed to use Lyapunov's stability analysis method. On the other hand, the numerical calculation of the system does not diverge to infinity. One can analyze the image of the dynamics of the system by solving the PDE for given parameters. Such analysis will allow to find a function $g(T_u, BCG, time) \to \{0, 1\}$ when the source space contained in $\mathbb{R}^3$ and the image space is exactly $\{0, 1\}$. This allows to set the start condition of the problem and to find whether the treatment will succeed or not without the need to solve the PDE from scratch each time.

Calculating an approximation to the function g first requires to sample the parameter's space. There are six parameters affecting the system: t, BCG, T_u, C_1, C_2, C_3 when $C_1, C_2, C_3 \in \mathbb{R}^+$ are thresholds of the three population sizes T, B, E, respectively, depending if the treatment succeeded or not. Assume there are lower and upper boundaries from biological experiments for the parameters yielding a compact parameter's set. This is because the set is complete as a sub-set of $\mathbb{R}^6$ and bounded. Assuming the solution is continuous and can be restored from discrete sampling, define $h \in, 1 \gg h$ such that h is the size of the sampling step.

Using the output of the algorithm (3), one can extract the border pixels. In this case, a border pixel is a pixel with neighbor pixels from in the case the treatment

Algorithm 3 Sample the PDE's image space of function g

```
procedure PDEImageSampling(PDEmodel, B, C, h)    ▷ B is an array of boundries, C is an
array of thresholds and h is the sample step size
  output ← zeros(B[0][0], B[0][1], B[1][0], B[1][1], B[2][0], B[2][1])
  while i ∈ [B[0][0], B[0][1]] do
    while j ∈ [B[1][0], B[1][1]] do
      while k ∈ [B[2][0], B[2][1]] do
        s ←solve(PDEmodel(i, j, k))
        if s[0] < c1 and s[1] < c2 and s[2] < c3 then
        output[i][j][k] ← 1
        EndIf
  return output
```

succeeded and in the case is did not succeeded. Boundary following algorithm [19] for the three-dimensional case performs such a task.

One can take advantage again of the *least squares* analysis method to find an approximation to the function describing the border between the two cases. We assume that the model is as follows:

$$F(x, y) = a_1 sin(x) + a_2 cos(x) + a_3 sin(y) + a_4 cos(y) + a_5 sin(x)cos(y) + a_6 sin(y)cos(x) \\ + a_7 sin(x)sin(y) + a_8 cos(x)cos(y). \quad (18)$$

This produces the following models for both the PDE and the ODE models, respectively

$$F_{pde}(x, y) = 2.644 sin(x) + 3.904 cos(x) + 9.636 sin(y) + 8.931 cos(y) - 8.544 sin(x)cos(y) \\ -2.607 sin(y)cos(x) - 1.266 sin(x)sin(y) - 9.393 cos(x)cos(y) \quad (19)$$

Using Eq. (19), it is possible to predict the needed time (if it exists) so that the tumor cell population size is small enough $T(t, r) < C_1$ on one hand and the effector, BCG-infected, cell populations sizes are not growing to large $B(t, r) < C_2$, $E(t, r) < C_3$ on the other hand, yielding a successful treatment, given only the tumor's initial cell population size and the BCG injection rate.

5 Conclusions and Future Work

It is safe to claim that mathematical modeling is a useful tool for studying the mechanism of tumor growth and response to therapy. The use of numerical simulation of complex mathematical models that is not yet analytically solvable can help predict the outcome of treatment and determine better therapeutic protocols. As population analysis is a common way of describing such systems [8–10], it is important to add the geometrical configuration of the problem into the dynamics since the system parameter values vary across different geometries.

Bifurcation analysis of the mathematical model considered in this paper was not previously available because the numerical methods developed for bifurcation analysis require continuous vector fields. We found that PDE representation in bladder cancer treatment with BCG provides more accurate predictions to observations done in vitro in mice and humans than the ODE representation. As can be observed from Figure (2), the delta between the ODE and PDE model in all cell population sizes are in a factor of 100, where the PDE model's predictions better fits previous observations with respect to the ODE model's predictions [16].

On the other hand, after five weeks of treatment, the delta between the models converges to a constant for each population function (E, B, T_i, T_u) and basically indicates a complete linear correlation between the ODE and PDE models ($R^2 \to 1$).

The difference between the models is initially associated with the introduction of the geometry reflected in the diffusion coefficients introduced into the dynamics of the system. In fact, from the very beginning, there is a disagreement between the models: for PDE there is diffusion dynamics, and for ODE there is an instant reaction to the introduction of BCG. After diffusion spreads throughout the space, it behaves like an instantaneous response, and therefore, the ODE and PDE models ultimately work identically, as can be seen from the calculation of the delta between the models.

This study develops a numerical method for the stability analysis of PDE's solutions of a mathematical model with pulsed BCG immunotherapy based on well-known algorithms from the field of computer vision. We can make a few clinical conclusions based on analysis of function (19): (1) BCG injected with a rate smaller than sixty thousand cells almost does not have an effect. (2) In the case of bladder cancer, when there are 10% or less cancer cells from the overall population and BCG is injected at a rate of eighty thousand cells, then the cancer can be cured in ninety percent of the cases for the treatment that is given between eight and ten weeks. (3) There is a strong linear correlation between the amount of BCG injected and the time of the treatment in the successful cases when the cancer cell population is around five percent of the overall cell population at the beginning of the treatment.

References

1. Jemal A., Bray F., Center M. M., Ferlay J., Ward E., Forman D., Global cancer statistics, ca: a cancer. J. Clin. **61**, 69–90 (2011)
2. Morales, A., Eidinger, D., Bruce, A.W.: Intracavity Bacillus Calmette-Guérin in the treatment of superficial bladder tumors. J. Urol. **116**, 180–183 (1976)
3. Guzev E., Halachmi S., Bunimovich-Mendrazitsky S.: Additional extension of the mathematical model for BCG immunotherapy of bladder cancer and its validation by auxiliary tool. Int. J. Nonlinear Sci. Numer. Simul. (2019)
4. Byrne, H.M.: Dissecting cancer through mathematics: from the cell to the animal model. Nat. Rev. Cancer **10**(3), 221–230 (2010)

5. Kuznetsov, V.A., Makalkin, I.A., Taylor, M.A., Perelson, A.S.: Nonlinear dynamics of immunogenic tumours: parameter estimation and global bifurcation analysis. Bull. Math. Biol. **56**, 295–321 (1994)
6. Kim, J.C., Steinberg, G.D.: The limits of bacillus Calmette-Guerin for carcinoma in situ of the bladder. J. Urol. **165**(3), 745–56 (2001)
7. Castiglione, F., Piccoli, B.: Cancer immunotherapy, mathematical modeling and optimal control. J. Theor. Biol. **247**, 723–732 (2007)
8. De Pillis, L.G., Gu, W., Radunskaya, A.E.: Mixed immunotherapy and chemotherapy of tumors: modeling, applications and biological interpretations. J. Theor. Biol. **238**, 841–862 (2006)
9. Kirschner, D., Panetta, J.C.: Modeling immunotherapy of the tumor-immune interaction. J. Math. Biol. **37**, 235–252 (1998)
10. Panetta, J.C.: A mathematical model of periodically pulsed chemotherapy: tumor recurrence and metastasis in a competitive environment. Bull. Math. Biol. **58**, 425–447 (1996)
11. De Pillis, L.G., Radunskaya, A.E., Wiseman, C.L.: A validated mathematical model of cell-mediated immune response to tumor growth. Cancer Res. **65**(17), 7950–7958 (2005)
12. Bunimovich-Mendrazitsky, S., Pisarev, V., Kashdan, E.: Modeling and simulation of a low-grade urinary bladder carcinoma. Comput. Biol. Med. **58**, 118–129 (2015)
13. Bunimovich-Mendrazitsky, S., Goltser, Y.: Use of quasi-normal form to examine stability of tumor-free equilibrium in a mathematical model of BCG treatment of bladder cancer. Math. Biosci. Eng. **8**, 529–547 (2011)
14. Nave, O., Hareli, S., Elbaz, M., Iluz, I.H., Bunimovich-Mendrazitsky, S.: BCG and IL2 model for bladder cancer treatment with fast and slow dynamics based on SPVF method-stability analysis. Math. Biosci. Eng. (MBE) **16**(5), 5346–5379 (2019)
15. Shaikhet L., Bunimovich-Mendrazitsky S.: Stability analysis of delayed immune response BCG infection in bladder cancer treatment model by stochastic perturbations. Comput. Math. Methods Med. (2018)
16. Bunimovich-Mendrazitsky, S., Shochat, E., Stone, L.: Mathematical model of BCG immunotherapy in superficial bladder cancer. Bull. Math. Biol. **69**(6), 1847–1870 (2007)
17. Skeel, R.D., Berzins, M.: A method for the spatial discretization of parabolic equations in one space variable. SIAM J. Sci. Stat. Comput. **11**, 1–32 (1990)
18. Lazebnik T., Yantez S.: A stable algorithm for matrx exponent calculation, Funct. Differ. Equ. (2016)
19. Jonghoon, S., Seungho, C., Jinwook, S., Dongchul, K., Cheolho, C., Tack-Don, H.: Fast Contour-Tracing Algorithm Based on a Pixel-Following Method for Image Sensors, MDPI (2016); Lazebnik, T., Yantez, S.: A stable algorithm for matrx exponent calculation, Funct. Differ. Equ. (2016)
20. Björck, Å.: Numerical methods for least squares problems. SIAM J. Sci. Stat. Comput. Book OT51 (1996)

Marchuk's Models of Infection Diseases: New Developments

Irina Volinsky, Alexander Domoshnitsky, Marina Bershadsky, and Roman Shklyar

Abstract We consider mathematical models of infectious diseases built by G. I. Marchuk in his well-known book on immunology. These models are in the form of systems of ordinary delay differential equations. We add a distributed control in one of the equations describing the dynamics of the antibody concentration rate. Distributed control looks here naturally since the change of this concentration rather depends on the corresponding average value of the difference of the current and normal antibody concentrations on the time interval than on their difference at the point t only. Choosing this control in a corresponding form, we propose some ideas of the stabilization in the cases, where other methods do not work. The main idea is to reduce the stability analysis of a given integro-differential system of the order n, to one of the auxiliary systems of the order $n + m$, where m is a natural number, which is "easy" for this analysis in a corresponding sense. Results for these auxiliary systems allow us to make conclusions for the given integro-differential system of the order n. We concentrate our attempts in the analysis of the distributed control in an integral form. An idea of reducing integro-differential systems to systems of ordinary differential equations is developed. We present results about the exponential stability of stationary points of integro-differential systems using the method based on the presentation of solution with the help of the Cauchy matrix. Various properties of integro-differential systems are studied by this way. Methods of the general theory of functional differential equations developed by N. V. Azbelev and his followers are used. One of them is the Azbelev W-transform. We propose ideas allowing to achieve faster convergence to stationary point using a distributed control. We obtain estimates of solutions using estimates of the Cauchy matrices.

Keywords Functional differential equations · Exponential stability · Cauchy matrix

I. Volinsky (✉) · A. Domoshnitsky · M. Bershadsky · R. Shklyar
Ariel University, Ariel, Israel
e-mail: irinav@ariel.ac.il

A. Domoshnitsky
e-mail: adom@ariel.ac.il

A. Domoshnitsky et al. (eds.), *Functional Differential Equations and Applications*, Springer Proceedings in Mathematics & Statistics 379,
https://doi.org/10.1007/978-981-16-6297-3_10

1 Introduction

Mathematical models in the form of systems of nonlinear ordinary differential equations are used in many fields of science and technology to describe various phenomena. In medicine, the purpose of mathematical modeling is the analysis and prediction of the development of diseases and their possible treatment. A comprehensive work on mathematical models in the field of immunology was summarized by Marchuk in his book [18]. The models constructed there reflect the most significant patterns of the immune system acting during these diseases. This model was studied in many works. Note, for example, the recent papers [19, 20] and the bibliography therein. The adding control was proposed, for example, in [6, 7, 9–11, 16, 17, 22, 23]. In the works [8, 21], the basic mathematical model that takes into account the discrete control of the immune response is proposed. See also the recent papers [7, 10, 17], where distributed control was considered. It can be noted that the use of information about the behavior of a disease and the immune system for a long time (defined by distributed control, for example, in the form of an integral term) looks very natural in choosing the strategy of a possible treatment. Optimal control in the basic model of the infectious disease was considered in the work [8], where the control function characterizing realization of an immunotherapy which includes in administration of immunoglobulin or donor antibodies is proposed. In the work [2], the model of influence of an immunotherapy on dynamics of an immune response which represents generalization of basic model was considered. On the basis of the proposed model, the problem of determination of coefficients on the basis of laboratory dates was considered and a suitable management was proposed in [5, 8]. Such task was called control in uncertain conditions [22]. A control algorithm in uncertain conditions was proposed in the work ([8], see pp. 71–73).

In the recent papers [7, 10, 17], we present new approach for the study of the model of infectious diseases. In this paper, we summarize their results and formulate mathematical problems which look very natural from the medical point of view.

Our contribution in the modeling is a distributed feedback control which is added to the equation describing the concentration of antibodies. This step transforms these systems into functional differential ones. As a result, we have to study the properties of solutions of these systems such as asymptotic behavior in the neighborhood of stationary points and stability of the stationary points. the importance of stationary points should be stressed. These points describe the conditions of the healthy body or the chronic disease. The aim of the treatment is to lead the process to one of the stationary points. Further, we try to obtain estimates of solutions of linear and nonlinear systems of functional differential equations. One of the ways to these estimates is the construction of the Cauchy matrix. First steps in this direction were proposed in the recent paper [10].

2 Description of Model

In this paper, we deal with the system of functional differential equations

$$x'(t) + (Ax)(t) = (\Phi T x)(t), \quad t \in [0, \infty), \quad x = \{x_1, , x_n\} \tag{2.1}$$

where the operators T and A are linear continuous. $T, A : C^n_{[0,\infty)} \to L^n_{\infty[0,\infty)}$ ($C^n_{[0,\infty)}$, $L^n_{\infty[0,\infty)}$, are the spaces of continuous, and essentially bounded vector functions $x : [0, \infty) \to R^n$ respectively), $F : L^n_{\infty[0,\infty)} \to L^n_{\infty[0,\infty)}$ can be a linear or nonlinear bounded operator. We could analyze various boundary value problems for Eq. (2.1). One of them is the initial value problem. One of the main questions is the stability of this system [3]. We consider the stationary points for corresponding operators in the spaces of continuous functions $C^n_{[0,\infty)}$ or essentially bounded functions $L^n_{\infty[0,\infty)}$. We use our theoretical results in application to Marchuk's model of infectious diseases. This model reflects the most significant patterns of the immune system functioning during infectious diseases and focuses on the interactions between antigens and antibodies at different levels. We try to investigate the stability of stationary points of the immune system and its response to the treatment. We propose the control in the distributed form and obtain stabilization in the neighborhood of the stationary point in the model of infectious diseases. From the applications' point of view, the goal of the control in the system can be interpreted as a possibility to provide a corresponding immune response. It is noted in [22] that the immune response mechanisms provide a key to understanding disease processes and methods of effective medical treatment [18]. We try to combine our theoretical results with possible applications. Let us start with a description of one of these applications. Consider, for example, the Marchuk model of infectious diseases

$$\begin{cases} \frac{dV}{dt} = \beta V(t) - \gamma F(t) V(t) \\ \frac{dC}{dt} = \zeta(m) \alpha F(t) V(t) - \mu_c (C(t) - C^*) \\ \frac{dF}{dt} = \rho C(t) - \eta\gamma F(t) V(t) - \mu_f F(t) \\ \frac{dm}{dt} = \sigma V(t) - \mu_m m(t) \end{cases} \tag{2.2}$$

where $V(t)$—the antigen concentration rate, $C(t)$—the plasma cell concentration rate, $F(t)$—the antibody concentration rate, $m(t)$—the relative features of the body. It is clear that system (2.2) can be presented in the form of general system (2.1). Let us describe the coefficients: β—coefficient describing the antigen activity, γ—the antigen neutralizing factor, μ_f—coefficient inversely proportional to the decay time of the antibodies, μ_m—coefficient inversely proportional to the organ recovery time, μ_c—coefficient of reduction of plasma cells due to aging (inversely proportional to the lifetime), σ—constant related with a particular disease, ρ—rate of production of antibodies by one plasma cell. Denote C^* and F^*—the plasma rate concentration and antibody concentration of the healthy body, respectively. It is assumed that during a certain period of time τ, the plasma is restored as a result of the interaction between the antigen and the antibody cells. The product $\zeta(m)\alpha F(t)V(t)$ includes the following coefficients: α is the stimulation factor of the immune system. The function

$$\zeta(m) = \begin{cases} 1, & 0 \le m < m^* \\ \frac{1-m}{1-m^*}, & m^* \le m \le 1 \end{cases},$$

is a continuous function, characterizing the health of the organ, which depends on the relative characteristics m of the body, where m^* is the maximum proportion of cells destroyed by antigens in the case that the normal functioning of the immune system is still possible. This function is non-negative and does not increase. The function $m(t)$ can be described as $1 - \frac{1-M(t)}{1-M^*(t)}$, where $M(t)$ is the characteristic of a healthy organ (mass or area) and $M^*(t)$ is the corresponding characteristic of a healthy part of the affected organ. Let us discuss now every equation in the model (2.2) in a more detailed form. The first equation $\frac{dV}{dt} = \beta V(t) - \gamma F(t)V(t)$ presents the block of the virus dynamics. It describes the changes in the antigen concentration rate and includes the amount of the antigen in the blood. The antigen concentration decreases as a result of the interaction with the antibodies. The immune process characterizes the antibodies, whose concentration changes with time (destruction rate), is described by the equation $\frac{dF}{dt} = \rho C(t) - \eta\gamma F(t)V(t) - \mu_f F(t)$. The amount of the antibody cells also decreases as a result of the natural destruction. However, the plasma restores the antibodies, and therefore, the plasma state plays an important role in the immune process. Thus, the change in concentration rate of the plasma cell is included in several differential equations describing this system. Taking into account the healthy body level of plasma cells and their natural aging, the term $\mu_c(C(t) - C^*)$ is included in the second equation of the system (2.2). The second and third equations present the humoral immune response dynamics. Concerning the last equation of system (2.2) $\frac{dm}{dt} = \sigma V(t) - \mu_m m(t)$. The following can be noted: (1) the value of m increases with the antigen's concentration rate $V(t)$; (2) the maximum value of m is unity, in the case of 100% organ damage or zero for a fully healthy organ. The coefficient μ_m describes the rate of generation of the target organ. This model was considered in the recent work of Skvortsova [23]. Adding the control in the model introduced in Marchuk's book [18] is proposed, for example, in the works by Rusakov and Chirkov [21, 22], where the importance of this development is explained.

3 Stabilization Through a Support of the Immune System

Our first goal is the stabilization of the process in the neighborhood of a suitable stationary solution. We make a corresponding linearization and then use the concepts of the stability theory proposed by N. V. Azbelev and his followers in the well-known books [1, 3, 4] for linear functional differential systems. The main idea is to choose "close" in a corresponding sense auxiliary linear system, to solve it and to construct its Cauchy matrix (see, for example, [7, 10, 15]). Then the scheme of the Azbelev W-transform is used. We propose new ideas in choosing "close" systems. For a system of the order n, a corresponding "close" system can be of the order $n + m$. Our main idea here is to reduce the analysis of a given system of the order n to one of the auxiliary systems of the order $n + m$, which is "easy" in a corresponding sense. Results for the auxiliary system allows us to make conclusions for the given system of the order n. We essentially concentrate our attempts in the analysis of the distributed control in an integral form. The integral terms reflect an orientation on average values in the construction of the control. Another reason for the appearance of the integral terms is in the use of the "history of the process" to choose a strategy of a possible treatment. In our model, we demonstrate among other ideas that observation on the process of diseases can be very important in treatment. It should be also noted that the proposed control can be realized practically. To sum up all these consequences, we can conclude that the control in the integral form is reasonable from the medical point of view. Stability properties of integro-differential systems are studied.

Modifying model (2.2), we propose the control in the following form

$$u(t) = -b \int_0^t (F(s) - F^* - \varepsilon)e^{-k(t-s)} ds. \tag{3.1}$$

Adding this control in the third equation of (2.2), we obtain the following system

$$\begin{cases} \frac{dV}{dt} = \beta V(t) - \gamma F(t) V(t) \\ \frac{dC}{dt} = \zeta(m) \alpha F(t) V(t) - \mu_c (C(t) - C^*) \\ \frac{dF}{dt} = \rho C(t) - \eta\gamma F(t) V(t) - \mu_f F(t) + u(t), \\ \frac{dm}{dt} = \sigma V(t) - \mu_m m(t) \end{cases} \tag{3.2}$$

where $u(t)$ is defined by (3.1). Let F^* be the value of the antibody concentration rate for a healthy body. While the case of $F^* > \frac{\beta}{\gamma}$ is considered by G. I. Marchuk in the book [18]. We try to consider the "bad" case where $F^* < \frac{\beta}{\gamma}$. It is clear that system (2.2) could not be stable in this case in the neighborhood of the stationary point $(0, C^*, F^*, 0)$. Consider the following system of five equations

$$\begin{cases} \frac{dV}{dt} = \beta V(t) - \gamma F(t) V(t) \\ \frac{dC}{dt} = \zeta(m) \alpha F(t) V(t) - \mu_c (C(t) - C^*) \\ \frac{dF}{dt} = \rho C(t) - \eta\gamma F(t) V(t) - \mu_f F(t) + u(t) \\ \frac{dm}{dt} = \sigma V(t) - \mu_m m(t) \\ \frac{du}{dt} = -b(F(t) - F^* - \varepsilon) - ku(t) \end{cases} \tag{3.3}$$

The following assertion allows us to reduce analysis of system (3.2) to one of system (3.3).

Lemma 3.1 *The components of the solution-vector* $y(t) = col(v(t), s(t), f(t), m(t))$ *of system (3.2) and four first components of the solution-vector* $x(t) = col(v(t), s(t), f(t), m(t), \tilde{u}(t))$ *of system (3.3) satisfying the initial condition* $u(0) = 0$ *coincide.*

Theorem 3.1 *Let the inequality* $\varepsilon\gamma > \beta - \gamma F^*$, $k > 0$, $b > 0$ *be fulfilled, then the stationary solution* $(0, C^*, F^* + \varepsilon, 0, 0)$ *of system (3.3) is exponentially stable.*

To prove Theorem 3.1, we reduce the analysis of system (3.2) to one of system (3.3) by Lemma 3.1, linearize in the neighborhood of the stationary point $(0, C, F + \varepsilon, 0, 0)$ and then the negativity of roots to the characteristic polynomial of system (3.3) is demonstrated (see, for example, [17]).

Thus, we can stabilize the process at the point $(0, C^*, F^* + \varepsilon, 0, 0)$. It means that we have to support the immune system for a long time and to hold it on the level $F^* + \varepsilon$, where $\varepsilon > \frac{\beta - \gamma F^*}{\gamma}$.

4 Distributed Control and the Lyapunov Characteristic Exponents

To use the control in order to make convergence to set stationary state faster is the second goal. Note that the stationary points present the condition of the healthy body or at least chronical process of disease which we try to reach. This problem is directly related to the duration of a possible treatment. In many cases, this may have an important influence on the choice of treatment method and on the decision on the acceptability of such treatment in principle.

The goal of this part is to obtain faster tending to set stationary state.

Consider the system

$$\begin{cases} \frac{dv}{dt} = \beta v(t) - \gamma F^* f(t) v(t) \\ \frac{ds}{dt} = \alpha V_m \frac{F^*}{C^*} \zeta(m) f(t) v(t) - \mu_c (s(t) - 1) \\ \frac{df}{dt} = \frac{\rho C^*}{F^*} s(t) - \eta\gamma V_m f(t) v(t) - \mu_f f(t) - b\tilde{u}(t), \\ \frac{dm}{dt} = \sigma V_m v(t) - \mu_m m(t) \\ \frac{d\tilde{u}}{dt} = f(t) - 1 - k\tilde{u}(t) \end{cases} \tag{4.1}$$

where $\tilde{u} = \int_0^t (f(s) - 1) e^{-k(t-s)}$. Denoting in (4.1)

$$\alpha_1 = \beta, \alpha_2 = \gamma F^*, \alpha_3 = \alpha V_m \frac{F^*}{C^*}, \alpha_4 = \mu_f = \frac{\rho C^*}{F^*}, \alpha_5 = \mu_c, \alpha_6 = \sigma V_m, \alpha_7 = \mu_m, \alpha_8 = \eta\gamma V_m, \tag{4.2}$$

and linearizing system (4.1) in the neighborhood of stationary point $v = m = \tilde{u} = 0$, $s = f = 1$, we can write system (4.1) in the form

$$\begin{cases} \frac{dx_1}{dt} = (\alpha_1 - \alpha_2) x_1 \\ \frac{dx_2}{dt} = \alpha_3 x_1 - \alpha_5 x_2 \\ \frac{dx_3}{dt} = -\alpha_8 x_1 + \alpha_4 x_2 - \alpha_4 x_3 - b x_5, \\ \frac{dx_4}{dt} = \alpha_6 x_1 - \alpha_7 x_4 \\ \frac{dx_5}{dt} = x_3 - k x_5 \end{cases} \tag{4.3}$$

and to linearize system (2.2) and write it in the form

$$\begin{cases} \frac{dx_1}{dt} = (\alpha_1 - \alpha_2)x_1 \\ \frac{dx_2}{dt} = \alpha_3 x_1 - \alpha_5 x_2 \\ \frac{dx_3}{dt} = -\alpha_8 x_1 + \alpha_4 x_2 - \alpha_4 x_3 \\ \frac{dx_4}{dt} = \alpha_6 x_1 - \alpha_7 x_4 \end{cases}, \quad (4.4)$$

Denote $\lambda_i, i = \overline{1,4}$ the roots of the characteristic polynomial of systems (4.4), and $\tilde{\lambda} = \max_{1 \le i \le 4} \lambda_i$, $\tilde{\lambda^*} = \max_{1 \le j \le 5} \mathbf{Re}(\lambda_j^*)$ of (4.3).

Theorem 4.1 *If $\beta < \gamma F^*$, $b > 0$ and $k > 0$, then integro-differential system (4.3) is exponentially stable and if in addition, the inequality $k > \alpha_4$ is fulfilled, then $\tilde{\lambda} \ge \tilde{\lambda^*}$.*

To prove Theorem 4.1, after reducing analysis of system (3.2) to one of system (3.3) by Lemma 3.1 and linearizing in the neighborhoods of the stationary points $(0, C^-, F^-, 0)$ and $(0, C^-, F^-, 0, 0)$ of system (2.2) and (3.3), respectively, we compare the roots of characteristic polynomials of system (4.3) and (4.4) (see, for example, [7]).

In Figs. 1, 2, 3, and 4, the solution of model of pneumonia with the natural flow of data without the control of disease are presented by curves of red color, disease in the case of considered distributed control-by curves of green color.

Figure 1 demonstrates the dynamics in antigen concentration during the course of the disease. The insert detailing the process in the first two days was performed on a different scale and demonstrates the fact that the management transfers the disease from the "acute" form to the "subclinical" one (the antigen concentration only decreases after injection). Figure 2 demonstrates the dynamics in plasma cell concentration during the disease process. It can be seen from the figure that control leads to a faster increase in the concentration of plasma cells, which in this case ensures a transition to the "subclinical" form of the disease. In addition, it is necessary to note a fourfold increase in the maximum concentration of plasma cells in the case

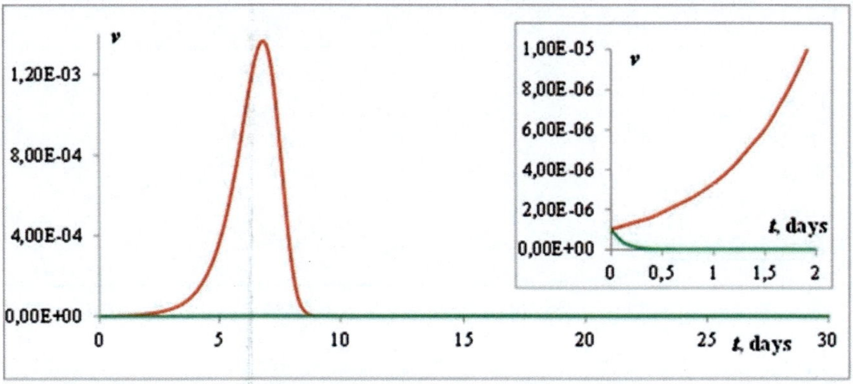

Fig. 1 Dynamics of the immune response: antigen

of control, compared with the option without control. Figure 3 demonstrates the dynamics in antibody concentration during the disease process. The graph shows that the concentration of antibodies in the solution with control practically does not change, because in this case, they are replaced by donor antibodies, which is what the control actually consists of. The dynamics in the proportion of target organ cells destroyed by antigen during the disease process is presented in Fig. 4. The values for the variant with control are given with an increase of 10^4 times. Thus, control allows

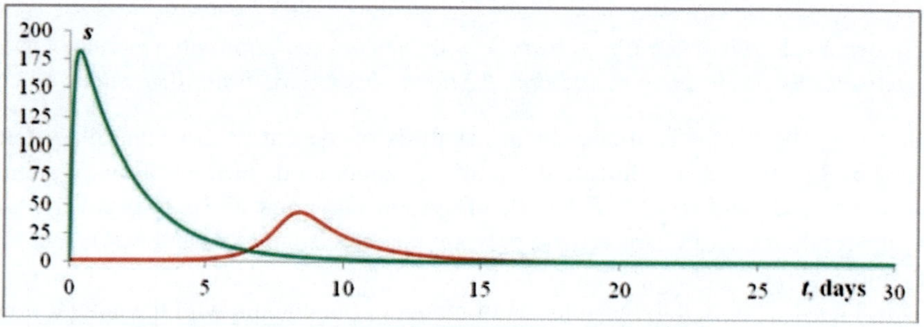

Fig. 2 Dynamics of the immune response: plasma

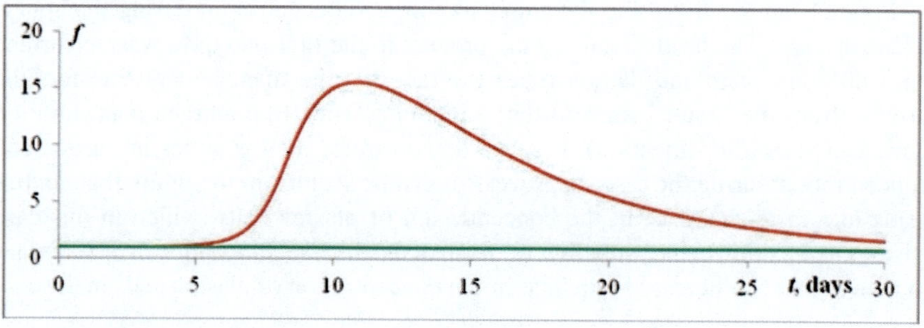

Fig. 3 Dynamics of the immune response: antibodies

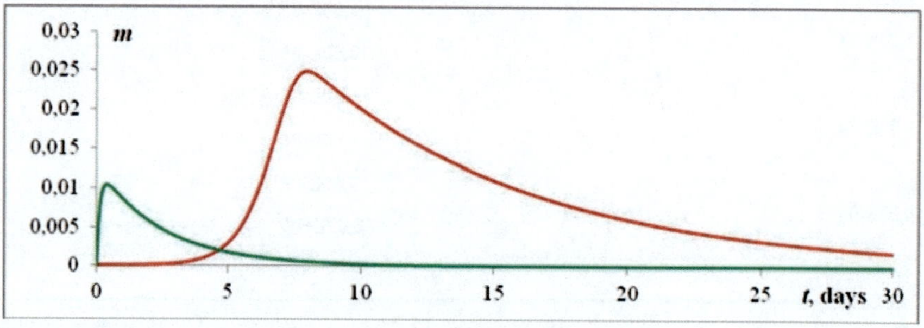

Fig. 4 Dynamics of the immune response: rate of the destroyed cells

to reduce the maximum proportion of affected cells of the target organ by more than 2.5×10^4 times.

5 Cauchy Matrix

To estimate the size of the neighborhood of the stationary solutions which usually describe the states of the healthy body is the third goal of our research. In practical problems, it is necessary since we have to hold the process in a corresponding zone. The process going beyond a certain admissible neighborhood of a stationary solution may be dangerous for patients.

In constructing every model, the influences of various additional factors that have seemed to be nonessential were neglected. The influence effect of choosing nonlinear terms by their linearization in the neighborhood of stationary solution is also neglected. Even in the frame of linearized model, only approximate values of coefficients instead of exact ones are used. Changes of these coefficients with respect to time are not usually taken into account. It looks important to estimate the influence of all these factors.

In order to make this, we have to obtain estimates of the elements of the Cauchy matrix of corresponding linearized (in a neighborhood of a stationary point) system. Consider the system

$$x'(t) = P(t)x(t) + G(t),$$

where $P(t)$ is a $(n \times n)$-matrix, $G(t)$ is n-vector. Its general solution $x(t) = col\{x_1(t), \dots x_n(t)\}$ can be represented in the form (see, for example [1])

$$x(t) = \int_0^t C(t,s)G(s)ds + C(t,0)x(0),$$

where $n \times n$-matrix $C(t,s)$ is called the Cauchy matrix. Its jth column ($j = 1, \dots, n$) for every fixed s as a function of t, is a solution of the corresponding homogeneous system

$$x'(t) = P(t)x(t),$$

satisfying the initial conditions $x_i(s) = \delta_{ij}$, where

$$\delta_{ij} = \begin{cases} 1, & i = j, \\ 0, & i \neq j, \end{cases} \quad i = 1, \dots, n,$$

This Cauchy matrix $C(t,s)$ satisfies the following symmetric properties $C(t,s) = X(s)X^{-1}(s)$, where $X(t)$ is a fundamental matrix, $C(t,0) = C(t,s)C(s,0)$, and in the case of constant matrix $P(t) = P$, we have $X(t-s) = C(t,s)$ is a fundamental matrix for every $s \geq 0$. These definitions and properties allow us to construct and estimate $C(t, s)$. The construction of the Cauchy matrix of system (4.3) can be found, for example, in [10].

6 Stabilization with the Use of Uncertain Coefficient in the Control

Consider the following system of equations with uncertain coefficient in the control

$$\begin{cases} \frac{dV}{dt} = \beta V(t) - \gamma F(t) V(t) \\ \frac{dC}{dt} = \zeta(m(t)) \alpha F(t) V(t) - \mu_c (C(t) - C^*) \\ \frac{dF}{dt} = \rho C - \eta \gamma F(t) V(t) - \mu_f F(t) - (b + \Delta b(t)) u(t), \\ \frac{dm}{dt} = \sigma V(t) - \mu_m m(t) \\ \frac{du}{dt} = F(t) - F^* - ku(t) \end{cases} \tag{6.1}$$

where $u(t) = \int_0^t (F(s) - F^*) e^{-k(t-s)} ds$

This system can be rewritten in the form

$$\begin{cases} x_1' = (a_1 - a_2) x_1 + g_1(x_1(t), x_3(t)) \\ x_2' = a_3 x_1 - a_5 x_2 + g_2(x_1(t), x_3(t)) \\ x_3' = -a_8 x_1 + a_4 x_2 - a_4 x_3 - (b + \Delta b(t)) x_5 + g_3(x_1(t), x_3(t)), \\ x_4' = a_6 x_1 - a_7 x_4 \\ x_5' = x_3 - k x_5 \end{cases} \tag{6.2}$$

where $g_i(x_1(t), x_3(t))(t)$, $1 \leq i \leq 3$ results of "mistakes" we made in the process of the linearization.

Consider the system

$$X' = AX + \Delta B(t) X + F(t), \tag{6.3}$$

where

$$X(t) = \begin{pmatrix} x_1(t) \\ x_2(t) \\ x_3(t) \\ x_4(t) \\ x_5(t) \end{pmatrix}, \quad \Delta B(t) = \begin{pmatrix} 0 & 0 & 0 & 0 & 0 \\ 0 & 0 & 0 & 0 & 0 \\ 0 & 0 & 0 & 0 & -\Delta b(t) \\ 0 & 0 & 0 & 0 & 0 \\ 0 & 0 & 0 & 0 & 0 \end{pmatrix}.$$

On the basis of the estimates of the elements of the Cauchy matrix, we obtain the following assertions on the stability of system (6.2). Denoting $Q_j = ess\ \sup_{t\geq 0}\int_0^t \sum_{i=1}^5 \left|(\Delta B\,(t)\, C(t,s))_{ij}\right| ds$ and $\Delta b^* = ess\ \sup_{t\geq 0} |\Delta b(t)|$, we obtain the estimates:

$$
\begin{aligned}
Q_1 &\leq \Delta b^* \left[\begin{array}{c} \left|\frac{\alpha_{24}(\alpha_{32}-\alpha_{35})-\alpha_{25}(\alpha_{32}-\alpha_{34})}{\alpha_{15}\alpha_{24}(\alpha_{31}-\alpha_{32})}\right| \frac{1}{|\lambda_1|} + \\ \left|\frac{\alpha_{24}(\alpha_{31}-\alpha_{35})-\alpha_{25}(\alpha_{31}-\alpha_{34})}{\alpha_{15}\alpha_{24}(\alpha_{31}-\alpha_{32})}\right| \frac{1}{|\lambda_2|} + \left|\frac{\alpha_{25}}{a_5\alpha_{15}\alpha_{24}}\right| \end{array} \right], \\
Q_2 &\leq \Delta b^* \left[\left|\frac{\alpha_{32}-\alpha_{34}}{\alpha_{24}(\alpha_{31}-\alpha_{32})}\right| \frac{1}{|\lambda_1|} + \left|\frac{\alpha_{31}-\alpha_{34}}{\alpha_{24}(\alpha_{31}-\alpha_{32})}\right| \frac{1}{|\lambda_2|} + \frac{1}{|a_5\alpha_{24}|} \right], \\
Q_3 &\leq \Delta b^* \left[\frac{1}{|\alpha_{31}-\alpha_{32}|}\frac{1}{|\lambda_1|} + \frac{1}{|\alpha_{31}-\alpha_{32}|}\frac{1}{|\lambda_2|} \right], \\
Q_4 &= 0, \\
Q_5 &\leq \Delta b^* \left[\left|\frac{\alpha_{32}}{\alpha_{31}-\alpha_{32}}\right| \frac{1}{|\lambda_1|} + \left|\frac{\alpha_{31}}{\alpha_{31}-\alpha_{32}}\right| \frac{1}{|\lambda_2|} \right].
\end{aligned}
\tag{6.4}
$$

Theorem 6.1 ([10]) *Let* $k > 0$, $b > 0$ *and* a_i, $1 \leq i \leq 8$, *are real positive and different,* $a_1 < a_2$, $(a_4 - k)^2 > 4b$ *and the inequality* $\max_{1\leq j\leq 5}\left\{\left|Q_j\right|\right\} < 1$ *be true. Then system* (6.2) *is exponential stable.*

Denoting $P_j = ess\ \sup_{t\geq 0}\int_0^t \sum_{i=1}^5 \left|(\Delta B\,(t)\, C(t,s))_{ij}\right| ds$, we obtain the estimates

$$
\begin{aligned}
P_1 &\leq \Delta b^* \left[\begin{array}{c} \left|\frac{\beta_{24}\beta_{35}-\beta_{25}\beta_{34}}{\beta_{31}\beta_{15}\beta_{24}}\right| \frac{4}{(a_4+k)^2} + \left|\frac{\beta_{24}(\beta_{31}-\beta_{35})-\beta_{25}(\beta_{31}-\beta_{34})}{\beta_{31}\beta_{52}\beta_{24}\beta_{15}}\right| \frac{4}{|a_4^2-k^2|} \\ + \left|\frac{\beta_{25}}{\beta_{15}\beta_{24}}\right| \frac{1}{|a_5|} + \frac{1}{|\beta_{15}|}\frac{1}{|a_1-a_2|} \end{array} \right], \\
P_2 &\leq \Delta b^* \left[\left|\frac{\beta_{34}}{\beta_{24}\beta_{31}}\right| \frac{4}{(a_4+k)^2} + \left|\frac{\beta_{31}-\beta_{34}}{\beta_{31}\beta_{24}\beta_{52}}\right| \frac{4}{|a_4^2-k^2|} + \frac{1}{|\beta_{24}|}\frac{1}{|a_5|} \right], \\
P_3 &\leq \Delta b^* \left[\frac{1}{|\beta_{31}|}\frac{4}{(a_4+k)^2} + \frac{1}{|\beta_{31}\beta_{52}|}\frac{4}{|a_4^2-k^2|} \right], \\
P_4 &= 0, \\
P_5 &\leq \Delta b^* \frac{1}{|\beta_{52}|}\frac{4}{|a_4^2-k^2|}.
\end{aligned}
\tag{6.5}
$$

Theorem 6.2 ([10]) *Let* $k > 0$, $b > 0$ *and* a_i, $1 \leq i \leq 8$, *are real positive and different,* $a_1 < a_2$, $(a_4 - k)^2 = 4b$ *and the inequality* $\max_{1\leq j\leq 5}\left\{\left|P_j\right|\right\} < 1$ *be true. Then system (6.2) is exponential stable.*

Denoting $R_j = ess\ \sup_{t\geq 0} \int\limits_0^t \sum\limits_{i=1}^{5} \left|(\Delta B(t)\, C(t,s))_{ij}\right| ds$ we obtain estimates

$$
\begin{gathered}
R_1 \leq \Delta b^* \left[\begin{array}{c} \left|\frac{\gamma_{24}-\gamma_{25}}{\gamma_{15}\gamma_{24}}\right| \frac{2}{|a_4+k|} + \left|\frac{\gamma_{24}(2\gamma_{35}-a_4+k)+\gamma_{25}(a_4-2a_5+3k)}{\gamma_{32}\gamma_{15}\gamma_{24}}\right| \frac{1}{|a_4+k|} \\ + \left|\frac{\gamma_{25}}{\gamma_{15}\gamma_{24}}\right| \frac{1}{|a_5|} + \left|\frac{1}{\gamma_{15}}\right| \frac{1}{|a_1-a_2|} \end{array} \right], \\
R_2 \leq \Delta b^* \left[\frac{1}{|\gamma_{24}|}\frac{2}{|a_4+k|} + \left|\frac{a_4-3k+2a_5}{\gamma_{24}\gamma_{32}}\right| \frac{1}{|a_4+k|} + \frac{1}{|\gamma_{24}|}\frac{1}{|a_5|}\right], \\
R_3 \leq \Delta b^* \frac{1}{|\gamma_{32}|}\frac{2}{|a_4+k|}, \\
R_4 = 0, \\
R_5 \leq \Delta b^* \left[\frac{2}{|a_4+k|} + \left|\frac{a_4-k}{\gamma_{32}}\right| \frac{1}{|a_4+k|}\right].
\end{gathered}
\tag{6.6}
$$

Theorem 6.3 ([10]) *Let $k > 0$, $b > 0$ and a_i, $1 \leq i \leq 8$, are real positive and different, $a_1 < a_2$, $(a_4 - k)^2 < 4b$ and the inequality* $\max_{1\leq j\leq 5}\{|R_j|\} < 1$ *be true. Then system (6.2) is exponential stable.*

Remark 6.1 Note that the approach presented here can be used in the model of testosterone regulation (see, for example [11–14]).

Acknowledgements This paper is part of the third and fourth author's Ph.D. thesis which is being carried out in the Department of Mathematics at Ariel University.

References

1. Agarwal, R., Berezansky, L., Braverman, E., Domoshnitsky, A.: Nonoscillation Theory of Functional Differential Equations with Applications. Springer, Berlin (2012)
2. Asachenkov, A.L., Marchuk, G.I.: The specified mathematical model of an infectious disease. In: Marchuk, G.I. (ed.) Mathematical Modeling in Immunology and Medicine, pp. 44–59. Science, Novosibirsk (1982). (in Russian)
3. Azbelev, N.V., Simonov, P.M.: Stability of differential equations with aftereffect. Reference - 240 Pages, ISBN 9780415269575 - CAT# TF1327 (2002)
4. Azbelev, N.V., Maksimov, V.P., Rakhmatullina, L.F.: Introduction to Theory of Functional-Differential Equations. Nauka, Moscow (1991)
5. Bard, Y.: Nonlinear estimation of parameters. Statistics (1979) (in Russian)
6. Belih, L.N.: O the numerical solution of models of diseases. In: Marchuk, G.I., Belih, L.N. (eds.) Mathematical Models in Immunology and Medicine, pp. 291–297. World (1986). (in Russian)
7. Bershadsky, M., Chirkov, M., Domoshnitsky, A., Rusakov, S., Volinsky, I.: Distributed control and the lyapunov characteristic exponents in the model of infectious diseases. Complexity **2019**, Article ID 5234854. Published 13 November 2019. https://doi.org/10.1155/2019/5234854
8. Chirkov, M.V.: Parameter Identification and Control in Mathematical Models of the Immune Response. Thesis. Perm State University, Perm (2014)
9. Domoshnitsky A., Volinsky I., Polonsky A.: Stabilization of third order differential equation by delay distributed feedback control. Math. Slovaca **69**(5), 1165–1175 (2019). https://doi.org/10.1515/ms-2017-0298

10. Domoshnitsky, A., Volinsky, I., Bershadsky, M.: Around the model of infection disease: the cauchy matrix and its properties. Symmetry **11**(8), 1016 (2019). https://doi.org/10.3390/sym11081016 (Published: August 6, 2019)
11. Domoshnitsky, A., Volinsky, I., Pinhasov, O., Bershadsky, M.: Questions of stability of functional differential systems around the model of testosterone regulation. Bound. Value Prob. **2019**(1), 1–13 (2019). https://doi.org/10.1186/s13661-019-01295-2
12. Domoshnitsky, A., Volinsky, I., Pinhasov, O.: Some developments in the model of testosterone regulation. AIP Conf. Proc. **2159**, 030010 (2019). https://doi.org/10.1063/1.5127475
13. Domoshnitsky, A., Volinsky, I., Shklyar, R.: About Green's functions for impulsive differential equations. Funct. Differ. Equ. **20**(1–2), 55–81 (2013)
14. Domoshnitsky, A., Volinsky, I.: About differential inequalities for nonlocal boundary value problems with impulsive delay equations. Math. Bohem. **140**(2), 121–128 (2015)
15. Domoshnitsky, A., Volinsky, I., Polonsky, A., Sitkin, A.: Practical constructing the Cauchy function of integro-differential equation. Funct. Differ. Equ. **23**(3–4), 109–118 (2016)
16. Domoshnitsky, A., Volinsky, I., Polonsky, A., Sitkin, A.: Stabilization by delay distributed feedback control. Math. Model. Nat. Phenom. **12**(6), 91–105 (2017)
17. Domoshnitsky, A., Bershadsky, M., Volinsky, I.: Distributed control in stabilization of model of infection diseases. Russ. J. Biomech. **23**(4), 494–499 (2019). https://doi.org/10.15593/RJBiomech/2019.4.08
18. Marchuk, G.I.: Mathematical Modelling of Immune Response in Infection Diseases. Mathematics and Its Applications. Springer, Berlin (1997)
19. Martsenyuk, V.P., Andrushchak, I.Ye., Gvozdetska, I.S.: Qualitiative analysis of the antineoplastic immunitty system on the basis of a decision tree. Cybern. Syst. Anal. **51**(3) (2015)
20. Martsenyuk, V.P.: Construction and study of stability of an antitumor immunity model. Cybern. Syst. Anal. **40**(5) (2004)
21. Rusakov, S.V., Chirkov, M.V.: Mathematical model of influence of immunotherapy on dynamics of immune response. Probl. Control **6**, 45–50 (2012)
22. Rusakov, S.V., Chirkov, M.V.: Identification of parameters and control in mathematical models of immune response. Russ. J. Biomech. **18**(2), 259–269 (2014)
23. Skvortsova, M.: Asymptotic properties of solutions in Marchuk's basic model of disease. Funct. Differ. Equ. **24**(3–4), 127–135 (2017)

The Second Lyapunov Method for Time-Delay Systems

G. V. Demidenko and I. I. Matveeva

Abstract Some classes of systems of delay differential equations are considered. We give a review of methods for the study of the stability of solutions in the case of constant and periodic coefficients in linear terms. Special attention is paid to the development of the second Lyapunov method. A number of authors' results for linear and nonlinear delay differential equations obtained by using various Lyapunov–Krasovskii functionals are presented. The application of discrete analogs of the constructed functionals to the study of the stability of solutions to delay difference equations is discussed.

Keywords Time-delay systems · Periodic coefficients · Exponential stability · Lyapunov-Krasovskii functionals · Estimates for solutions

1 Introduction

The theory of stability of solutions to delay equations started to develop in the middle of the previous century. The foundations of the theory were laid in the works of Andronov and Mayer [3], Bellman [8], Zverkin [122], Krasovskii [70], Myshkis [90], Pontryagin [95], Razumikhin [96], Chebotarev and Meyman [13], El'sgol'ts [40], and others. The great interest of mathematicians to delay equations in those years was associated with the necessity to solve important applied problems in which time-delay effect played a significant role. Delay equations appear in a large amount of automatic regulation and control theory problems, in problems of automation and mechanics, radiophysics, when modeling immunology processes, when studying

G. V. Demidenko (✉) · I. I. Matveeva
Sobolev Institute of Mathematics, Acad. Koptyug avenue 4, 630090 Novosibirsk, Russia
e-mail: demidenk@math.nsc.ru

I. I. Matveeva
e-mail: matveeva@math.nsc.ru

Novosibirsk State University, Pirogov street 1, 630090 Novosibirsk, Russia

A. Domoshnitsky et al. (eds.), *Functional Differential Equations and Applications*, Springer Proceedings in Mathematics & Statistics 379,
https://doi.org/10.1007/978-981-16-6297-3_11

gene networks, economics, etc. (e.g. [6, 10, 11, 15, 20, 43, 49–51, 72, 74, 76, 86, 89, 92, 93, 101, 110, 112, 114, 116, 118, 121]).

Nowadays, there are a huge number of works devoted to various problems for delay equations, in particular, stability problems. A number of monographs are dedicated to this subject (e.g. Myshkis [90] (1951), [91] (1972), El'sgol'ts [40] (1955), [41] (1964), Krasovskii [71] (1959), Pinney [94] (1958), Bellman and Cooke [9] (1963), Rubanik [102] (1969), Halanay and Wexler [57] (1971), El'sgol'ts and Norkin [42] (1971), Mitropol'skii and Martynuk [87] (1979), Kolmanovskii and Nosov [65] (1981), Shimanov [104] (1983), Razumikhin [97] (1977), Hale [55] (1984), Korenevskii [68] (1989), [69] (2008), Azbelev, Maksimov and Rakhmatullina [4] (1991), Györi and Ladas [52] (1991), Dolgii [39] (1996), Kolmanovskii and Myshkis [66] (1999), Azbelev and Simonov [5] (2001), Vlasov and Medvedev [115] (2008), Agarwal, Berezansky, Braverman and Domoshnitsky [1] (2012), Kharitonov [60] (2013), Gil' [46] (2014), Michiels and Niculescu [85] (2014), Romanovskii, Belgart, Dobrovolskii, Rogozin and Trotsenko [100] (2015) and others). However, despite the rapid development of the stability theory, there are many unsolved questions. For example, how to describe the maximal attraction domain for nonlinear equations, how to specify the maximal stabilization rate of solutions at infinity, what algorithm should be used in numerical studies of stability of solutions with the guaranteed accuracy, etc.? These questions are especially actual for nonautonomous equations.

In this work, we consider some classes of time-delay systems of the form

$$\frac{d}{dt}y(t) = F\left(t, y(t), y(t-\tau), \frac{d}{dt}y(t-\tau)\right), \quad t > 0. \tag{1}$$

Here we give a brief overview of our results on the asymptotic stability of the zero solution to these classes obtained in recent years. We also present some of our results for delay differential equations.

By now, the most investigated problems are the problems on the stability of stationary solutions to autonomous delay equations, herewith spectral methods are widely used. They are based on the spectral criteria of the asymptotic stability. According to such criterion, the asymptotic stability of the zero solution to the linear time-delay system

$$\frac{d}{dt}y(t) = Ay(t) + By(t-\tau), \quad t > 0, \tag{2}$$

is equivalent to the location of the roots of the quasipolynomial

$$\det\left(A + e^{-\lambda\tau}B - \lambda I\right) = 0 \tag{3}$$

in the left half-plane $C_- = \{\lambda \in C : \operatorname{Re}\lambda < 0\}$ (for example, see [41, 55]). In this case, the asymptotic stability of solutions implies the exponential stability. For linear time-delay systems of the form

$$\frac{d}{dt}(y(t) + Dy(t-\tau)) = Ay(t) + By(t-\tau), \quad t > 0, \tag{4}$$

the necessary condition for the asymptotic stability of the zero solution is the requirement that the roots of the quasipolynomial

$$\det\,(A + e^{-\lambda\tau}B - \lambda I - \lambda e^{-\lambda\tau}D) = 0 \tag{5}$$

belong to the left half-plane C_-. A sufficient condition for the exponential stability of the zero solution to (4) is the requirement that the roots of (5) belong to the left half-plane $C_{-,\gamma} = \{\lambda \in C : \mathrm{Re}\,\lambda \le \gamma < 0\}$ (for instance, see [41, 55]). If $D \neq 0$ systems of the form (4) are called systems of *neutral type*. Such equations were introduced in the book [40].

According to the theorems on stability in the first approximation, a sufficient condition for the asymptotic stability of the zero solution for a wide class of nonlinear systems of the form (1) is also the requirement that the roots of the corresponding quasipolynomials belong to the left half-plane (for example, see [9, 41, 55]). However, investigating the stability of the zero solution to systems of the form (1), the verification of this condition can be a rather complicated task. On the one hand, the quasipolynomials can have a countable number of roots, on the other hand, the problem of finding roots of the quasipolynomials is, generally speaking, *ill-conditioned*. This can be a serious obstacle in studying the stability of solutions by using computer software. Therefore, as in the case of systems of ordinary differential equations, when studying the asymptotic stability of solutions to time-delay systems, various criteria on the location of quasipolynomial roots in the left half-plane become important. For this purpose, researchers often use the *method of D-decompositions* (for instance, see [113]), *amplitude-phase method* (for example, see [41]), *Chebotarev–Meyman method* [13], and also methods based on *analogs* of Lyapunov's theorems [70, 96].

2 Lyapunov–Krasovskii Functionals

One of the most common is the method based on the use of *Lyapunov–Krasovskii functionals* [71], which is a development of the second Lyapunov method. The advantage of this method is the simplicity of formulated statements and the reduction of studying asymptotic stability to solving well-conditioned problems. Below we formulate the well-known result obtained by this method for the linear system (2).

Theorem 1 (N. N. Krasovskii) *Suppose that there exist matrices $H = H^* > 0$ and $K = K^* > 0$ such that the matrix*

$$\begin{pmatrix} HA + A^*H + K & HB \\ B^*H & -K \end{pmatrix}$$

is negative definite. Then the zero solution to (2) is asymptotically stable.

In the proof of this assertion, the following Lyapunov–Krasovskii functional was used

$$\langle Hy(t), y(t)\rangle + \int_{t-\tau}^{t} \langle Ky(s), y(s)\rangle ds. \tag{6}$$

The functional (6) is an analog of the Lyapunov function $\langle Hy, y\rangle$ for the system of ordinary differential equations

$$\frac{dy}{dt} = Ay + F(y), \quad t > 0, \tag{7}$$

which is constructed by using the solution to the Lyapunov equation

$$HA + A^*H = -C, \quad C = C^* > 0. \tag{8}$$

It is well known that this equation plays a significant role in the study of the asymptotic stability of solutions to systems of the forms (7). In particular, using the solution to (8), we can obtain an estimate for solutions to the linear system

$$\frac{dy}{dt} = Ay, \quad t > 0,$$

which characterizes the decay rate at infinity. For example, in the case of $C = I$ the following Krein's estimate is valid [14]:

$$\|y(t)\| \le \sqrt{2\|A\|\|H\|} \exp\left(-\frac{t}{2\|H\|}\right) \|y(0)\|, \quad t > 0. \tag{9}$$

Using the solution to the Lyapunov equation (8), we can also estimate the attraction domain of the zero solution to the nonlinear system (7) and establish estimates of the exponential decay of its solutions without finding the spectrum of the matrix A. We emphasize that, in contrast to the problem of finding the matrix spectrum, the construction of the solution to (8) is a well-conditioned problem (see [47]). That is why the approach based on the use of the Lyapunov equation became a basis for the development of numerical methods for studying asymptotic stability of solutions to ordinary differential equations with *guaranteed accuracy* [48].

It should be noted that the Lyapunov–Krasovskii functional (6) can also be used in the study of the asymptotic stability of solutions to nonlinear time-delay systems. However, in contrast to the Lyapunov function used for proving Krein's estimate (9), the use of the functional (6) does not allow us to obtain analogs of Krein's estimate for time-delay systems. A review of some results obtained by means of Lyapunov–Krasovskii functionals until 2003 is given in [98].

The problem of obtaining analogs of Krein's estimate for solutions to the linear time-delay systems (2) by using some Lyapunov–Krasovskii type functionals without finding the roots of the quasipolynomials has been solved relatively recently [21, 58, 62, 88]. (It is interesting to note that all these articles were published almost simultaneously!). The methods proposed in these works are based on the use of various Lyapunov–Krasovskii functionals.

In particular, in [21] the following Lyapunov–Krasovskii functional was proposed

$$V(t, y) = \langle Hy(t), y(t)\rangle + \int_{t-\tau}^{t} \langle K(t-s)y(s), y(s)\rangle ds, \tag{10}$$

where $H = H^* > 0$ and $K(s) = K^*(s) > 0$. Note that, in contrast to the functional (6), here the matrix K is variable. Below we present the result of [21] for the linear system (2).

Consider the initial value problem for (2)

$$\begin{cases} \dfrac{d}{dt}y(t) = Ay(t) + By(t-\tau), \quad t > 0, \\ y(t) = \varphi(t), \quad t \in [-\tau, 0], \quad y(+0) = \varphi(0), \end{cases} \tag{11}$$

where $\varphi(t) \in C([-\tau, 0])$ is a given vector-function. The following theorem is valid.

Theorem 2 ([21]) *Suppose that there exist matrices $H = H^* > 0$ and $K(s) \in C^1([0, \tau])$ such that*

$$K(s) = K^*(s) > 0, \quad \frac{d}{ds}K(s) < 0, \quad s \in [0, \tau],$$

and the matrix

$$\mathbf{C} = -\begin{pmatrix} HA + A^*H + K(0) & HB \\ B^*H & -K(\tau) \end{pmatrix}$$

is positive definite. Then, for the solution $y(t)$ to the initial value problem (11), the following estimate holds

$$\|y(t)\| \le \sqrt{\|H^{-1}\| V(0, \varphi)}\, e^{-\varepsilon t/2}, \tag{12}$$

where

$$\varepsilon = \min\left\{\frac{c_{\min}}{\|H\|}, k\right\},$$

$c_{\min}$ are the minimal eigenvalues of the matrix $\mathbf{C}$, $k > 0$ is the maximal number such that

$$\frac{d}{ds}K(s) + kK(s) \le 0, \quad s \in [0, \tau].$$

The estimate (12) is an analog of Krein's estimate (9) for solutions to ordinary differential equations. We emphasize that this estimate was received without using the information about the location of the roots of (3).

In [21], it is also shown that the use of the Lyapunov–Krasovskii functional (10) allows us to obtain analogs of Krein's estimate for solutions to **nonlinear** time-delay systems and to determine attraction regions of the zero solution, thereby establishing the exponential stability of the zero solution. In this context, we present a brief extract from [84]: "To the best of the authors' knowledge, there is no similar constructive method for delay systems of the form $\dot{x}(t) = Ax(t) + Bx(t-h) + f(x(t), x(t-h))$. This is to provide estimates of the attraction region by using explicitly a quadratic Lyapunov–Krasovskii functional associated to the exponentially stable linear system $\dot{x}(t) = Ax(t) + Bx(t-h)$." We should add to this that our work [21] was published in 2005, in Russian, while [84] was published in 2007, in English ...In addition to the mentioned results for autonomous time-delay systems, in [21, 22], similar results were established for the first time for time-delay systems with periodic coefficients.

For linear time-delay systems of neutral type (4), in [16], it was proposed to use the following Lyapunov–Krasovskii functional

$$V(t, y) = \langle H(y(t) + Dy(t-\tau)), (y(t) + Dy(t-\tau))\rangle + \int_{t-\tau}^{t} \langle K(t-s)y(s), y(s)\rangle ds, \qquad (13)$$

where $H = H^* > 0$ and $K(s) = K^*(s) > 0$. Using this functional, in [16] estimates characterizing the exponential decay of solutions to (4) at infinity were obtained. Below we present one of the results from [16].

Consider the initial value problem

$$\begin{cases} \dfrac{d}{dt}(y(t) + Dy(t-\tau)) = Ay(t) + By(t-\tau), \quad t > 0, \\ y(t) = \varphi(t), \quad t \in [-\tau, 0], \quad y(+0) = \varphi(0), \end{cases} \qquad (14)$$

where $\varphi(t) \in C^1([-\tau, 0])$ is a given vector-function. The following theorem is valid.

Theorem 3 ([16]) *Let* $\|D\| < 1$. *Suppose that there exist matrices*

$$H = H^* > 0, \quad K(s) \in C^1([0, \tau])$$

such that

$$K(s) = K^*(s) > 0, \quad \frac{d}{ds}K(s) < 0, \quad s \in [0, \tau],$$

and

$$\mathbf{C} = -\begin{pmatrix} HA + A^*H + K(0) & HB + A^*HD \\ B^*H + D^*HA & D^*HB + B^*HD - K(\tau) \end{pmatrix} > 0.$$

Then the zero solution to (4) is exponentially stable.

In [16], there were also established estimates for solutions to (14), similar to Krein's estimate (9). From these estimates, it follows that all solutions decrease with an exponential rate and the rate significantly depends on the matrix D. We emphasize that the estimates are obtained without any information about the roots of the quasipolynomial (5). In [24], the results from [16] were generalized to the case when the spectrum of the matrix D belongs to the unit disk $\{\lambda \in C : |\lambda| < 1\}$. In [23, 26, 28, 32] similar results were obtained for nonlinear time-delay systems of the form

$$\frac{d}{dt}(y(t) + Dy(t-\tau)) = Ay(t) + By(t-\tau) + F(t, y(t), y(t-\tau)), \quad t > 0.$$

Somewhat different types of estimates for solutions to time-delay systems of neutral type (4) were obtained in [7, 59, 62].

In contrast to autonomous delay equations, the problem of stability of solutions to nonautonomous delay equations is essentially less studied. Basic studies for nonautonomous equations were carried out for linear delay equations with periodic coefficients

$$\frac{d}{dt}y(t) = A(t)y(t) + B(t)y(t-\tau), \quad t > 0, \tag{15}$$

$$A(t+T) \equiv A(t), \quad B(t+T) \equiv B(t).$$

The foundations of stability theory for equations of the form (15) were laid in the works of Zverkin [123], Stokes [111], Halanay [56], Hahn [53], Hale [55], Shimanov [104]. The main approach in these studies is the development of *Floquet theory* and the use of *monodromy operator*. This approach was also used to study the stability of solutions to linear time-delay systems of neutral type with periodic coefficients of the form

$$\frac{d}{dt}(y(t) + D(t)y(t-\tau)) = A(t)y(t) + B(t)y(t-\tau), \quad t > 0, \tag{16}$$

$$D(t+T) \equiv D(t), \quad A(t+T) \equiv A(t), \quad B(t+T) \equiv B(t).$$

Further development of the Floquet theory for delay equations is described, for example, in the papers [39, 44, 45, 67, 73], and others.

In addition to this approach to the stability problem of solutions to systems of the form (15), (16), the following methods have been developed in the literature: the *method of generating functions* (e.g. [75, 103]), the *method of monotone operators* (e.g. [4, 12]), the *method of Lyapunov–Krasovskii functionals* (e.g. [38, 63–65]). There are also some generalizations to the case of delay equations with almost periodic coefficients (for example, see [2, 99]).

It should be noted that it is often difficult to verify conditions for the asymptotic stability of solutions to nonautonomous delay equations. Difficulties also arise when describing attraction domains in the case of nonlinear delay equations and when obtaining asymptotic estimates of solutions as $t \to \infty$.

For the first time, analogs of Krein's estimate for solutions to time-delay systems of the form

$$\frac{d}{dt}y(t) = A(t)y(t) + B(t)y(t-\tau) + F(t, y(t), y(t-\tau)), \quad t > 0, \tag{17}$$

$$A(t+T) \equiv A(t), \quad B(t+T) \equiv B(t).$$

were obtained in [21, 22]. A Lyapunov–Krasovskii functional was proposed in order to obtain these estimates. To describe this functional, we first give the criterion for the asymptotic stability of solutions to linear systems of ordinary differential equations with periodic coefficients of the form

$$\frac{d}{dt}y(t) = A(t)y(t), \quad A(t+T) \equiv A(t), \quad t \geq 0, \tag{18}$$

established in [17].

Consider the boundary value problem for the Lyapunov differential equation

$$\begin{cases} \frac{d}{dt}H + HA(t) + A^*(t)H = -Q(t), & t \in [0, T], \\ H(0) = H(T), \end{cases} \tag{19}$$

where $Q(t) = Q^*(t) > 0$ is a matrix with continuous entries on $[0, T]$. The following criterion for the asymptotic stability of the zero solution to (18) is valid:

Theorem 4 ([17])

I. If the zero solution to (18) is asymptotically stable, then there exists a unique solution $H(t) = H^*(t) > 0$ to the boundary value problem (19).

II. If the boundary value problem (19) has a solution $H(t) = H^*(t)$ such that $H(0) > 0$, then the zero solution to (18) is asymptotically stable.

Using this criterion, an estimate for solutions to the linear system (18) was obtained in [17]. Let $Q(t) \equiv I$, then, for the solutions to (18), we have

$$\|y(t)\| \leq \sqrt{\max_{\xi\in[0,T]} \|H(\xi)\| \|H(0)\|} \exp\left(- \int_0^t \frac{1}{2\|H(\xi)\|} d\xi \right) \|y(0)\|, \quad t > 0. \tag{20}$$

Here the symbol $H(t)$ denotes T-periodic continuation of the solution to (19) on the whole semi-axis $\{t \geq 0\}$. The estimate (20) is an analog of Krein's estimate.

Note that, by the use of the solution $H(t)$ to the boundary value problem (19), in [18, 19] the asymptotic stability of stationary solutions of nonlinear systems of ordinary differential equations was studied. In particular, an independent proof of Bogolyubov's theorem on the stability of an inverted pendulum with a vibrating point of suspension was obtained.

The advantage of using the solution $H(t)$ to (19) in comparison, for example, with the calculation of multipliers, is the fact that the problem of constructing the solution to (19) is well-conditioned.

On the base of the mentioned criterion, the authors of [21] introduced the following Lyapunov–Krasovskii functional

$$V(t, y) = \langle H(t)y(t), y(t)\rangle + \int_{t-\tau}^{t} \langle K(t-s)y(s), y(s)\rangle ds \tag{21}$$

and proposed to use it for the study of the asymptotic stability of solutions to the time-delay systems (17).

Below we give one of the results for the linear time-delay systems (17) ($F(t, u, v) \equiv 0$). Consider the initial value problem

$$\begin{cases} \dfrac{d}{dt}y(t) = A(t)y(t) + B(t)y(t-\tau), \quad t > 0, \\ y(t) = \varphi(t), \quad t \in [-\tau, 0], \quad y(+0) = \varphi(0), \end{cases} \tag{22}$$

where $\varphi(t) \in C([-\tau, 0])$ is a given vector-function. The following theorem is valid.

Theorem 5 ([21]) *Suppose that there exist matrices*

$$H(t) = H^*(t) \in C^1([0, T]) \quad \text{and} \quad K(s) = K^*(s) \in C^1([0, \tau])$$

such that

$$H(0) = H(T) > 0, \quad K(s) > 0, \quad \frac{d}{ds}K(s) < 0, \quad s \in [0, \tau],$$

and the composite matrix

$$\mathbf{C}(t) = -\begin{pmatrix} \frac{d}{dt}H(t) + H(t)A(t) + A^*(t)H(t) + K(0) & H(t)B(t) \\ B^*(t)H(t) & -K(\tau) \end{pmatrix}$$

is positive definite for $t \in [0, T]$. Then for the solution $y(t)$ to the initial value problem (22) the following estimate holds

$$\|y(t)\| \le \sqrt{h_{\min}^{-1}(t)V(0, \varphi)} \exp\left(-\int_0^t \frac{\varepsilon(\xi)}{2} d\xi\right), \tag{23}$$

where

$$\varepsilon(t) = \min\left\{\frac{c_{\min}(t)}{\|H(t)\|}, k\right\},$$

$h_{\min}(t) > 0$ and $c_{\min}(t) > 0$ are the minimal eigenvalues of T-periodic continuations of the matrices $H(t)$ and $\mathbf{C}(t)$, respectively, $k > 0$ is the maximal number such that

$$\frac{d}{ds}K(s) + kK(s) \leq 0, \quad s \in [0, \tau].$$

The inequality (23) is an analog of Krein's estimate (9) for ordinary differential equations, it implies the exponential stability of the zero solution to (17). Studies of the asymptotic stability of the zero solution to the nonlinear time-delay system (17) were carried out in [22, 30, 77].

The use of the Lyapunov–Krasovskii functionals (13) and (21) when obtaining analogs of Krein's estimate led us to the idea to introduce the functional

$$V(t, y) = \langle H(t)(y(t) + Dy(t-\tau)), (y(t) + Dy(t-\tau))\rangle + \int_{t-\tau}^{t} \langle K(t-s)y(s), y(s)\rangle ds \tag{24}$$

in order to study the exponential stability of the zero solution to the systems of neutral type with periodic coefficients in linear terms

$$\frac{d}{dt}(y(t) + Dy(t-\tau)) = A(t)y(t) + B(t)y(t-\tau) + F(t, y(t), y(t-\tau)). \tag{25}$$

Below we formulate one of the results of [25].

Theorem 6 ([25]) *Suppose that there exist matrices*

$$H(t) = H^*(t) \in C^1([0, T]), \quad K(s) = K^*(s) \in C^1([0, \tau])$$

$$H(0) = H(T) > 0, \quad K(s) > 0, \quad \frac{d}{ds}K(s) < 0, \quad s \in [0, \tau],$$

such that the matrix

$$\mathbf{C}(t) = \begin{pmatrix} C_{11}(t) & C_{12}(t) \\ C_{12}^*(t) & C_{22}(t) \end{pmatrix} \tag{26}$$

is positive definite for $t \in [0, T]$, where

$$C_{11}(t) = -\frac{d}{dt}H(t) - H(t)A(t) - A^*(t)H(t) - K(0),$$

$$C_{12}(t) = -\frac{d}{dt}H(t)D - H(t)B(t) - A^*(t)H(t)D,$$

$$C_{22}(t) = -D^* \frac{d}{dt} H(t) D - D^* H(t) B(t) - B^*(t) H(t) D + K(\tau).$$

Then the zero solution to the linear time-delay system (25) ($F(t, u, v) \equiv 0$) is exponentially stable.

Some estimates for solutions to (25), which are analogs to Krein's inequality (9), were also obtained in [25]. Following the reasoning scheme from [21, 22] and using the functional (24), similar results on the stability of the zero solution to nonlinear time-delay systems of the form (25) were also obtained (for instance, see [31, 35]). Below we formulate some results of [35].

We assume that the conditions of Theorem 6 are fulfilled, $F(t, u, v)$ is a real-valued continuous vector-function satisfying locally Lipschitz condition with respect to u and the estimate

$$\|F(t, u, v)\| \le q_1 \|u\|^{1+\omega_1} + q_2 \|v\|^{1+\omega_2}, \quad q_i, \ \omega_i \ge 0, \ i = 1, 2.$$

Here and further we consider T-periodic continuation of the matrix $H(t)$ on the whole semi-axis $\{t \ge 0\}$, keeping the same notation. Introduce the following matrices

$$S_{11}(t) = -\frac{d}{dt} H(t) - H(t) A(t) - A^*(t) H(t) - K(0),$$

$$S_{12}(t) = H(t) A(t) D - H(t) B(t) + K(0) D, \quad S_{22} = K(\tau) - D^* K(0) D,$$

$$P(t) = S_{11}(t) - S_{12}(t) S_{22}^{-1} S_{12}^*(t).$$

It should be noted that the matrix $\mathbf{C}(t)$ in (26) is positive definite if and only if the matrices $P(t)$ and S_{22} are positive definite. Denote by $h_{\min}(t)$, $p_{\min}(t)$, and $s_{\min}$ the minimal eigenvalues of $H(t)$, $P(t)$, and S_{22}, respectively. Obviously, $h_{\min}(t) \ge h_{\min} > 0$, $p_{\min}(t) \ge p_{\min} > 0$, $s_{\min} > 0$. Introduce the following notations

$$s_{\max} = \max_{t \in [0,T]} \|S_{22}^{-1} S_{12}^*(t)\|,$$

$$\delta(s) = q_1 \left(1 + (\varepsilon_1)^{-1}\right)^{\omega_1} \|D\|^{1+\omega_1} s^{\omega_1} + q_2 s^{\omega_2}, \quad \varepsilon_1 > 0, \quad s \ge 0,$$

$$p = \min_{t \in [0,T]} \frac{p_{\min}(t)}{\|H(t)\|}, \qquad h = \min_{t \in [0,T]} \frac{1}{\|H(t)\|},$$

$$\varepsilon_2 = \frac{s_{\min}}{2 p_{\min}} \left(s_{\max} + \sqrt{s_{\max}^2 + \frac{p_{\min}}{s_{\min}}} \right),$$

$\kappa > 0$ is the maximal real such that

$$\frac{d}{ds} K(s) + \kappa K(s) \le 0, \quad s \in [0, \tau],$$

$\rho > 0$ is such that

$$\delta(\rho) < \frac{s_{\min} h}{2\varepsilon_2},$$

$$Q = \max_{t\in[0,T]} \left(\|H(t)\| \|H^{-1}(t)\|^{1+\omega_1/2} q_1 (1+\varepsilon_1)^{\omega_1}\right).$$

Put

$$\gamma(t) = \min\left\{\left(\frac{p_{\min}(t)}{\|H(t)\|} - 2(s_{\max} + (4\varepsilon_2)^{-1})\delta(\rho)\right), \kappa\right\},$$

$$r^{-\omega_1/2} = Q\omega_1 \left(\int_0^T \exp\left(-\frac{\omega_1}{2}\int_0^\eta \gamma(\xi)d\xi\right)d\eta\right)\left(1 - \exp\left(-\frac{\omega_1}{2}\int_0^T \gamma(\xi)d\xi\right)\right)^{-1},$$

$$\beta(t) = \frac{\gamma(t)}{2}, \quad \beta^+ = \max_{t\in[0,T]} \beta(t), \quad \beta^- = \min_{t\in[0,T]} \beta(t),$$

$$\Phi_1 = \max_{s\in[-\tau,0]} \|\varphi(s)\|, \quad \Phi_2 = \frac{\sqrt{h_{\min}^{-1} V(0,\varphi)}}{\left(1 - \left(r^{-1}V(0,\varphi)\right)^{\omega_1/2}\right)^{1/\omega_1}}.$$

By the definitions of $\varepsilon_2 > 0$ and $\rho > 0$, it is easy to verify that $\gamma(t) \geq \gamma_{\min} > 0$.

If the conditions of Theorem 6 hold, it is not difficult to show that the spectrum of D belongs to the unit disk $\{\lambda \in C : |\lambda| < 1\}$ and so $\|D^j\| \to 0$ as $j \to \infty$. Let l be the minimal positive integer such that $\|D^l\| < 1$. Consider the initial value problem for (25)

$$\begin{cases} \frac{d}{dt}(y(t) + Dy(t-\tau)) = A(t)y(t) + B(t)y(t-\tau) + F(t, y(t), y(t-\tau)), \\ y(t) = \varphi(t), \quad t \in [-\tau, 0], \quad y(+0) = \varphi(0), \end{cases} \tag{27}$$

where $\varphi(t) \in C^1([-\tau, 0])$ is a given vector-function. Below we establish estimates for solutions to (27) in the cases

$$\|D^l\| < e^{-l\beta^+\tau}, \quad e^{-l\beta^+\tau} \leq \|D^l\| \leq e^{-l\beta^-\tau}, \quad e^{-l\beta^-\tau} < \|D^l\| < 1.$$

Theorem 7 ([35]) *Let the conditions of Theorem 6 hold and let*

$$\|D^l\| < e^{-l\beta^+\tau}.$$

Assume that the initial function $\varphi(t)$ in (27) belongs to $\mathscr{E}_1$, where

$$\mathscr{E}_1 = \Big\{\varphi(s) \in C^1[-\tau, 0] : \Phi_1 < \rho, \; V(0,\varphi) < r,$$

$$\Phi_2(1 - \|D^l\|e^{l\beta^+\tau})^{-1}\sum_{j=0}^{l-1}\|D^j\|e^{j\beta^+\tau} + \max\{\|D\|, \ldots, \|D^l\|\}\Phi_1 < \rho\Bigg\}.$$

Then a solution to (27) is defined on the whole semi-axis $\{t \geq 0\}$ *and satisfies the estimate*

$$\|y(t)\| \leq \left(\Phi_2\big(1 - \|D^l\|e^{l\beta^+\tau}\big)^{-1}\sum_{j=0}^{l-1}\|D^j\|e^{j\beta^+\tau}\right.$$

$$\left. + \max\left\{\|D\|e^{\beta^+\tau}, \ldots, \|D^l\|e^{l\beta^+\tau}\right\}\Phi_1\right)\exp\left(-\int_0^t \beta(\xi)d\xi\right), \quad t > 0.$$

Theorem 8 ([35]) *Let the conditions of Theorem 6 hold and let*

$$e^{-l\beta^+\tau} \leq \|D^l\| \leq e^{-l\beta^-\tau}.$$

Assume that the initial function $\varphi(t)$ *in (27) belongs to* $\mathscr{E}_2$*, where*

$$\mathscr{E}_2 = \Bigg\{\varphi(s) \in C^1[-\tau, 0] : \quad \Phi_1 < \rho, \quad V(0, \varphi) < r,$$

$$\Phi_2 B_l\sum_{j=0}^{l-1}\|D^j\|e^{j\beta^+\tau} + \max\{\|D\|, \ldots, \|D^l\|\}\Phi_1 < \rho\Bigg\},$$

$$B_l = \max_{t \geq 0}\left(1 + \frac{t}{l\tau}\right)\exp\left(-\int_0^t \hat{\beta}(\xi)d\xi\right),$$

$$\hat{\beta}(t) = \min\left\{\beta(t), -\frac{1}{l\tau}\ln\|D^l\|\right\}.$$

Then a solution to (27) is defined on the whole semi-axis $\{t \geq 0\}$ *and satisfies the estimate*

$$\|y(t)\| \leq \left(\Phi_2\left(1 + \frac{t}{l\tau}\right)\sum_{j=0}^{l-1}\|D^j\|e^{j\beta^+\tau}\right.$$

$$\left. + \max\left\{1, \|D\|e^{\beta^+\tau}, \ldots, \|D^{(l-1)}\|e^{(l-1)\beta^+\tau}\right\}\Phi_1\right)\exp\left(-\int_0^t \hat{\beta}(\xi)d\xi\right), \quad t > 0.$$

Theorem 9 ([35]) *Let the conditions of Theorem 6 hold and let*

$$e^{-l\beta^- \tau} < \|D^l\| < 1.$$

Assume that the initial function $\varphi(t)$ *in* (27) *belongs to* $\mathscr{E}_3$, *where*

$$\mathscr{E}_3 = \Big\{\varphi(s) \in C^1[-\tau, 0] : \quad \Phi_1 < \rho, \quad V(0, \varphi) < r,$$

$$\Phi_2\big(1 - (\|D^l\|e^{l\beta^-\tau})^{-1}\big)^{-1} \sum_{j=0}^{l-1} \|D^j\|e^{j\beta^-\tau} + \max\{\|D\|, \dots, \|D^l\|\}\Phi_1 < \rho\Big\}.$$

Then a solution to (27) *is defined on the whole semi-axis* $\{t \geq 0\}$ *and satisfies the estimate*

$$\|y(t)\| \leq \Bigg(\Phi_2\big(1 - (\|D^l\|e^{l\beta^-\tau})^{-1}\big)^{-1} \sum_{j=0}^{l-1} \|D^j\|e^{j\beta^-\tau}$$

$$+\|D^l\|^{-1}\|D\| \max\left\{1, \|D\|, \dots, \|D^{l-1}\|\right\} \Phi_1\Bigg) \exp\left(\frac{t}{l\tau} \ln \|D^l\|\right), \quad t \geq l\tau.$$

Theorems 7–9 give us estimates for attraction domains of the zero solution to (25) and estimates characterizing the decay rate of the solutions to (27) at infinity.

We now consider a more general class of time-delay systems of neutral type

$$\frac{d}{dt}y(t) + D(t)\frac{d}{dt}y(t - \tau) = A(t)y(t) + B(t)y(t - \tau)$$
$$+ F\left(t, y(t), y(t - \tau), \frac{d}{dt}y(t - \tau)\right), \tag{28}$$

where all the matrices have T-periodic continuous entries, including the matrix $D(t)$, and nonlinear terms can contain $\frac{d}{dt}y(t - \tau)$. We suppose that $F(t, u, v, w)$ is a real-valued continuous vector-function satisfying locally Lipschitz condition with respect to u and the estimate

$$\|F(t, u, v, w)\| \leq q_1\|u\|^{1+\omega_1} + q_2\|v\|^{1+\omega_2} + q_3\|v\|^{1+\omega_3}, \quad q_i, \ \omega_i \geq 0, \ i = 1, 2, 3.$$

Using the Lyapunov–Krasovksii functional

$$V(t, y) = \langle H(t)y(t), y(t)\rangle + \int_{t-\tau}^{t} \langle K(t - s)y(s), y(s)\rangle ds$$

$$+\int_{t-\tau}^{t}\left\langle L(t-s)\frac{d}{ds}y(s),\frac{d}{ds}y(s)\right\rangle ds, \tag{29}$$

conditions for the exponential stability of the zero solution to (28) were obtained and estimates of exponential decay of solutions to (28) were established (for example, see [78–80]). We now formulate one of the results.

Let the matrices $H(t)\in C^1[0,T]$, $K(s)$, and $L(s)\in C^1[0,\tau]$ be such that

$$H(t)=H^*(t),\quad t\in[0,T],\quad H(0)=H(T)>0, \tag{30}$$

$$K(s)=K^*(s)\geq 0,\quad \frac{d}{ds}K(s)\leq 0,\quad s\in[0,\tau], \tag{31}$$

$$L(s)=L^*(s)\geq 0,\quad \frac{d}{ds}L(s)\leq 0,\quad s\in[0,\tau]. \tag{32}$$

Define the matrix

$$\mathbf{C}(t)=\begin{pmatrix} C_{11}(t) & C_{12}(t) & C_{13}(t)\\ C_{12}^*(t) & C_{22}(t) & C_{23}(t)\\ C_{13}^*(t) & C_{23}^*(t) & C_{33}(t)\end{pmatrix}$$

with the entries

$$C_{11}(t)=-\frac{d}{dt}H(t)-H(t)A(t)-A^*(t)H(t)-K(0)-A^*(t)L(0)A(t),$$

$$C_{12}(t)=-H(t)B(t)-A^*(t)L(0)B(t),$$

$$C_{13}(t)=H(t)D(t)+A^*(t)L(0)D(t),$$

$$C_{22}(t)=K(\tau)-B^*(t)L(0)B(t),$$

$$C_{23}(t)=B^*(t)L(0)D(t),$$

$$C_{33}(t)=L(\tau)-D^*(t)L(0)D(t).$$

Theorem 10 ([78]) *Assume that there exist matrices $H(t)$, $K(s)$, and $L(s)$, satisfying the conditions (30)–(32), such that the inequality holds*

$$\left\langle \mathbf{C}(t)\begin{pmatrix}u\\v\\w\end{pmatrix},\begin{pmatrix}u\\v\\w\end{pmatrix}\right\rangle \geq \langle P(t)u,u\rangle,\quad u,v,w\in C^n,\quad t\in[0,T],$$

where $P(t)>0$ is a positive definite Hermitian matrix with continuous entries. If

$$\frac{d}{ds}K(s)+kK(s)\leq 0,\quad \frac{d}{ds}L(s)+lL(s)\leq 0,\quad s\in[0,\tau],$$

for some k, $l > 0$ then the zero solution to the linear time-delay system (28) ($F(t, u, v, w) \equiv 0$) is exponentially stable.

As we see, the choice of a suitable Lyapunov–Krasovskii functional makes it possible to establish the exponential stability and obtain sharper estimates of the decay rate of solutions at infinity for a wider class of systems. In [81], we introduced the following class of Lyapunov–Krasovskii functionals

$$V(t, y) = \left\langle \mathscr{H}(t) \begin{pmatrix} y(t) \\ y(t-\tau) \end{pmatrix}, \begin{pmatrix} y(t) \\ y(t-\tau) \end{pmatrix} \right\rangle$$

$$+ \int_{t-\tau}^{t} \left\langle \mathscr{K}(t, t-s) \begin{pmatrix} y(s) \\ \frac{d}{ds} y(s) \end{pmatrix}, \begin{pmatrix} y(s) \\ \frac{d}{ds} y(s) \end{pmatrix} \right\rangle ds. \tag{33}$$

In particular, this class contains the functionals mentioned above; so, we have the functionals (10), (21) for

$$\mathscr{H}(t) = \begin{pmatrix} H(t) & 0 \\ 0 & 0 \end{pmatrix}, \qquad \mathscr{K}(t, s) = \begin{pmatrix} K(s) & 0 \\ 0 & 0 \end{pmatrix},$$

the functionals (13), (24) for

$$\mathscr{H}(t) = \begin{pmatrix} H(t) & H(t)D \\ D^* H(t) & D^* H(t) D \end{pmatrix}, \qquad \mathscr{K}(t, s) = \begin{pmatrix} K(s) & 0 \\ 0 & 0 \end{pmatrix},$$

the functional (29) for

$$\mathscr{H}(t) = \begin{pmatrix} H(t) & 0 \\ 0 & 0 \end{pmatrix}, \qquad \mathscr{K}(t, s) = \begin{pmatrix} K(s) & 0 \\ 0 & L(s) \end{pmatrix}.$$

Using this class of functionals, more exact estimates for solutions to (28) were obtained in [81, 83]. These estimates are analogs of Krein's estimate and characterize the decay rate at infinity.

It is also possible to define functionals of Lyapunov–Krasovskii type to study the exponential stability of solutions to nonlinear time-delay systems with several constant and variable delays (including distributed delays). Moreover, one can obtain estimates characterizing the exponential decay of the solutions at infinity and apply these results to some models of population dynamics (for example, see [27, 29, 33, 79, 82, 105–109, 119, 120]).

It should be noted that the proposed approach can be extended on delay difference equations and functional difference equations with periodic coefficients (for instance, see [34, 36, 37]). Below we formulate some results of [36].

Consider systems of delay difference equations with periodic coefficients of the following form

$$x_{n+1} = A(n)x_n + B(n)x_{n-\tau(n)}, \quad n = 0, 1, 2, \dots, \tag{34}$$

where $\{A(n)\}$, $\{B(n)\}$ are sequences of N-periodic $(m \times m)$-matrices, $\tau(n) \in \mathbf{N}$ is a delay parameter. Using the discrete analog of the functional (21)

$$v(n, x) = \langle H(n)x_n, x_n\rangle + \sum_{j=n-\tau}^{n-1} \langle K_{n-j-1}x_j, x_j\rangle,$$

one can establish conditions for the asymptotic stability of the zero solution to (34) and obtain estimates for solutions to (34), characterizing the decay rate as $n \to \infty$.

Assume that the delay parameter is bounded

$$1 \le \tau(n) \le \tau < \infty.$$

Then (34) can be rewritten as the following system of linear difference equations with variable coefficients

$$x_{n+1} = A(n)x_n + \sum_{j=1}^{\tau} B_j(n)x_{n-j}, \quad n = 0, 1, \dots,$$

where

$$B_j(n) = \begin{cases} B(n) & \text{for} \quad \tau(n) = j, \\ 0 & \text{for} \quad \tau(n) \neq j. \end{cases}$$

Theorem 11 ([36]) *Assume that there exist Hermitian positive definite matrices*

$$H(n), \quad K_j, \quad j = 0, 1, \dots, \tau,$$

such that

$$H(0) = H(N), \quad \Delta_j = K_{j-1} - K_j > 0, \quad j = 1, \dots, \tau,$$

and the matrices

$$\mathbf{C}(n) = -\begin{pmatrix} C_{00}(n) & A^*(n)H(n+1)B_1(n) & \dots & A^*(n)H(n+1)B_\tau(n) \\ B_1^*(n)H(n+1)A(n) & C_{11}(n) & \dots & B_1^*(n)H(n+1)B_\tau(n) \\ \vdots & \vdots & \ddots & \vdots \\ B_\tau^*(n)H(n+1)A(n) & B_\tau^*(n)H(n+1)B_1(n) & \dots & C_{\tau\tau}(n) \end{pmatrix}$$

with

$$C_{00}(n) = A^*(n)H(n+1)A(n) - H(n) + K_0,$$

$$C_{jj}(n) = B_j^*(n)H(n+1)B_j(n) - \frac{1}{2}\Delta_j, \quad j = 1, \dots, \tau - 1,$$

$$C_{\tau\tau}(n) = B_\tau^*(n)H(n+1)B_\tau(n) - K_\tau$$

are also positive definite for $n = 0, \ldots, N-1$. *Then the zero solution to (34) is asymptotically stable.*

Obviously, there exists a constant $c_1 > 0$ such that

$$\left\langle \mathbf{C}(n) \begin{pmatrix} x_n \\ \vdots \\ x_{n-\tau} \end{pmatrix}, \begin{pmatrix} x_n \\ \vdots \\ x_{n-\tau} \end{pmatrix} \right\rangle \geq c_1 \sum_{i=0}^{\tau} \|x_{n-i}\|^2.$$

The following result holds.

Theorem 12 ([36]) *Assume that the conditions of Theorem 11 hold. Let* $æ_j \in (0,1)$, $j = 1, 2, \ldots, \tau$, *be such that*

$$-\frac{1}{2}\Delta_i + æ_i K_{i-1} \leq 0, \quad i = 1, \ldots, \tau - 1, \quad -\Delta_\tau + æ_\tau K_{\tau-1} \leq 0.$$

Then for a solution to (34) with the initial data $x_0, x_{-1}, \ldots, x_{-\tau}$ *the inequality holds*

$$\|x_n\|^2 \leq (h_1(n))^{-1} \prod_{j=0}^{n-1} (1 - \varepsilon_j) \left(\langle H(0)x_0, x_0 \rangle + \sum_{l=-\tau}^{-1} \langle K_{-l-1} x_l, x_l \rangle \right),$$

where $h_1(n) > 0$ *is the minimal eigenvalue of the matrix* $H(n)$

$$\varepsilon_j = \min \left\{ æ_1, \ldots, æ_\tau, \frac{c_1}{\|H(j)\|} \right\}.$$

Using discrete analogs of the functionals (10), (13), (21), (24), (29), (33), one can study the exponential stability of solutions to more wide classes of delay difference equations.

Acknowledgements This work was supported by the Russian Foundation for Basic Research (project no. 19-01-00754).

References

1. Agarwal, R.P., Berezansky, L., Braverman, E., Domoshnitsky, A.: Nonoscillation Theory of Functional Differential Equations with Applications. Springer, New York (2012)
2. Aleksenko, N.V., Romanovskii, R.K.: The method of Lyapunov functionals for linear difference-differential systems with almost-periodic coefficients. Differ. Equ. **37**, 159–165 (2001)
3. Andronov, A.A., Mayer, A.G.: Simplest linear delay systems (in Russian). Autom. Remote. Control. **7**, 95–106 (1946)

4. Azbelev, N.V., Maksimov, V.P., Rakhmatullina, L.F.: Introduction to the Theory of Functional Differential Equations. Methods and Applications. Hindawi Publishing Corporation, New York (2007)
5. Azbelev, N.V., Simonov, P.M.: Stability of Solutions to Equations with Ordinary Derivatives. Izd. Perm University, Perm (2001).(in Russian)
6. Arutyunyan, N.Kh., Kolmanovskii, V.B.: Creep Theory of Inhomogeneous Bodies. Nauka, Moscow (1983).(in Russian)
7. Baštinec, J., Diblík, J., Khusainov, D.Ya., Ryvolová, A.: Exponential stability and estimation of solutions of linear differential systems of neutral type with constant coefficients. Bound. Value Probl. Art. ID 956121, 1–20 (2010)
8. Bellman, R.: Stability Theory of Differential Equations. McGraw-Hill Book Company, New York, Toronto, London (1953)
9. Bellman, R., Cooke, K.L.: Differential-Difference Equations. Academic Press, New York, London (1963)
10. Belotserkovskii, S.M., Kochetkov, Yu.A., Krasovskii, A.A., Novitskii, V.V.: Introduction to Aeroautoelasticity. Nauka, Moscow (1980).(in Russian)
11. Belykh, L.N.: Analysis of Mathematical Models in Immunology. Nauka, Moscow (1988).(in Russian)
12. Berezanskii, L.M.: Development of N.V. Azbelev's W-method in problems of the stability of solutions of linear functional-differential equations. Differ. Equ. **22**, 521–529 (1986)
13. Chebotarev, N.G., Meyman, N.N.: The Routh-Hurwitz problem for polynomials and entire functions (in Russian). Tr. Mat. Inst. Steklova. **26**, 3–331 (1949)
14. Daleckii, Ju.L., Krein, M.G.: Stability of Solutions to Differential Equations in Banach Space. American Mathematical Society, Providence (1974)
15. Day, W.A.: The Thermodynamics of Simple Materials with Fading Memory. Springer, New York, Heidelberg (1972)
16. Demidenko, G.V.: Stability of solutions to linear differential equations of neutral type. J. Anal. Appl. **7**, 119–130 (2009)
17. Demidenko, G.V., Matveeva, I.I.: On stability of solutions to linear systems with periodic coefficients. Siberian Math. J. **42**, 282–296 (2001)
18. Demidenko, G.V., Matveeva, I.I.: On asymptotic stability of solutions to nonlinear systems of differential equations with periodic coefficients. Selcuk J. Appl. Math. **3**, 37–48 (2002)
19. Demidenko, G.V., Matveeva, I.I.: On stability of solutions to quasilinear periodic systems of differential equations. Siberian Math. J. **45**, 1041–1052 (2004)
20. Demidenko, G.V., Kolchanov, N.A., Likhoshvai, V.A., Matushkin, Yu.G., Fadeev, S.I.: Mathematical modeling of regulatory circuits of gene networks. Comput. Math. Math. Phys. **44**, 2166–2183 (2004)
21. Demidenko, G.V., Matveeva, I.I.: Asymptotic properties of solutions to delay differential equations (in Russian). Vestnik Novosib. Gos. Univ. Ser. Mat. Mekh. Inform. **5**, 20–28 (2005)
22. Demidenko, G.V., Matveeva, I.I.: Stability of solutions to delay differential equations with periodic coefficients of linear terms. Siberian Math. J. **48**, 824–836 (2007)
23. Demidenko, G.V., Kotova, T.V., Skvortsova, M.A.: Stability of solutions to differential equations of neutral type. J. Math. Sci. **186**, 394–406 (2012)
24. Demidenko, G.V., Vodop'yanov, E.S., Skvortsova, M.A.: Estimates of solutions to the linear differential equations of neutral type with several delays of the argument. J. Appl. Indust. Math. **7**, 472–479 (2013)
25. Demidenko, G.V., Matveeva, I.I.: On estimates of solutions to systems of differential equations of neutral type with periodic coefficients. Siberian Math. J. **55**, 866–881 (2014)
26. Demidenko, G.V., Matveeva, I.I.: On exponential stability of solutions to one class of systems of differential equations of neutral type. J. Appl. Indust. Math. **8**, 510–520 (2014)
27. Demidenko, G.V., Matveeva, I.I.: Estimates for solutions to linear systems of neutral type with several delays. J. Anal. Appl. **12**, 37–52 (2014)
28. Demidenko, G.V., Matveeva, I.I.: Estimates for solutions to a class of nonlinear time-delay systems of neutral type. Electron. J. Diff. Equ. **2015**, 1–14 (2015)

29. Demidenko, G.V., Matveeva, I.I.: Estimates for solutions to a class of time-delay systems of neutral type with periodic coefficients and several delays. Electron. J. Qual. Theory Differ. Equ. **2015**, 1–22 (2015)
30. Demidenko, G.V., Matveeva, I.I.: Asymptotic stability of solutions to a class of linear time-delay systems with periodic coefficients and a large parameter. J. Ineq. Appl. **2015**, 1–10 (2015)
31. Demidenko, G.V., Matveeva, I.I.: On the robust stability of solutions to linear differential equations of neutral type with periodic coefficients (in Russian). Sib. Zh. Ind. Mat. **18**, 18–29 (2015)
32. Demidenko, G.V., Matveeva, I.I.: Exponential stability of solutions to nonlinear time-delay systems of neutral type. Electron. J. Diff. Equ. **2016**, 1–20 (2016)
33. Demidenko, G.V., Matveeva, I.I.: Estimates for solutions of one class of nonlinear neutral type systems with several delays. J. Math. Sci. **213**, 811–822 (2016)
34. Demidenko, G.V., Baldanov, D.S.: Asymptotic stability of solutions to delay difference equations. J. Math. Sci. **221**, 815–825 (2017)
35. Demidenko, G.V., Matveeva, I.I., Skvortsova, M.A.: Estimates for solutions to neutral differential equations with periodic coefficients of linear terms. Siberian Math. J. **60**, 828–841 (2019)
36. Demidenko, G.V., Baldanov, D.S.: Exponential stability of solutions to delay difference equations with periodic coefficients. In: Demidenko, G.V., Romenski, E., Toro, E., Dumbser, M.(eds.) Continuum mechanics, applied mathematics and scientific computing: Godunov's legacy – A liber amicorum to Professor Godunov. Springer Nature, Cham, Switzerland, pp. 93–100 (2020)
37. Demidenko, G.V., Matveeva, I.I.: On estimates of solutions to one class of functional difference equations with periodic coefficients. In: Demidenko, G.V., Romenski, E., Toro, E., Dumbser, M.(eds.) Continuum mechanics, applied mathematics and scientific computing: Godunov's legacy - A liber amicorum to Professor Godunov. Springer Nature, Cham, Switzerland, pp. 101–109 (2020)
38. Dolgii, Yu.F., Kim, A.V.: On the method of Lyapunov functionals for systems with aftereffect. Differ. Equ. **27**, 918–922 (1991)
39. Dolgii, Yu.F.: Stability of Periodic Differential-Difference Equations. Izd. Ural University, Ekaterinburg (1996).(in Russian)
40. El'sgol'ts, L.E.: Qualitative Methods in Mathematical Analysis. American Mathematical Society, Providence (1964)
41. El'sgol'ts, L.E.: Introduction to the Theory of Differential Equations with Deviating Argument. Nauka, Moscow (1964).(in Russian)
42. El'sgol'ts, L.E., Norkin, S.B.: Introduction to the Theory and Application of Differential Equations with Deviating Arguments. Academic, Academic Press, New York, London (1973)
43. Erneux, T.: Applied Delay Differential Equations. Surveys and Tutorials in the Applied Mathematical Sciences, vol. 3. Springer, New York (2009)
44. Gasilov, G.L.: On the characteristic equation of a system of linear differential equations with periodic coefficients and delays (in Russian). Izv. Vyssh. Uchebn. Zaved. Matematika (4), 60–66 (1972)
45. Germanovich, O.P.: Linear Periodic Equations of Neutral Type and Their Applications. Izd. Leningrad University, Leningrad (1986).(in Russian)
46. Gil', M.I.: Stability of Neutral Functional Differential Equations. Atlantis Studies in Differential Equations, vol. 3. Atlantis Press, Paris (2014)
47. Godunov, S.K.: Ordinary Differential Equations with Constant Coefficients. Izd. Novosibirsk University, Novosibirsk (1994).(in Russian)
48. Godunov, S.K.: Modern Aspects of Linear Algebra. Translations of Mathematical Monographs, vol. 175. American Mathematical Society, Providence (1998)
49. Gopalsamy K. Stability and Oscillations in Delay Differential Equations of Population Dynamics. Mathematics and its Applications, vol. 74. Kluwer Academic Publishers, Dordrecht (1992)

50. Goryachenko, V.D.: Methods of Research of Stability of Nuclear Reactors. Atomizdat, Moscow (1977).(in Russian)
51. Góreckii, H.: Analysis and Synthesis of Control Systems with Delay. Mashinostroenie, Moscow (1974).(in Russian)
52. Györi, I., Ladas, G.: Oscillation Theory of Delay Differential Equations: With Applications. Oxford Mathematical Monographs. Oxford Science Publications. The Clarendon Press, Oxford University Press, New York (1991)
53. Hahn, W.: On difference differential equations with periodic coefficients. J. Math. Anal. Appl. **3**, 70–101 (1961)
54. Hartman, Ph.: Ordinary Differential Equations. Wiley, New York, London, Sydney (1964)
55. Hale, J.K.: Theory of Functional Differential Equations. Springer, New York, Heidelberg (1977)
56. Halanay, A.: Stability theory of linear periodic systems with delay. Acad. Repub. Popul. Roum., Rev. Math. Pures Appl. **6**, 633–653 (1961)
57. Halanay, A., Wexler, D.: Qualitative Theory of Impulse Systems. Mir, Moscow (1971).(in Russian)
58. Kharitonov, V.L., Hinrichsen, D.: Exponential estimates for time delay systems. Syst. Control Lett. **53**, 395–405 (2004)
59. Kharitonov, V., Mondié, S., Collado, J.: Exponential estimates for neutral time-delay systems: an LMI approach. IEEE Trans. Automat. Control **50**, 666–670 (2005)
60. Kharitonov, V.L.: Time-Delay Systems. Lyapunov Functionals and Matrices. Control Engineering. Birkhäuser/Springer, New York (2013)
61. Khusainov, D.Ya., Shatyrko, A.V.: The Method of Lyapunov Functions and the Investigation of the Stability of Functional-Differential Systems. Izd. Kiev University, Kiev (1997). (in Russian)
62. Khusainov, D.Ya., Ivanov, A.F., Kozhametov, A.T.: Convergence estimates for solutions of linear stationary systems of differential-difference equations with constant delay. Differ. Equ. **41**, 1196–1200 (2005)
63. Khusainov, D.Ya., Kozhametov, A.T.: Convergence of solutions of the neutral type nonautonomous systems. Russ. Math. **50**, 65–69 (2006)
64. Kim, A.V.: Direct Lyapunov's Method in the Stability Theory of Systems with Delays. Izd. Ural University, Ekaterinburg (1992). (in Russian)
65. Kolmanovskii, V.B., Nosov, V.R.: Stability and Periodic Regimes of Control Systems with Aftereffect. Nauka, Moscow (1981).(in Russian)
66. Kolmanovskii, V.B., Myshkis, A.D.: Introduction to the Theory and Applications of Functional-Differential Equations. Mathematics and its Applications, vol. 463. Kluwer Academic Publishers, Dordrecht (1999)
67. Komlenko, Yu.V., Tonkov, E.L.: The Lyapunov-Floquet representation for differential equations with aftereffect. Russ. Math. **39**, 38–43 (1995)
68. Korenevskii, D.G.: Stability of Dynamical Systems under Random Perturbations of Parameters. Algebraic Criteria. Naukova Dumka, Kiev (1989). (in Russian)
69. Korenevskii, D.G.: The Destabilizing Effect of Parametric White Noise in Continuous and Discrete Dynamical Systems. Akademperiodika, Kiev (2008).(in Russian)
70. Krasovskii, N.N.: On application of second Lyapunov's method for systems with time delays. Prikl. Mat. Mekh. **20**, 315–327 (1956)
71. Krasovskii, N.N.: Stability of Motion. Applications of Lyapunov's Second Method to Differential Systems and Equations with Delay. Stanford University Press, Stanford (1963)
72. Kuang, Y.: Differential Equations with Applications in Population Dynamics. Mathematics in Science and Engineering, vol. 191. Academic Press, Boston (1993)
73. Lyubich, Yu.I., Tkachenko, V.A.: On a Floquet theory for equations with retarded argument (in Russian). Differ. Uravn. **5**, 648–656 (1969)
74. MacDonald, N.: Biological Delay Systems: Linear Stability Theory. Cambridge Studies in Mathematical Biology, vol. 8. Cambridge University Press, Cambridge (1989)

75. Malygina, V.V.: On the stability of equations with periodic parameters (in Russian). In: Functional-Differential Equations, Interuniversity Collection Science Works, pp. 41–43. Perm (1987)
76. Marchuk, G.I.: Mathematical Models in Immunology. Nauka, Moscow (1980).(in Russian)
77. Matveeva, I.I.: Estimates of solutions to a class of systems of nonlinear delay differential equations. J. Appl. Indust. Math. **7**, 557–566 (2013)
78. Matveeva, I.I.: On exponential stability of solutions to periodic neutral-type systems. Siberian Math. J. **58**, 264–270 (2017)
79. Matveeva, I.I.: On the exponential stability of solutions of periodic systems of the neutral type with several delays. Differ. Equ. **53**, 725–735 (2017)
80. Matveeva, I.I.: On the robust stability of solutions to periodic systems of neutral type. J. Appl. Indust. Math. **12**, 684–693 (2018)
81. Matveeva, I.I.: Estimates of the exponential decay of solutions to linear systems of neutral type with periodic coefficients. J. Appl. Indust. Math. **13**, 511–518 (2019)
82. Matveeva, I.I.: On exponential stability of solutions to linear periodic systems of neutral type with time-varying delay. Siberian Electron. Math. Reports. **16**, 748–756 (2019)
83. Matveeva, I.I.: Estimates of exponential decay of solutions to one class of nonlinear systems of neutral type with periodic coefficients. Comput. Math. Math. Phys. **60**, 601–609 (2020)
84. Melchor-Aguilar D., Niculescu, S.-I.: Estimates of the attraction region for a class of nonlinear time-delay systems. IMA J. Math. Control Inform. **24**, 523–550 (2007)
85. Michiels, W., Niculescu, S.-I.: Stability, Control, and Computation for Time-Delay Systems. An Eigenvalue-Based Approach. Advances in Design and Control, vol. 27. Society for Industrial and Applied Mathematics, Philadelphia (2014)
86. Mikhalevich, V.S., Kozorez, V.V., Rashkovan, V.M., Khusainov, D.Ya., Cheborin, O.G.: "Magnetic Potential Well" — Stabilization Effect of Superconducting Dynamical Systems. Naukova Dumka, Kiev (1991). (in Russian)
87. Mitropol'skij, Yu.A., Martynyuk, D.I.: Periodic and Quasi-Periodic Oscillations of Systems with Delay. Vishcha Shkola, Kiev (1979). (in Russian)
88. Mondié, S., Kharitonov, V.L.: Exponential estimates for retarded time-delay systems: an LMI approach. IEEE Trans. Automat. Control. **50**, 268–273 (2005)
89. Murray, J.D.: Lectures on Nonlinear Differential-Equation Models in Biology. Clarendon Press, Oxford (1977)
90. Myshkis, A.D.: Linear Differential Equations with Retarded Argument. Gostekhizdat, Moscow, Leningrad (1951).(in Russian)
91. Myshkis, A.D.: Linear Differential Equations with Retarded Argument. Nauka, Moscow (1972).(in Russian)
92. Pertsev, N.V.: Application of the monotone method and of M-matrices to the analysis of the behavior of solutions of some models of biological processes (in Russian). Sib. Zh. Ind. Mat. **5**, 110–122 (2002)
93. Pertsev, N.V.: Global solvability and estimates of solutions to the Cauchy problem for the retarded functional differential equations that are used to model living systems. Siberian Math. J. **59**, 113–125 (2018)
94. Pinney, E.: Ordinary Difference-Differential Equations. University of California Press, Berkeley, Los Angeles (1958)
95. Pontryagin, L.S.: On zeros of some transcendental functions (in Russian). Izv. Akad. Nauk SSSR, Ser. Mat. **6** 115–134 (1942)
96. Razumikhin, B.S.: On stability of systems with time lag (in Russian). Prikl. Mat. Mekh. **20**, 500–512 (1956)
97. Razumikhin, B.S.: Direct Method of Investigation of Stability of Systems with Aftereffect (in Russian). Preprint. VNIISI, Moscow, 75 p. (1984)
98. Richard, J.P.: Time-delay systems: an overview of some recent advances and open problems. Automatica. **39**, 1667–1694 (2003)
99. Romanovskii, R.K., Trotsenko, G.A.: The method of Lyapunov functionals for neutral type linear difference-differential systems with almost periodic coefficients. Siberian Math. J. **44**, 355–362 (2003)

100. Romanovskii, R.K., Bel'gart, L.V., Dobrovol'skii, S.M., Rogozin, A.V., Trotsenko, G.A.: Method of Lyapunov Functions for Almost Periodic Systems. Publishing House SB RAS, Novosibirsk (2015). (in Russian)
101. Romanyukha, A.A., Rudnev, S.G.: A variational principle for modelling infection immunity by the example of pneumonia (in Russian). Mat. Model. **13**, 65–84 (2001)
102. Rubanik, V.P.: Oscillations of Quasilinear Systems with Delay. Nauka, Moscow (1969).(in Russian)
103. Shil'man, S.V.: The Method of Generating Functions in the Theory of Dynamical Systems. Nauka, Moscow (1978).(in Russian)
104. Shimanov, S.N.: Stability of Linear Systems with Periodic Coefficients and Delay. Izd. Ural University, Sverdlovsk (1983). (in Russian)
105. Skvortsova, M.A.: Estimates of solutions to a system describing birds' migration. J. Anal. Appl. **13**, 15–27 (2015)
106. Skvortsova, M.A.: Asymptotic properties of solutions to a system describing the spread of avian influenza. Siberian Electron. Math. Reports. **13**, 782–798 (2016)
107. Skvortsova, M.: Asymptotic properties of solutions in Marchuk's basic model of disease. Func. Diff. Equ. **24**, 127–135 (2017)
108. Skvortsova, M.A.: Asymptotic properties of solutions in a model of antibacterial immune response (in Russian). Siberian Electron. Math. Reports. **15**, 1198–1215 (2018)
109. Skvortsova, M.A.: On estimates of solutions in a predator-prey model with two delays (in Russian). Siberian Electron. Math. Reports. **15**, 1697–1718 (2018)
110. Solodov, A.V., Solodova, E.A.: Systems with Variable Delay. Nauka, Moscow (1980).(in Russian)
111. Stokes, A.P.: A Floquet theory for functional differential equations. Proc. Nat. Acad. Sci. USA **48**, 1330–1334 (1962)
112. Svirezhev, Yu.M., Pasekov, V.P.: Fundamentals of Mathematical Genetics. Nauka, Moscow (1982).(in Russian)
113. Tsypkin, Ya.Z.: Theory of Linear Impulse Systems. Gos. Izd. Fiz.-Mat. Lit., Moscow (1963). (in Russian)
114. Vielle, B., Chauvet, G.: Delay equation analysis of human respiratory stability. Math. Biosci. **152**, 105–122 (1998)
115. Vlasov, V.V., Medvedev, D.A.: Functional-differential equations in Sobolev spaces and related problems of spectral theory. J. Math. Sci. **164**, 659–841 (2010)
116. Volterra, V.: Lessons on the Mathematical Theory of Struggle for Life. Nauka, Moscow (1976).(in Russian)
117. Wilkinson, J.H.: The Algebraic Eigenvalue Problem. Clarendon Press, Oxford (1965)
118. Wolkowicz, G.S.K., Xia, H.: Global asymptotic behavior of a chemostat model with discrete delays. SIAM. J. Appl. Math. **57**, 1019–1043 (1997)
119. Yskak, T.: Stability of solutions to systems of differential equations with distributed delay. Func. Diff. Equ. **25**, 97–108 (2018)
120. Yskak, T.: On the stability of systems of linear differential equations of neutral type with distributed delay. J. Appl. Indust. Math. **13**, 575–583 (2019)
121. Zubov, V.I.: Analytical Dynamics of a System of Solids. Izd. Leningrad University, Leningrad (1983). (in Russian)
122. Zverkin, A.M.: On the theory of linear delay differential equations with periodic coefficients (in Russian). Dokl. Akad. Nauk SSSR **128**, 882–885 (1959)
123. Zverkin, A.M.: Differential difference equations with periodic coefficients (in Russian). In: Bellman, R.E., Cooke, K.L. Differential-Difference Equations. pp. 498–535. Supplement to the Russian translation of the book. Mir, Moscow (1967)

Part III
Qualitative Study of Solutions of Differential Equations

Some Extremal Problems for Solutions of the Modified Helmholtz Equation in the Half-Space

Gershon Kresin and Tehiya Ben Yaakov

Abstract Representations for the sharp coefficients in pointwise estimates involving the gradient of the solution to the modified Helmholtz equation $(\Delta - c^2)u = 0$ in the half-space $\mathbb{R}^n_+$ are described. It is assumed that the boundary data of the Dirichlet and Neumann problems in $\mathbb{R}^n_+$ belong to the space L^p. Each of these representations includes an extremal problem with respect to a vector parameter inside of an integral over the unit sphere in $\mathbb{R}^n$. Explicit formulas for solutions to the extremal problems are indicated for $p \in [2, \infty]$ and $p \in [2, (n+2)/2]$ in the cases of Dirichlet and Neumann boundary data, respectively.

Keywords Extremal problems · Modified Helmholtz equation · Half-space

1 Background

In the present paper, we describe representations for the sharp coefficients in pointwise estimates for solutions to the modified Helmholtz equation $(\Delta - c^2)u = 0$, $c > 0$, in the half-space $\mathbb{R}^n_+ = \{x = (x', x_n) : x' \in \mathbb{R}^{n-1}, x_n > 0\}$. It is assumed that the boundary data of the Dirichlet and Neumann problems in $\mathbb{R}^n_+$ for the modified Helmholtz equation belong to the space $L^p(\mathbb{R}^{n-1})$. These representations include some extremal problems on the unit sphere $\mathbb{S}^{n-1}$ in $\mathbb{R}^n$, finding solutions of which we arrive at explicit formulas for the sharp coefficients in estimates to solutions to the boundary value problems.

We use the term *sharp estimate* if the coefficient in front of a function characteristic in the majorant part of an inequality can't be diminished. This best coefficient we call also *sharp*.

Previous results of similar nature were obtained in [1–3], where solutions of the Laplace equation in $\mathbb{R}^n_+$ were considered.

G. Kresin (✉) · T. B. Yaakov
Department of Mathematics, Ariel University, 40700 Ariel, Israel
e-mail: kresin@ariel.ac.il

A. Domoshnitsky et al. (eds.), *Functional Differential Equations and Applications*, Springer Proceedings in Mathematics & Statistics 379,
https://doi.org/10.1007/978-981-16-6297-3_12

In particular, in [3] the explicit formula for the sharp coefficient $\mathcal{A}_{n,p}(x)$ in the inequality

$$\left|\nabla\left\{\frac{u(x)}{x_n}\right\}\right| \le \mathcal{A}_{n,p}(x)\,\left\|u(\cdot,0)\right\|_p \tag{1}$$

was derived, where x is an arbitrary point $\mathbb{R}^n_+$, u is a harmonic function in $\mathbb{R}^n_+$, represented by the Poisson integral with boundary values in $L^p(\mathbb{R}^{n-1})$, and $||\cdot||_p$ is the norm in $L^p(\mathbb{R}^{n-1})$, $1 \le p \le \infty$. It was shown that

$$\mathcal{A}_{n,p}(x) = \frac{A_{n,p}}{x_n^{2+(n-1)/p}},$$

where

$$A_{n,p} = \frac{2n}{\omega_n}\left\{\frac{\pi^{\frac{n-1}{2}}\Gamma\left(\frac{3p+n-1}{2(p-1)}\right)}{\Gamma\left(\frac{(n+2)p}{2(p-1)}\right)}\right\}^{1-\frac{1}{p}}$$

for $1 < p < \infty$, and $A_{n,1} = 2n/\omega_n$, $A_{n,\infty} = 1$. Here and henceforth, we denote by $\omega_n = 2\pi^{n/2}/\Gamma(n/2)$ the area of the unit sphere in $\mathbb{R}^n$.

Another sharp estimate for the modulus of the gradient of harmonic function u in $\mathbb{R}^n_+$ was obtained in [2]:

$$|\nabla u(x)| \le \mathcal{N}_{n,p}(x)\left\|\frac{\partial u}{\partial \nu}(\cdot,0)\right\|_p, \tag{2}$$

where ν is the unit normal vector to $\partial\mathbb{R}^n_+$, $p \in [1, n]$, $x \in \mathbb{R}^n_+$. The sharp coefficient $\mathcal{N}_{n,p}(x)$ in (2) is given by

$$\mathcal{N}_{n,p}(x) = \frac{N_{n,p}}{x_n^{(n-1)/p}},$$

where

$$N_{n,p} = \frac{2^{1/p}}{\omega_n}\left\{\frac{2\pi^{(n-1)/2}\Gamma\left(\frac{n+p-1}{2p-2}\right)}{\Gamma\left(\frac{np}{2p-2}\right)}\right\}^{1-\frac{1}{p}}$$

for $1 < p \le n$, and $N_{n,1} = 2/\omega_n$.

In what follows, we will give analogues of the sharp estimates (1), (2) for solutions of the modified Helmholtz equation in the half-space $\mathbb{R}^n_+$. The proofs of the statements below are contained in [4].

2 An Extremal Problem for Integral over $\mathbb{S}^{n-1}$ with Vector Parameter

Let $\boldsymbol{e}_\sigma$ stand for the n-dimensional unit vector joining the origin to a point σ on the sphere $\mathbb{S}^{n-1}$. In what follows by $\boldsymbol{e}_i$, we mean the unit vector of the ith coordinate axis. We denote by $\boldsymbol{e}$ and $\boldsymbol{z}$ the n-dimensional unit vectors and assume that $\boldsymbol{e}$ is a fixed vector.

The following assertion plays an important role in the solution to extremal problems which arise in the next two sections.

Proposition 1 *Let*

$$G_\alpha(\boldsymbol{z}) = \int_{\mathbb{S}^{n-1}} \omega\big((\boldsymbol{e}_\sigma, \boldsymbol{e})\big)\big|(\boldsymbol{e}_\sigma, \boldsymbol{e})\big|^\alpha \big|(\boldsymbol{e}_\sigma, \boldsymbol{z})\big|^{2-\alpha} d\sigma,$$

where ω *is a continuous non-negative even function on* $[-1, 1]$ *with continuous positive derivative on* $(0, 1)$. *Then for any* $\alpha \in [0, 2)$, *the equality*

$$\max_{|\boldsymbol{z}|=1} G_\alpha(\boldsymbol{z}) = G_\alpha(\boldsymbol{e}) = \int_{\mathbb{S}^{n-1}} \omega\big((\boldsymbol{e}_\sigma, \boldsymbol{e})\big)(\boldsymbol{e}_\sigma, \boldsymbol{e})^2 d\sigma$$

holds.

3 Sharp Weighted Estimate for the Gradient of Solution to the Dirichlet Problem

We denote by $||\cdot||_p$ the norm in the space $L^p(\mathbb{R}^{n-1})$, that is,

$$||f||_p = \left\{\int_{\mathbb{R}^{n-1}} |f(x')|^p\, dx'\right\}^{1/p},$$

if $1 \le p < \infty$, and $||f||_\infty = \text{ess sup}\{|f(x')| : x' \in \mathbb{R}^{n-1}\}$.

Solution to the Dirichlet problem in $\mathbb{R}^n_+$ for the modified Helmholtz equation,

$$(\Delta - c^2)u = 0 \ \text{ in } \ \mathbb{R}^n_+, \quad u\big|_{x_n=0} = f(x') \tag{3}$$

with continuous and bounded function f on $\mathbb{R}^{n-1}$, is given by (e.g. [6])

$$u(x) = \frac{c^n x_n}{2^{(n-2)/2}\pi^{n/2}} \int_{\mathbb{R}^{n-1}} \frac{K_{n/2}\big(c|y-x|\big)}{\big(c|y-x|\big)^{n/2}} f(y')dy', \tag{4}$$

where $y = (y', 0)$, $y' \in \mathbb{R}^{n-1}$ and K_ν denotes the modified Bessel function of the third kind (Macdonald function).

We consider the solution to problem (3) with $f \in L^p(\mathbb{R}^{n-1})$ represented by (4), where $p \in [1, \infty]$. A related theory of harmonic functions in $\mathbb{R}^n_+$ with boundary values from $L^p(\mathbb{R}^{n-1})$ is described, for instance, in [7] (Chap. 2, Sect. 2).

Theorem 1 *Let x be an arbitrary point in $\mathbb{R}^n_+$. The sharp coefficient $C_p(x)$ in the inequality*

$$\left|\nabla\left\{\frac{u(x)}{x_n}\right\}\right| \le C_p(x)||f||_p$$

is given by

$$C_1(x) = \left(\frac{c^{n+2}}{2^{n-2}\pi^n}\right)^{1/2} \frac{K_{(n+2)/2}(cx_n)}{x_n^{n/2}}$$

for $p = 1$ and

$$C_p(x) = \frac{2^{1/p}}{\pi^{n/2} x_n^{2+\frac{n-1}{p}}} \max_{|z|=1} \left\{\int_{\mathbb{S}^{n-1}} \rho_n^{\frac{p}{p-1}}\big((\boldsymbol{e}_\sigma, \boldsymbol{e}_n)\big)|(\boldsymbol{e}_\sigma, \boldsymbol{e}_n)|^{\frac{n+p}{p-1}}|(\boldsymbol{e}_\sigma, z)|^{\frac{p}{p-1}} d\sigma\right\}^{\frac{p-1}{p}}$$

for $p \in (1, \infty]$, where

$$\rho_m(t) = \int_0^\infty \xi^{m/2} e^{-\xi - \frac{c^2 x_n^2}{4\xi t^2}} d\xi.$$

In particular, the solution to the extremal problem with respect to $z \in \mathbb{S}^{n-1}$ for $p \in [2, \infty]$ is given by

$$C_p(x) = \frac{2^{1/p}}{\pi^{n/2} x_n^{2+\frac{n-1}{p}}} \left\{\int_{\mathbb{S}^{n-1}} \rho_n^{\frac{p}{p-1}}\big((\boldsymbol{e}_\sigma, \boldsymbol{e}_n)\big)|(\boldsymbol{e}_\sigma, \boldsymbol{e}_n)|^{\frac{n+2p}{p-1}} d\sigma\right\}^{\frac{p-1}{p}}$$

$$= \frac{\omega_{n-1}^{1-\frac{1}{p}} c^{\frac{n+2}{2}} x_n^{\frac{n-2}{2}-\frac{n-1}{p}}}{\pi^{n/2} 2^{(n-2)/2}} \left\{\int_0^{\pi/2} K_{(n+2)/2}^{p/(p-1)}\left(\frac{cx_n}{\cos\vartheta}\right) \cos^{\frac{2p-n(p-2)}{2(p-1)}}\vartheta \sin^{n-2}\vartheta d\vartheta\right\}^{\frac{p-1}{p}}.$$

As a special case, one has

$$C_\infty(x) = \frac{c^{(n+2)/2} x_n^{(n-2)/2}}{2^{\frac{n-4}{2}}\sqrt{\pi}\,\Gamma\left(\frac{n-1}{2}\right)} \int_0^{\pi/2} K_{(n+2)/2}\left(\frac{cx_n}{\cos\vartheta}\right) \frac{\sin^{n-2}\vartheta}{\cos^{(n-2)/2}\vartheta} d\vartheta\,.$$

4 Sharp Estimate for the Gradient of Solution to the Neumann Problem

Solution to the Neumann problem in $\mathbb{R}^n_+$ for the modified Helmholtz equation,

$$(\Delta - c^2)u = 0 \ \text{ in } \ \mathbb{R}^n_+, \qquad \left.\frac{\partial u}{\partial x_n}\right|_{x_n=0} = g(x') \tag{5}$$

with continuous and bounded function g on $\mathbb{R}^{n-1}$, is given by

$$u(x) = -\frac{2c^{(n-2)/2}}{(2\pi)^{n/2}} \int_{\mathbb{R}^{n-1}} \frac{K_{(n-2)/2}\big(c|y-x|\big)}{\big(c|y-x|\big)^{(n-2)/2}} g(y')dy'. \tag{6}$$

Here, as before, $y = (y', 0)$, $y' \in \mathbb{R}^{n-1}$ and K_ν means the Macdonald function. Formula (6) with any integer $n \geq 2$ can be obtained by the Fourier transform. This formula is well known for the cases $n = 2$ and $n = 3$ (e.g. [5], Sects. 7.3 and 8.3).

We consider the solution to problem (5) with $g \in L^p(\mathbb{R}^{n-1})$ represented by (6), where $p \in [1, \infty]$.

Theorem 2 *Let x be an arbitrary point in $\mathbb{R}^n_+$. The sharp coefficient $\mathcal{K}_p(x)$ in the inequality*

$$|\nabla u(x)| \leq \mathcal{K}_p(x)||g||_p$$

is given by

$$\mathcal{K}_1(x) = \frac{c}{2^{(n-2)/2}\pi^{n/2}} \frac{K_{n/2}(cx_n)}{x_n^{(n-2)/2}}$$

for $p = 1$ and

$$\mathcal{K}_p(x) = \frac{2^{(1-p)/p}}{\pi^{\frac{n}{2}} c^{\frac{n-2}{2}} x_n^{\frac{n-1}{p-1}}} \max_{|z|=1} \left\{ \int_{\mathbb{S}^{n-1}} \rho_{n-2}^{\frac{p}{p-1}}\big((\boldsymbol{e}_\sigma, \boldsymbol{e}_n)\big) |(\boldsymbol{e}_\sigma, \boldsymbol{e}_n)|^{\frac{n-p}{p-1}} |(\boldsymbol{e}_\sigma, \boldsymbol{z})|^{\frac{p}{p-1}} d\sigma \right\}^{\frac{p-1}{p}}$$

for $p \in (1, \infty]$, where the function $\rho_m(t)$ is defined as before.

In particular, the solution to the extremal problem with respect to $z \in \mathbb{S}^{n-1}$ for $p \in [2, (n+2)/2]$ is given by

$$\mathcal{K}_p(x) = \frac{2^{(1-p)/p}}{\pi^{n/2} c^{(n-2)/2} x_n^{\frac{n-1}{p-1}}} \left\{ \int_{\mathbb{S}^{n-1}} \rho_{n-2}^{\frac{p}{p-1}}\big((\boldsymbol{e}_\sigma, \boldsymbol{e}_n)\big) |(\boldsymbol{e}_\sigma, \boldsymbol{e}_n)|^{\frac{n}{p-1}} d\sigma \right\}^{\frac{p-1}{p}},$$

$$= \frac{c\, \omega_{n-1}^{(p-1)/p} x_n^{\frac{n}{2} - \frac{n-1}{p-1}}}{\pi^{n/2} 2^{(n-2)/2}} \left\{ \int_0^{\pi/2} K_{n/2}^{p/(p-1)} \left(\frac{cx_n}{\cos\vartheta} \right) \cos^{\frac{(2-p)n}{2(p-1)}} \vartheta \sin^{n-2} \vartheta d\vartheta \right\}^{\frac{p-1}{p}}.$$

As a special case, one has

$$\mathcal{K}_2(x) = \frac{c x_n^{1/2}}{\pi^{(n+1)/4} 2^{(n-3)/2} \sqrt{\Gamma\left(\frac{n-1}{2}\right)}} \left\{ \int_0^{\pi/2} K_{n/2}^2 \left(\frac{c x_n}{\cos \vartheta} \right) \sin^{n-2} \vartheta \, d\vartheta \right\}^{1/2}.$$

References

1. Kresin, G., Maz'ya, V.: Optimal estimates for the gradient of harmonic functions in the multidimensional half-space. Discr. Cont. Dyn. Syst. series B **28**, 425–440 (2010)
2. Kresin, G., Maz'ya, V.: Sharp real-part theorems in the upper half-plane and similar estimates for harmonic functions. J. Math. Sci. (New York) **179**, 144–163 (2011)
3. Kresin, G., Maz'ya, V.: Generalized Poisson integral and sharp estimates for harmonic and biharmonic functions in the half-space. Math. Model. Nat. Phen. (2018). https://doi.org/10.1051/mmnp/2018032
4. Kresin, G., Ben Yaakov, T.: Sharp pointwise estimates for solutions of the modified Helmholtz equation. Pure Appl. Funct. Anal. (to appear)
5. Polyanin, A.D., Nazaikinskii, V.E.: Handbook of Linear Partial Differential Equations for Engineers and Scientists, 2nd edn. Chapman and Hall/CRC Press, Boca Raton (2016)
6. Schot, S.H.: The first boundary value problem for the iterated Helmholtz equation in a half-space. Appl. Anal. **54**, 151–161 (1994)
7. Stein, E.M., Weiss, G.: Introduction to Fourier Analysis on Euclidean Spaces. Princeton University Press, Princeton (1971)

Delay Optimization Problem for One Class of Functional Differential Equation

Medea Iordanishvili, Tea Shavadze, and Tamaz Tadumadze

Abstract For the nonlinear optimization problem with delays necessary optimality conditions are proved: for delays in the phase coordinates and controls in the form of equality; for control functions in the form of linearized integral maximum principle.

Keywords Controlled functional differential equation · Necessary conditions of optimality

1 Introduction

As is known, the real controlled dynamical systems contain effects with delayed action and are described by differential equations with delay in controls [1]. To illustrate this, here we consider a model of marketing relation below.

Let $t_1 > t_0$, $\theta_2 > \theta_1 > 0$, $p_2 > p_1 \geq 0$ and $q_2 > q_1 \geq 0$ be given numbers. Let market relation demand and supply are described by the functions $D(t, p, q)$ and $S(t, \varsigma, \upsilon)$, which are continuous and continuously differentiable with respect to (p, q) and (ς, υ).

Let the function $p(t) \in P = [p_1, p_2], t \in I_1 = [t_0 - \theta_2, t_1]$ is price of a good i_1 changing over time and $q(t) \in Q = [q_1, q_2], t \in I_1$ is price of a good i_2. Suppose that at time $t \in I_2 = [t_0, t_1]$ will be satisfied demand of consumer on the good i_1 which

M. Iordanishvili
Department of Computer Sciences, I. Javakhishvili Tbilisi State University,
13 University Str., 0186 Tbilisi, Georgia
e-mail: medea.iordanishvili@tsu.ge

T. Shavadze (✉)
I. Vekua Institute of Applied Mathematics, I. Javakhishvili Tbilisi State University,
2 University Str., 0186 Tbilisi, Georgia
e-mail: Tea.shavadze@gmail.com

T. Tadumadze
Department of Mathematics, I. Vekua Institute of Applied Mathematics,
I. Javakhishvili Tbilisi State University, 2 University Str., 0186 Tbilisi, Georgia
e-mail: tamaz.tadumadze@tsu.ge

A. Domoshnitsky et al. (eds.), *Functional Differential Equations and Applications*,
Springer Proceedings in Mathematics & Statistics 379,
https://doi.org/10.1007/978-981-16-6297-3_13

has been ordered at time $t-h$, where $h \in I_3 = [\theta_1, \theta_2]$ is a fixed delay parameter and will be satisfied demand of consumer on the good i_2, which has been ordered at time $t-\theta$, where $\theta \in [\theta_1, \theta_2]$, in general, is non fixed delay.

The function

$$E(t) = D(t, p(t), q(t)) - S(t, p(t-h), q(t-\theta)), t \in I_2,$$

we call the disbalance index.

If $E(t) = 0$, then at the moment t, we do not have a disbalance between supply and demand, and the customer will buy exactly the quantity of goods he needs.

It is clear that at various time moment t, the disbalance index E(t) is possible to be not positive as well as positive. At time t, if $E(t) > 0$, then the demand exaggerates the supply, If $E(t) < 0$, then the supply exaggerates the demand. To describe the development of marketing relation process in time, i.e., create dynamical model, we consider the integral index of disbalance

$$y(t) = E(t_0) + \int_{t_0}^{t} E(s)ds. \tag{1}$$

The function $y(t)$ gives complete information about the disbalance from the initial time t_0 to any time t. From (1), we get the differential equation

$$\dot{y}(t) = D(t, p(t), q(t)) - S(t, p(t-h), q(t-\theta)), t \in I_2 \tag{2}$$

with the initial condition

$$y(t_0) = y_0 = E(t_0).$$

In the paper, for the nonlinear functional differential equation

$$\dot{x}(t) = f(t, x(t), x(t-\tau), u(t), u(t-h), v(t), v(t-\theta)), x(t) \in \mathbb{R}^n, \ t \in I_2,$$

where $\tau > 0$, a delay optimization problem is considered. The necessary optimality conditions are proved: for delays τ and θ in the form of equality; for control functions $u(t)$ and $v(t)$ in the form of linearized integral maximum principle.

2 Statement of the Problem. Necessary Optimality Conditions

Let $U \subset \mathbb{R}^{r_1}$ and $V \subset \mathbb{R}^{r_2}$ be convex and compact sets. Let the n-dimensional function $f(t, x, x_1, u, u_1, v, v_1)$ be continuous on $I_2 \times \mathbb{R}^n \times \mathbb{R}^n \times U^2 \times V^2$ and continuously differentiable with respect to (x, x_1, u, u_1, v, v_1); there exists a number $M > 0$ such that for all $(t, x, x_1, u, u_1, v, v_1) \in I_2 \times \mathbb{R}^n \times \mathbb{R}^n \times U^2 \times V^2$ we have

$$|f(t,x,x_1,u,u_1,v,v_1)|+|f_x(\cdot)|+|f_{x_1}(\cdot)|+|f_u(\cdot)|+|f_{u_1}(\cdot)|$$

$$+|f_v(\cdot)|+|f_{v_1}(\cdot)|\le M.$$

Furthermore, let $x_0\in\mathbb{R}^n$ be fixed points and $\tau_2>\tau_1>0$ be given numbers. Let $\Omega_0(I_1,U)$ be a set of piecewise-continuous control functions $u(t)\in U, t\in I_1$, with $\| u \|=\sup\{|u(t)|: t\in I_1$ and $\Omega_1(I_1,V)$ be a set of absolutely continuous control functions $v(t)\in V, t\in I_1$, with $\| v \|=\sup\{|v(t)|: t\in I_1$.

To each element

$$w=(\tau,\theta,u(\cdot),v(\cdot))\in W:=[\tau_1,\tau_2]\times I_3\times\Omega_0(I_1,U)\times\Omega_1(I_1,V)$$

we assign the functional differential equation

$$\dot{x}(t)=f(t,x(t),x(t-\tau),u(t),u(t-h),v(t),v(t-\theta)),\ t\in I_2 \tag{3}$$

with the initial condition

$$x(t)=\varphi(t), t\in[t_0-\tau_2,t_0),\ x(t_0)=x_0, \tag{4}$$

where $\varphi(t)\in\mathbb{R}^n, t\in[t_0-\tau_2,t_0]$ is a given absolutely continuous initial function.

Definition 1 Let $w=(\tau,\theta,u(\cdot),v(\cdot))\in W$. A function $x(t)=x(t;w)\in\mathbb{R}^n, t\in I_2$, is called a solution of Eq. (3) with the initial condition (4) or a solution corresponding to the element w and defined on the interval I_2 if it satisfies condition (4) and is absolutely continuous and satisfies Eq. (3) almost everywhere on I_2. end definition. By the step method and Gronwall inequality can be proved that for every element $w\in W$ there exists unique solution $x(t;w)$ defined on the interval $[\hat{\tau},t_1]$ and continuous with respect to w.

Let the scalar-valued functions $\vartheta^i(\tau,\theta,x), i=\overline{0,l}$, be continuously differentiable on $[\tau_1,\tau_2]\times I_3\times\mathbb{R}^n$.

Definition 2 An element $w\in W$ is said to be admissible if the corresponding solution $x(t)=x(t;w)$ satisfies the conditions

$$\vartheta^i(\tau,\theta,x(t_1))=0, i=\overline{1,l}. \tag{5}$$

Denote by W_0 the set of admissible elements.

Definition 3 An element $w_0=(\tau_0,\theta_0,u_0(\cdot),v_0(\cdot))\in W_0$ is said to be optimal if for an arbitrary element $w\in\Lambda_0$ we have

$$\vartheta^0(\tau_0,\theta_0,x_0(t_1))\le\vartheta^0(\tau,\theta,x(t_1)), \tag{6}$$

where $x_0(t)=x(t;w_0)$.

Equations (3)–(6) is called the delay optimization problem.

Theorem 1 *Let w_0 be an optimal element with $\tau_0 \in (\tau_1, \tau_2)$, $\theta_0 \in (\theta_1, \theta_2)$ and $t_0 + \tau_0 < t_1$. Moreover, the functions $u_0(t)$ and $u_0(t-h)$ is continuous at the point $t_0 + \tau_0$. Then there exist a vector $\pi = (\pi_0, \dots, \pi_l) \neq 0$ and a solution $\psi(t)$ of the equation*

$$\dot{\psi}(t) = -\psi(t) f_x[t] - \psi(t+\tau_0) f_{x_1}[t+\tau_0], t \in I_2 \tag{7}$$

with the initial condition

$$\psi(t_0) = \pi \Theta_{0x}, \ \psi(t) = 0, t > t_1, \tag{8}$$

where

$$f_x[t] = f_x(t, x_0(t), x_0(t-\tau_0), u_0(t), u_0(t-h), v_0(t), v_0(t-\theta_0)),$$

such that the following conditions hold;
(1) the condition for the delay τ_0

$$\pi\Theta_{0\tau} = \psi(t_0+\tau_0)\hat{f} + \int_{t_0}^{t_0+\tau_0} \psi(t) f_{x_1}[t]\dot{\varphi}(t-\tau_0)dt$$

$$+ \int_{t_0+\tau_0}^{t_1} \psi(t) f_{x_1}[t]\dot{x}_0(t-\tau_0)dt,$$

where $\Theta = (\vartheta^0, \dots, \vartheta^l)^T$, $\Theta_{0\tau} = \Theta_\tau(\tau_0, \theta_0, x_0(t_1))$ and

$$\hat{f} = f(t_0+\tau_0, x_0(t+\tau_0), x_0, u_0(t_0+\tau_0), u_0(t_0+\tau_0-h), v_0(t_0+\tau_0), v_0(t_0+\tau_0-\theta_0))$$

$$-f(t_0+\tau_0, x_0(t_0+\tau_0), \varphi(t_0), u_0(t_0+\tau_0), u_0(t_0+\tau_0-h), v_0(t_0+\tau_0), v_0(t_0+\tau_0-\theta_0));$$

(2) the condition for the delay θ_0

$$\pi\Theta_{0\theta} = \int_{t_0}^{t_1} \psi(t) f_{v_1}[t]\dot{v}_0(t-\theta_0)dt;$$

(3) the integral maximum principle for the optimal control $u_0(t)$

$$\int_{t_0}^{t_1} \psi(t)\Big[f_u[t]u_0(t) + f_{u_1}[t]u_0(t-h)\Big]dt$$

$$= \max_{u(\cdot)\in\Omega_0(I_1,U)} \int_{t_0}^{t_1} \psi(t)\Big[f_u[t]u(t) + f_{u_1}[t]u(t-h)\Big]dt.$$

(4) the integral maximum principle for the optimal control $v_0(t)$

$$\int_{t_0}^{t_1} \psi(t)\Big[f_v[t]v_0(t) + f_{v_1}[t]v_0(t-\theta_0)\Big]dt$$

$$= \max_{v(\cdot)\in\Omega_1(I_1,V)} \int_{t_0}^{t_1} \psi(t)\Big[f_v[t]v(t) + f_{v_1}[t]v(t-\theta_0)\Big]dt.$$

3 Optimization Problem for the Eq. (2). Necessary Optimality Conditions

To each element

$$\varrho = (\theta, p(\cdot), q(\cdot)) \in \Pi := I_3 \times \Omega_0(I_1, P) \times \Omega_1(I_1, Q)$$

we assign the differential equation

$$\dot{y} = D(t, p(t), q(t)) - S(t, p(t-h), q(t-\theta)), t \in I_2$$

with the initial condition

$$y(t_0) = y_0.$$

Definition 4 Let y_1 be a fixed number. An element $\varrho \in \Pi$ is said to be admissible if the corresponding solution $y(t) = y(t; \varrho)$ satisfies the condition

$$y(t_1) = y_1.$$

Denote by Π_0 the set of admissible elements.

Definition 5 An element $\varrho_0 = (\theta_0, p_0(\cdot), q_0(\cdot)) \in \Pi_0$ is said to be optimal if for an arbitrary element $\varrho \in \Pi_0$ we have

$$\int_{t_0}^{t_1} g(t, p_0(t), q_0(t))dt \le \int_{t_0}^{t_1} g(t, p(t), q(t))dt,$$

where the function $g(t, p, q)$ is continuous and continuously differentiable with respect to (p, q).

It is clear that the above formulated problem is equivalent to the following problem

$$\begin{cases} \dot{y}^0(t) = g(t, p(t), q(t)), \\ \dot{y}(t) = D(t, p(t), q(t)) - S(t, p(t-h), q(t-\theta)) \end{cases}$$

$$y^0(t_0) = 0, \, y(t_0) = y_0, \, y(t_1) = y_1,$$

$$y^0(t_1) \to \min,$$

which is particular case of the problem (3)–(6) considered in the previous section. Therefore, from Theorem 1 it follows

Theorem 2 Let ϑ_0 be an optimal element. Then there exists a vector $(\psi_0, \psi) \neq$, $\psi_0 \leq 0$ such that the following conditions hold:
(1) the integral condition for the optimal delay parameter θ_0

$$\psi \int_{t_0}^{t_1} S_v(t, p_0(t-h), q_0(t-\theta_0))\dot{q}_0(t-\theta_0)dt = 0;$$

(2) the integral maximum principle for the optimal control $p_0(t)$

$$\int_{t_0}^{t_1} \Big[\Big(\psi_0 g_p(t, p_0(t)) + \psi D_p(t, p_0(t), q_0(t)\Big) p_0(t)$$

$$-\psi S_\varsigma(t, p_0(t-h), q_0(t-\theta_0)) p_0(t-h)\Big] dt$$

$$= \max_{p(\cdot)\in\Omega_0(I_1,P)} \int_{t_0}^{t_1} \Big[\Big(\psi_0 g_p(t, p_0(t)) + \psi D_p(t, p_0(t), q_0(t)\Big) p(t)$$

$$-\psi S_\varsigma(t, p_0(t-h), q_0(t-\theta_0)) p(t-h)\Big] dt.$$

(3) the integral maximum principle for the optimal control $q_0(t)$

$$\int_{t_0}^{t_1} \Big[\Big(\psi_0 g_q(t, p_0(t), q_0(t)) + \psi D_q(t, p_0(t), q_0(t)\Big) q_0(t)$$

$$-\psi S_v(t, p_0(t-h), q_0(t-\theta_0)) q_0(t-\theta_0)\Big] dt$$

$$= \max_{q(\cdot)\in\Omega_1(I_1,Q)} \int_{t_0}^{t_1} \Big[\Big(\psi_0 g_q(t, p_0(t), q_0(t)) + \psi D_q(t, p_0(t), q_0(t)\Big) q(t)$$

$$-\psi S_v(t, p_0(t-h), q_0(t-\theta_0)) q(t-\theta_0)\Big] dt.$$

Analogous problem for the marketing relation model with the fixed θ is investigated in [2].

4 Proof of Theorem 1

Theorem 1 proved by the scheme given in [3, 4].

Proof On the convex set $Z = \mathbb{R}_+ \times W$, where $\mathbb{R}_+ = [0, \infty)$, let us define the mapping

$$\Phi : Z \to \mathbb{R}^{1+l} \tag{9}$$

by the formula

$$\Phi(z) = (\Phi^0(z), \ldots, \Phi^l(z))^T = \Theta(\tau, \theta, x(t_1; w)) + (\xi, 0 \ldots, 0)^T, z = (\xi, w) \in Z.$$

It is clear that

$$\Phi^0(z_0) \le \Phi^0(z), \Phi^i(z) = 0, i = \overline{1, l}, \forall z \in \mathbb{R}_+ \times W_0 \subset Z,$$

where $z_0 = (0, w_0)$.

Thus, the point $z_0 = (0, w_0) \in Z$ is a critical (see [3, 4]), since $Q(z_0) \in \partial Q(Z)$. Moreover, the mapping (9) is continuous.

There exists a small $\varepsilon_0 > 0$ such that for an arbitrary $\varepsilon \in (0, \varepsilon_0)$ and $\delta z = (\delta\xi, \delta w) \in B_{z_0} := [0, \alpha) \times B_{w_0} \subset Z - z_0$, where $\delta w = (\delta\tau, \delta\theta, \delta u, \delta v)$ and $B_{w_0} = [(\tau_0 - \alpha, \tau_0 + \alpha) - \tau_0] \times [(\theta_0 - \alpha, \theta_0 + \alpha) - \theta_0] \times [\Omega_0(I_1, U) - u_0(\cdot)] \times [\Omega_1(I_1, V) - v_0(\cdot)]$ we get $z_0 + \varepsilon\delta z \in Z$.

On the basis of the variation formula of solution [5, 6], we have

$$\Delta x(t_1; \varepsilon\delta w) := x(t_1; w_0 + \varepsilon\delta w) - x_0(t_1) = \varepsilon\delta x(t_1; \delta w) + o(\varepsilon\delta w),$$

$$(\varepsilon, \delta w) \in (0, \varepsilon_1) \times B_{w_0}$$

where

$$\delta x(t_1; \delta w) = -\Big[Y(t_0 + \tau_0; t_1)\hat{f} + \int_{t_0}^{t_0+\tau_0} Y(t; t_1) f_{x_1}[t]\dot{\varphi}(t - \tau_0)dt \tag{10}$$

$$+ \int_{t_0+\tau_0}^{t_1} Y(t; t_1) f_{x_1}[t]\dot{x}_0(t - \tau_0)dt\Big]\delta\tau - \Big[\int_{t_0}^{t_1} Y(t; t_1) f_{v_1}[t]\dot{v}_0(t - \theta_0)dt\Big]\delta\theta$$

$$+ \int_{t_0}^{t_1} Y(t; t_1)\Big[f_u[t]\delta u(t) + f_{u_1}[t]\delta u(t - h) + f_v[t]\delta v(t) + f_{v_1}[t]\delta v(t - \theta_0)\Big]dt$$

and

$$\lim_{\varepsilon\to 0} \frac{o(\varepsilon\delta w)}{\varepsilon} = 0 \text{ uniformly for } \delta w \in B_{w_0};$$

$Y(t;t_1)$ is the $n \times n-$matrix function satisfying the linear differential equation with advanced argument

$$\frac{d}{dt}Y(t;t_1) = -Y(t;t_1)f_x[t] - Y(t+\tau_0;t_1)f_{x_1}[t+\tau_0], t \in [t_{00}, t_1]$$

and the conditions

$$Y(t;t_1) = \begin{cases} E \ for \ t = t_1, \\ \hat{0} \ for \ t > t_1, \end{cases}$$

E is the identity matrix and $\hat{0}$ is the zero matrix.

Now we calculate a differential of the mapping (9) at the point z_0. We have

$$\Phi(z_0 + \varepsilon\delta z) - \Phi(z_0) = \Theta(\tau_0 + \varepsilon\delta\tau, \theta_0 + \varepsilon\delta\theta, x(t_1; w_0 + \varepsilon\delta w)) - \Theta(\tau_0, \theta_0, x_0(t_1))$$

$$+\varepsilon(\delta\xi, 0 \ldots, 0)^T, \varepsilon \in (0, \varepsilon_1), \delta w \in B_{z_0}.$$

We introduce the notation

$$\Theta[\varepsilon; s] = \Theta(\tau_0 + \varepsilon s\delta\tau, \theta_0 + \varepsilon s\delta\theta, x_0(t_1) + s\Delta x(t_1; \varepsilon\delta w))$$

Let us transform the difference

$$\Theta(\tau_0 + \varepsilon\delta\tau, \theta_0 + \varepsilon\delta\theta, x(t_1; w_0 + \varepsilon\delta w)) - \Theta(\tau_0, \theta_0, x_0(t_1))$$

$$= \int_0^1 \frac{d}{ds}\Theta[\varepsilon; s]ds = \int_0^1 \Big[\varepsilon\Big(\Theta_\tau[\varepsilon; s]\delta\tau + \Theta_\theta[\varepsilon; s]\delta\theta\Big) + \Theta_x[\varepsilon; s]\Delta x(t_1; \varepsilon\delta w)\Big]ds$$

$$= \int_0^1 \Big[\varepsilon\Big(\Theta_\tau[\varepsilon; s]\delta\tau + \Theta_\theta[\varepsilon; s]\delta\theta + \Theta_x[\varepsilon; s]\delta x(t_1; \varepsilon\delta w)\Big) + \Theta_x[\varepsilon; s]o(\varepsilon\delta w)\Big]ds$$

$$= \varepsilon\Big[\Theta_{0\tau}\delta\tau + \Theta_{0\theta}\delta\theta + \Theta_{0x}\delta x(t_1; \delta w)\Big] + \gamma(\varepsilon\delta w),$$

where

$$\gamma(\varepsilon\delta w) = \varepsilon\int_0^1 \Big\{[\Theta_\tau[\varepsilon; s] - \Theta_{0\tau}]\delta\tau + [\Theta_\theta[\varepsilon; s] - \Theta_{0\theta}]\delta\theta$$

$$+[\Theta_x[\varepsilon; s] - \Theta_{0x}]\delta x(t_1; \delta w) + \Theta_{0x}\frac{o(\varepsilon\delta w)}{\varepsilon}\Big\}ds.$$

It is easy to see that

$$\lim_{\varepsilon\to 0}[\Theta_\tau[\varepsilon; s] - \Theta_{0\tau}] = 0, [\Theta_\theta[\varepsilon; s] - \Theta_{0\theta}] = 0, [\Theta_x[\varepsilon; s] - \Theta_{0x}] = 0.$$

Therefore, $\gamma(\varepsilon\delta w) = o(\varepsilon\delta w)$. Thus,

$$\Phi(z_0 + \varepsilon\delta z) - \Phi(z_0) = \varepsilon d\Phi_{z_0}(\delta z) + o(\varepsilon\delta z),$$

where $o(\varepsilon\delta z) := o(\varepsilon\delta w)$ and differential $d\Phi_{z_0}(\delta z)$ of the mapping (9) has the form

$$d\Phi_{z_0}(\delta z) = \Theta_{0\tau}\delta\tau + \Theta_{0\theta}\delta\theta + \Theta_{0x}\delta x(t_1; \delta w) + (\delta\xi, 0, \dots, 0)^T.$$

Due to the relation (10), we get

$$d\Phi_{z_0}(\delta z) = \Big[\Theta_{0\tau} - \Theta_{0x}Y(t_0 + \tau_0; t_1)\hat{f} \tag{11}$$

$$-\int_{t_0}^{t_0+\tau_0} \Theta_{0x}Y(t; t_1)f_{x_1}[t]\dot{\varphi}(t - \tau_0)dt - \int_{t_0+\tau_0}^{t_1} \Theta_{0x}Y(t; t_1)f_{x_1}[t]\dot{x}_0(t - \tau_0)dt\Big]\delta\tau$$

$$+\Big[\Theta_{0\theta} - \int_{t_0}^{t_1} \Theta_{0x}Y(t; t_1)f_{v_1}[t]\dot{v}_0(t - \theta_0)dt\Big]\delta\theta + \int_{t_0}^{t_1} \Theta_{0x}Y(t; t_1)\Big\{f_u[t]\delta u(t)$$

$$+f_{u_1}[t]\delta u(t - h) + f_v[t]\delta v(t) + f_{v_1}[t]\delta v(t - \theta_0)\Big\}dt + (\delta\xi, 0, \dots, 0)^T.$$

From the necessary condition of criticality [3, 4], it follows that there exists a vector $\pi = (\pi_0, \dots, \pi_l) \neq 0$ such that

$$\pi d\Phi_{z_0}(\delta z) \le 0, \forall\ \delta z \in \mathbb{R}_+ \times \mathbb{R} \times \mathbb{R} \times [\Omega_0(I_1, U) - u_0(\cdot)] \tag{12}$$

$$\times[\Omega_1(I_1, V) - v_0(\cdot)].$$

Introduce the function

$$\psi(t) = \pi\Theta_{0x}Y(t; t_1) \tag{13}$$

as is easily seen, it satisfies the Eq. (7) and the condition (8).

Taking into account (11) and (13) from inequality (12), we obtain

$$\Big[\pi\Theta_{0\tau} - \psi(t_0 + \tau_0)\hat{f} - \int_{t_0}^{t_0+\tau_0} \psi(t)f_{x_1}[t]\dot{\varphi}(t - \tau_0)dt \tag{14}$$

$$-\int_{t_0+\tau_0}^{t_1} \psi(t)f_{x_1}[t]\dot{x}_0(t - \tau_0)dt\Big]\delta\tau + \Big[\pi\Theta_{0\theta} - \int_{t_0}^{t_1} \psi(t)f_{v_1}[t]\dot{v}_0(t - \theta_0)dt\Big]\delta\theta$$

$$+\int_{t_0}^{t_1} \psi(t)\Big\{f_u[t]\delta u(t) + f_{u_1}[t]\delta u(t - h) + f_v[t]\delta v(t) + f_{v_1}[t]\delta v(t - \theta_0)\Big\}dt$$

$$+\pi_0\delta\xi \le 0.$$

Let $\delta\tau = \delta\theta = 0$, $\delta u = 0$, $\delta v = 0$ in (14), then we obtain

$$\pi_0 \delta\xi \le 0, \forall \delta\xi \in \mathbb{R}_+.$$

This implies $\pi_0 \le 0$.

Let $\delta\xi = \delta\theta = 0$, $\delta u = 0$, $\delta v = 0$ then, taking into account that $\delta\tau \in \mathbb{R}$ from (14), we obtain the condition *(2.1)*.

Let $\delta\xi = \delta\tau = 0$, $\delta u = 0$, $\delta v = 0$ then, taking into account that $\delta\tau \in \mathbb{R}$ from (14), we obtain the condition *(2.2)*.

Let $\delta\xi = \delta\tau = \delta\theta = 0$, $\delta v = 0$ then, taking into account that $\delta u \in \Omega_0(I_0), U) - u_0(\cdot)$ from (14), we obtain the condition *(2.3)*.

Let $\delta\xi = \delta\tau = \delta\theta = 0$, $\delta u = 0$ then, taking into account that $\delta v \in \Omega_1(I_0), V) - v_0(\cdot)$ from (14), we obtain the condition *(2.4)*.

Acknowledgements This work was supported partly by the Sh. Rustaveli National Science Foundation (Georgia), Grant No. Ph.D.-F-17-89.

References

1. Kharatishvili, G., Nanetashvili, N., Nizharadze, T.: Dynamic control mathematical model of demand and satisfactions and problem of optimal satisfaction of demand. In: Proceedings of the International Scientific Conference "Problems of Control and Power Engineering", vol. 8, pp. 44–48 (2004)
2. Dvalishvili, Ph., Tadumadze, T.: Optimization of one marketing relation model with delay. J. Modern Technol. Eng. **4**(1), 5–10 (2019)
3. Kharatishvili, G.L., Tadumadze, T.A.: Formulas for the variation of a solution and optimal control problems for differential equations with retarded arguments. J. Math. Sci. (N. Y.) **140**(1), 1–175 (2007)
4. Tadumadze, T.: Variation formulas of solutions for functional differential equations with several constant delays and their applications in optimal control problems. Mem. Differ. Equ. Math. Phys. **70**, 7–97 (2017)
5. Shavadze, T.: Variation formulas of solutions for controlled functional differential equations with the continuous initial condition with regard for perturbations of the initial moment and several delays. Mem. Differ. Equ. Math. Phys. **74**, 125–140 (2018)
6. Iordanishvili, M.: Local variation formulas of solutions for the nonlinear controlled differential equation with the discontinuous initial condition and with delay in the phase coordinates and controls Transactions of A. Razmadze Mathematical Institute, vol. 174, pp. 10–16 (2019)

On Qualitative Research of Lattice Dynamical System of Two- and Three-Dimensional Biopixels Array

Vasyl Martsenyuk, Mikolaj Karpinski, Aleksandra Klos-Witkowska, and Andriy Sverstiuk

Abstract We consider the model of two- or three-dimensional biopixels array, which can be used for the design of biosensors. The model is based on the system of lattice differential equations with time delay, describing interactions of biological species of neighboring pixels. The qualitative analysis includes permanence and extinction of solutions, stability investigation, bifurcations, and transition to chaos. The stability conditions are obtained with help of the method of Lyapunov functionals. They are formulated in terms of the value of time necessary for immune response. Numerical research is presented with the help of phase portraits, square and hexagonal lattice plots, and bifurcation diagrams.

Keywords Lattice differnetial equations · Delay · Stability · Biopixel · Extinction · Phase plot

1 Introduction

Nowadays, reaction-diffusion models are used in designing and studies of a lot of detecting, measuring, and sensing devices. One of the examples is the immunosensor which is studied here. Such spatial-temporal models are described by the systems of partial or lattice differential equations.

V. Martsenyuk (✉) · M. Karpinski · A. Klos-Witkowska
University of Bielsko-Biala, 2 Willowa St, 43-309 Bielsko-Biala, Poland
e-mail: vmartsenyuk@ath.bielsko.pl

M. Karpinski
e-mail: mkarpinski@ath.bielsko.pl

A. Klos-Witkowska
e-mail: awitkowska@ath.bielsko.pl

A. Sverstiuk
Ternopil National Medical University, Voli Square 1, Ternopil 46001, Ukraine
e-mail: sverstyuk@tdmu.edu.ua

A. Domoshnitsky et al. (eds.), *Functional Differential Equations and Applications*, Springer Proceedings in Mathematics & Statistics 379,
https://doi.org/10.1007/978-981-16-6297-3_14

The biosensor models are traditionally studied from the viewpoint of their qualitative analysis. Even in case of a small number of spatial elements, they show complex behavior. In [1], it was shown that the model describes the chemical reaction of two morphogens (reactants), one of them diffusing within two compartments, results in "bi-chaotic" behavior. The origin of such chaotic phenomena[1] were also explained with the help of statistics of topological defects [2].

When considering continuously distributed reaction-diffusion models described by nonlinear partial differential equations, FeigenbaumSharkovskiiMagnitskii bifurcation theory can be applied, which results in a subharmonic cascade of bifurcations of stable limit cycles [3].

The lattice differential equations describe the systems with the discrete spatial structure, which is more consistent with pixel devices. These equations were also called earlier by a series of authors as spatially discrete differential equations [4].

Due to [5], a typical lattice differential equation takes the form

$$\dot{u_\xi} = g_\xi(\{u_\zeta\}_{\zeta \in \Lambda}), \xi \in \Lambda, \tag{1}$$

where we consider a lattice $\Lambda \subset \mathbb{R}^n$, which can be presented as a discrete subset of $\mathbb{R}^n$, consisting of either finite or infinite number of points, which are located in accordance with some regular spatial structure. The vectors u_ξ, $\xi \in \Lambda$ are the values of the state $u = \{u_\xi\}_{\xi \in \Lambda}$, calculated at the points of the lattice, and g_ξ are the right sides of the equations with the properties enabling us the solution existence.

As a rule, without loss of generality, they consider $\Lambda = \mathbb{Z}^n$, which is the integer lattice in $\mathbb{R}^n$. The methods developed can be easily applied to different types of lattices, namely, the planar rectangular and hexagonal lattice, the crystallographic lattices in $\mathbb{R}^3$.

They pay attention to the notion of delay in lattice differential equations, so-called delayed lattice differential equations. One of the applications dealing with them is the investigation of traveling wavefronts and their stability [5]. The main results are applied to the delayed and discretely diffusive models for the population (see, e.g., [6, 7]).

Lattice differential equations are used as models in a lot of applications, for example, cellular neural networks, image processing, chemical kinetics, material science, in particular, metallurgy and biology [5, 8]. Lattice models are extremely attractive from the viewpoint of population dynamics, especially in case of spatially separated populations [5, 6, 8–11].

There are few reasons requiring consideration of the hexagonal grid instead of rectangular ones (primarily in image and vision computing). Namely, the equal distances between neighboring pixels for hexagonal coordinate systems [12]; hexagonal points are packed more densely [13]; since the "hexagons are 'rounder' than squares", the presentation of curves are more consistent with help of hexagonal systems [13]; hence mathematical operations of edge detection and shape extraction are more successful when applying hexagonal lattices [14].

[1] They call it as "spiral turbulence" [2].

With the purpose of indexing hexagonal pixels, as a rule, they use two-[2] or three-element[3] coordinate systems [15]. Our reasoning will be based on the last one. In contrary to skewed axes, the use of the cubic coordinates enables us symmetries with respect to all three axes.

2 Lattice Model of Antibody–Antigen Interaction for Two-Dimensional Biopixels Array

Let $V_{i,j}(t)$ be the concentration of antigens, $F_{i,j}(t)$ be the concentration of antibodies in biopixel (i, j), $i, j = \overline{1, N}$.
The model is based on the following biological assumptions for arbitrary biopixel (i, j).

1. We have some constant birthrate $\beta > 0$ for antigen population.
2. Antigens are detected, binded, and finally neutralized by antibodies with some probability rate $\gamma > 0$.
3. We have some constant death rate of antibodies $\mu_f > 0$.
4. We assume that when the antibody colonies are absent, the antigen colonies are governed by the well-known delay logistic equation

$$\frac{dV_{i,j}(t)}{dt} = (\beta - \delta_v V_{i,j}(t-\tau))V_{i,j}(t), \tag{2}$$

   where β and δ_v are positive numbers and $\tau \geq 0$ denotes delay in the negative feedback of the antigen colonies.
5. The antibody decreases the average growth rate of antigen linearly with a certain time delay τ; this assumption corresponds to the fact that antibodies cannot detect and bind antigen instantly; antibodies have to spend τ units of time before they are capable of decreasing the average growth rate of the antigen colonies; these aspects are incorporated in the antigen dynamics by the inclusion of the term $-\gamma F_{i,j}(t-\tau)$, where γ is a positive constant which can vary depending on the specific colonies of antibodies and antigens.
6. In the absence of antigen colonies, the average growth rate of the antibody colonies decreases exponentially due to the presence of $-\mu_f$ in the antibody dynamics and so as to incorporate the negative effects of antibody crowding, we have included the term $-\delta_f F_{i,j}(t)$ in the antibody dynamics.
7. The positive feedback $\eta\gamma V_{i,j}(t-\tau)$ in the average growth rate of the antibody has a delay since mature adult antibodies can only contribute to the production of antibody biomass; one can consider the delay τ in $\eta\gamma V_{i,j}(t-\tau)$ as a delay in antibody maturation.

[2] So called "skewed-axis" coordinate system.

[3] It is also known as "cube hex coordinate system".

8. While the last delay need not be the same as the delay in the hunting term and in the term governing antigen colonies, we have retained this for simplicity. We remark that the delays in the antibody term, antibody replacement term, and antigen negative feedback term can be made different and a similar analysis can be followed.
9. We have some diffusion of antigens from four neighboring pixels $(i-1, j)$, $(i+1, j)$, $(i, j-1)$, $(i, j+1)$ (see Fig. 1) with diffusion $D > 0$. Here we consider only diffusion of antigens because the model describes the so-called "competitive" configuration of immunosensor [16]. When considering competitive configuration of immunosensor, the factors immobilized on the biosensor matrix are antigens, while the antibodies play the role of analytes or particles to be detected.
10. We consider surface lateral diffusion (movement of molecules on the surface on solid phase toward immobilizated molecules) [17]. Moreover, there are works [18, 19] which assume and consider surface diffusion as an entirely independent stage.
11. We extend the definition of the usual diffusion operator in case of surface diffusion in the following way. Let $n \in (0, 1]$ be a factor of diffusion disbalance. It means that only nth portion of antigens of the pixel (i, j) may be included in the diffusion process to any neighboring pixel as a result of surface diffusion.

For the reasoning given, we consider a very simple delayed antibody–antigen competition model for biopixels two-dimensional array, which is based on well-known Marchuk model [20–23] and using spatial operator $\hat{S}$ offered in [24] (Supplementary information, p. 10)

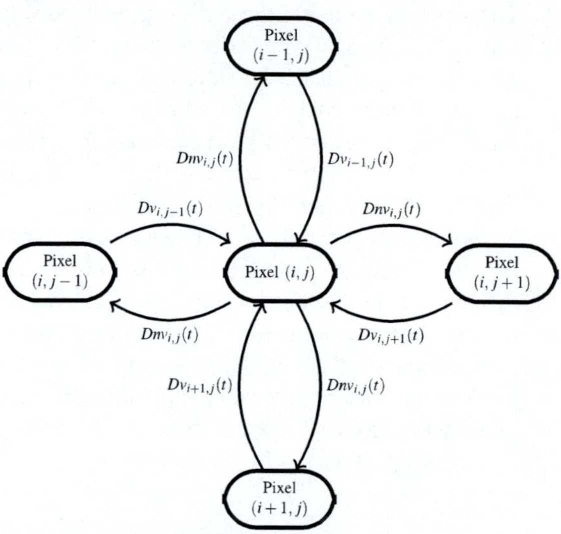

Fig. 1 Linear lattice interconnected four neighboring pixels model, $n > 0$ is disbalance constant

$$\begin{aligned} \frac{dV_{i,j}(t)}{dt} &= (\beta - \gamma F_{i,j}(t-\tau) - \delta_v V_{i,j}(t-\tau))V_{i,j}(t) + \hat{S}\{V_{i,j}\}, \\ \frac{dF_{i,j}(t)}{dt} &= \left(-\mu_f + \eta\gamma V_{i,j}(t-\tau) - \delta_f F_{ij}(t)\right) F_{i,j}(t) \end{aligned} \tag{3}$$

with given initial functions

$$\begin{aligned} &V_{i,j}(t) = V^0_{i,j}(t) \geq 0, \quad F_{i,j}(t) = F^0_{i,j}(t) \geq 0, \quad t \in [-\tau, 0), \\ &V_{i,j}(0), F_{i,j}(0) > 0. \end{aligned} \tag{4}$$

For a square $N \times N$ array of traps, we use the following discrete diffusion form of the spatial operator [24]

$$\hat{S}\{V_{i,j}\} = \begin{cases} D\Big[V_{1,2} + V_{2,1} - 2nV_{1,1}\Big] \;\; i, j = 1 \\ D\Big[V_{2,j} + V_{1,j-1} + V_{1,j+1} - 3nV_{i,j}\Big] \;\; i = 1, j \in \overline{2, N-1} \\ D\Big[V_{1,N-1} + V_{2,N} - 2nV_{1,N}\Big] \;\; i = 1, j = N \\ D\Big[V_{i-1,N} + V_{i+1,N} + V_{i,N-1} - 3nV_{i,N}\Big] \;\; i = \in \overline{2, N-1}, j = N \\ D\Big[V_{N-1,N} + V_{N,N-1} - 2nV_{N,N}\Big] \;\; i = N, j = N \\ D\Big[V_{N-1,j} + V_{N,j-1} + V_{N,j+1} - 3nV_{N,j}\Big] \;\; i = N, j \in \overline{2, N-1} \\ D\Big[V_{N-1,1} + V_{N,2} - 2nV_{N,1}\Big] \;\; i = N, j = 1 \\ D\Big[V_{i-1,1} + V_{i+1,1} + V_{i,2} - 3nV_{i,1}\Big] \;\; i \in \overline{2, N-1}, j = 1 \\ D\Big[V_{i-1,j} + V_{i+1,j} + V_{i,j-1} + V_{i,j+1} - 4nV_{i,j}\Big] \;\; i, j \in \overline{2, N-1} \end{cases} \tag{5}$$

Each colony is affected by the antigen produced in four neighboring colonies, two in each dimension of the array, separated by the equal distance Δ. We use the boundary condition $V_{i,j} = 0$ for the edges of the array $i, j = 0, N+1$. Further, we will use the following notation of the constant

$$k(i, j) = \begin{cases} 2 \;\; i, j = 1; \;\; i = 1, j = N; \;\; i = N, j = N; \;\; i = N, j = 1, \\ 3 \;\; i = 1, j \in \overline{2, N-1}; \;\; i \in \overline{2, N-1}, j = N; \;\; i = N, j \in \overline{2, N-1}; \\ \quad\; i \in \overline{2, N-1}, j = 1 \\ 4 \;\; i, j \in \overline{2, N-1} \end{cases} \tag{6}$$

which will be used in manipulations with the spatial operator (20).

Results of modeling (3) are presented further. It can be seen that the qualitative behavior of the system is determined mostly by the time of immune response τ (or time delay), diffusion D, and constant n.

2.1 Stability Investigation

2.1.1 Steady States

The steady states of the model (3) are the intersection of the null-clines $dV_{i,j}(t)/dt = 0$ and $dF_{i,j}(t)/dt = 0$, $i, j = \overline{1, N}$.

Antigen-free steady state. If $V_{i,j}(t) \equiv 0$, the free antigen equilibrium is at $\mathscr{E}^0_{i,j} \equiv \left(0, 0\right)$, $i, j = \overline{1, N}$ or $\mathscr{E}^0_{i,j} \equiv \left(0, -\frac{\mu_f}{\delta_f}\right)$, $i, j = \overline{1, N}$. The last solution does not have biological sense and cannot be reached for nonnegative initial conditions (19).

When considering endemic steady state $\mathscr{E}^*_{i,j} \equiv \left(V^*_{i,j}, F^*_{i,j}\right)$, $i, j = \overline{1, N}$ for (3) we get algebraic system:

$$\begin{aligned} &\left(\beta - \gamma F^*_{i,j} - \delta_v V^*_{i,j}\right) V^*_{i,j} + \hat{S}\left\{V^*_{i,j}\right\} = 0, \\ &\left(-\mu_f + \eta\gamma V^*_{i,j} - \delta_f F^*_{i,j}\right) F^*_{i,j} = 0, \quad i, j = \overline{1, N}. \end{aligned} \tag{7}$$

The solutions $\left(V^*_{i,j}, F^*_{i,j}\right)$ of (7) can be found as a result of solving lattice equation with respect to $V^*_{i,j}$, and using relation $F^*_{i,j} = \frac{-\mu_f + \eta\gamma V^*_{i,j}}{\delta_f}$

Thus we have to differ two cases.

Identical endemic state for all pixels. Let's assume there is the solution of (7) $V^*_{i,j} \equiv V^*$, $F^*_{i,j} \equiv F^*$, $i, j = \overline{1, N}$, i.e., $\hat{S}\left\{V^*_{i,j}\right\} \equiv 0$. Then $\mathscr{E}^*_{i,j} = \left(V^*, F^*\right)$, $i, j = \overline{1, N}$ can be calculated as

$$V^* = \frac{-\beta\delta_f - \gamma\mu_f}{\delta_v\delta_f - \eta\gamma^2}, \quad F^* = \frac{\delta_v\mu_f - \eta\gamma\beta}{\delta_v\delta_f - \eta\gamma^2}. \tag{8}$$

provided that $\delta_v\delta_f - \eta\gamma^2 < 0$.

Nonidentical endemic state for pixels. In the general case, we have an endemic steady state which is different from (8). It is shown numerically in Appendix B that it appears as a result of diffusion between pixels D.

At the absence of diffusion, i.e., $D = 0$, we have only an identical endemic state for pixels of external layer. At the presence of diffusion,i.e., $D > 0$, nonidentical endemic states tends to be identical ones (8) at internal pixels, which can be observed at numerical simulation. This phenomenon clearly appears at bigger amount of pixels.

Basic reproduction numbers. Here we define the basic reproduction number for antigen colony which is localized in pixel (i, j). When considering epidemic models, the basic reproduction number, $\mathscr{R}_0$, is defined as the expected number of secondary cases produced by a single (typical) infection in a completely susceptible population. It is important to note that $\mathscr{R}_0$ is a dimensionless number [25]. When applying this definition to the pixel (i, j), which is described by the Eq. (3), we get

$$\mathscr{R}_{0,i,j} = \mathscr{T}_{i,j}\overline{c}_{i,j}, d_{i,j}$$

where $\mathscr{T}_{i,j}$ is the transmissibility (i.e., probability of binding given constant between an antigen and antibody), $\overline{c}_{i,j}$ is the average rate of contact between antigens and antibodies, and $d_{i,j}$ is the duration of binding of antigen by antibody till deactivation.

Unfortunately, the lattice system (3) doesn't include all parameters, which allows to calculate the basic reproduction numbers in a clear form. Firstly, let's consider pixel $(i^\star, j^\star)$ without diffusion, i.e., $\hat{S}\left\{V_{i^\star,j^\star}\right\} \equiv 0$. In this case, the non-negative equilibria of (3) are

$$\mathscr{E}^0_{i^\star,j^\star} = \left(V^0, 0\right) := \left(\frac{\beta}{\delta_v}, 0\right), \quad \mathscr{E}^\star_{i^\star,j^\star} = \left(V^\star, F^\star\right).$$

Due to the approach which was offered in [26] (in pages 4 for ordinary differential equations, 5 for delay model), we introduce the basic reproduction number for pixel $(i^\star, j^\star)$ without diffusion, which is given by expression

$$\mathscr{R}_{0,i^\star,j^\star} := \frac{V^0}{V^\star} = \frac{\beta}{\delta_v V^\star} = \frac{\beta(\eta\gamma^2 - \delta_v\delta_f)}{\delta_v(\beta\delta_f + \gamma\mu_f)}.$$

Its biological meaning is given as being the average number of offsprings produced by a mature antibody in its lifetime when introduced in an antigen-only environment with antigen at carrying capacity.

According to the common theory, it can be shown that antibody-free equilibrium $\mathscr{E}^0_{i^\star,j^\star}$ is locally asymptotically stable if $\mathscr{R}_{0,i^\star,j^\star} < 1$ and it is unstable if $\mathscr{R}_{0,i^\star,j^\star} > 1$. It can be done with help of analysis of the roots of characteristic equation (similarly to [26], p. 5). Thus, $\mathscr{R}_{0,i^\star,j^\star} > 1$ is sufficient condition for existence of the endemic equilibrium $\mathscr{E}^\star_{i^\star,j^\star}$.

We can consider the expression mentioned above for the general case of the lattice system (3), i.e., when considering diffusion. In this case, we have the "lattice" of the basic reproduction numbers $\mathscr{R}_{0,i,j}$, $i, j = \overline{1, N}$ satisfying to

$$\mathscr{R}_{0,i,j} := \frac{V^0_{i,j}}{V^\star_{i,j}}, \quad i, j = \overline{1, N}, \tag{9}$$

where $\mathscr{E}^0_{i,j}$, $i, j = \overline{1, N}$ are nonidentical steady states, which are found as a result of solution of the algebraic system

$$\left(\beta - \delta_v V^0_{i,j}\right) V^0_{i,j} + \hat{S}\left\{V^0_{i,j}\right\} = 0, \quad i, j = \overline{1, N}, \tag{10}$$

endemic states $\mathscr{E}^\star_{i,j} = \left(V^\star_{i,j}, F^\star_{i,j}\right)$, $i, j = \overline{1, N}$ are found using (7).

It is worth to say that due to the common theory the conditions

$$\mathcal{R}_{0,i,j} > 1, \quad i, j = \overline{1, N} \tag{11}$$

are sufficient for the existence of endemic state $\mathcal{E}^{\star}_{i,j}$. We will check it only with help of numerical simulations.

2.2 Persistence of the Solutions

We will use the following definition which generalizes [27] for lattice differential equations.

Definition 1 System (3) is said to be uniformly persistent if for all $i, j = \overline{1, N}$ there exist compact regions $\mathcal{D}_{i,j} \subset \text{int}\, \mathbb{R}^2$ such that every solution $(V_{i,j}(t), F_{i,j}(t))$, $i, j = \overline{1, N}$ of (3) with the initial conditions (19) eventually enters and remains in the region $\mathcal{D}_{i,j}$.

Theorem 1 *Let* $(V_{i,j}(t), F_{i,j}(t))$, $i, j = \overline{1, N}$ *be the solutions of (3) with initials conditions (19). If*

$$\beta\eta\gamma - \mu_f\delta_v > 0, \tag{12}$$

then

$$0 < V_{i,j}(t) \le M_v, \quad 0 < F_{i,j}(t) \le M_f \tag{13}$$

for some large values of t. *Here*

$$M_v = \frac{\beta}{\delta_v} e^{\beta\tau}, \quad M_f = \frac{1}{\delta_f}\Big(\eta\gamma M_v - \mu_f\Big). \tag{14}$$

Proof Firstly, we can prove that there exists some large instant of time T_1 that $\hat{S}\{V_{i,j}(t)\} \le 0, i, j = \overline{1, N}, t > T_1$.

Let's assume the contrary, i.e., there is $i^\star, j^\star \in \overline{1, N}$, that $\hat{S}\{V_{i,j}(t)\} > 0$ at $t > T_1$, which is a contradiction with a balance principle.

Since the solutions of the system (3), (19) are positive, then

$$\frac{dV_{i,j}(t)}{dt} \le \Big(\beta - \delta_v V_{i,j}(t-\tau)\Big) V_{i,j}(t). \tag{15}$$

Further, we can apply the basic steps of proof of Lemma 3.1 [28] which is proved in nonlattice case (i.e., without spatial operator).

Remark 1 Conditions of uniform persistence of system (3) in nonlattice case were obtained in [29]. They resulted in inequality (12) provided that

$$\beta\delta_f + \mu_f\gamma > 0 \tag{16}$$

holds.

2.3 Extinction Research

The next result introduces a sufficient condition for the underlying grid size ensuring that the solution of (3) is non-vanishing.

Theorem 2 *Let for the system (3) the positive orthant Ω be positive invariant. Besides that, let N be such that $f_{extnc}(N) < 1$ holds, where*

$$f_{extnc}(N) = \max_{k,l=\overline{1,N}} \left| \beta - \frac{4D}{\Delta^2}\left(1 + \cos\frac{\pi(k+l)}{2(N+1)} \cos\frac{\pi(k-l)}{2(N+1)}\right)\right|. \tag{17}$$

Then $\lim_{t\to\infty} V_{i,j}(t) = 0$, $i, j = \overline{1, N}$.

Proof It requires a comparison principle for differential equations.

The following inequalities hold for $V_{i,j}(t)$

$$\frac{V_{i,j}(t)}{dt} < \beta V_{i,j}(t) + \hat{S}\left\{V_{i,j}(t)\right\}.$$

Consider N^2-vector of the form

$$V(n) = \Big(V_{1,1}(t), V_{1,2}(t), \ldots, V_{1,N}(t), V_{2,1}(t), \ldots, V_{2,N}(t), \ldots,$$
$$V_{N,1}(t), \ldots, V_{N,N}(t)\Big)^{\top}.$$

We compare $\frac{V(t)}{dt} \le CV(t)$, where $C = I_N \otimes A + B \otimes I_N$,

$$A = \begin{bmatrix} \beta - \frac{4D}{\Delta^2} & \frac{D}{\Delta^2} & & & & \\ \frac{D}{\Delta^2} & \beta - \frac{4D}{\Delta^2} & \frac{D}{\Delta^2} & & & \\ & \frac{D}{\Delta^2} & \ddots & & & \\ & & & \ddots & & \\ & & & & \beta - \frac{4D}{\Delta^2} & \frac{D}{\Delta^2} \\ & & & & \frac{D}{\Delta^2} & \beta - \frac{4D}{\Delta^2} \end{bmatrix} \in \mathbb{R}^{N\times N},$$

$$B = \begin{bmatrix} 0 & \frac{D}{\Delta^2} & & & & \\ \frac{D}{\Delta^2} & 0 & \frac{D}{\Delta^2} & & & \\ & \frac{D}{\Delta^2} & \ddots & & & \\ & & & \ddots & & \\ & & & & 0 & \frac{D}{\Delta^2} \\ & & & & \frac{D}{\Delta^2} & 0 \end{bmatrix} \in \mathbb{R}^{N\times N},$$

I_N is $N \times N$ identity matrix. The N^2 eigenvalues of C are of the form (see [30], Theorem 8.3.1) $\lambda_{k,l}(C) = \lambda_k(A) + \lambda_l(B)$, $k, l = \overline{1, N}$, where the eigenvalues of A

$$\lambda_k(A) = \beta - \frac{4D}{\Delta^2} - \frac{2D}{\Delta^2} \cos(\pi k/(N+1)), \quad k = \overline{1, N},$$

the eigenvalues of B

$$\lambda_l(B) = -\frac{2D}{\Delta^2} \cos(\pi l/(N+1)), \quad l = \overline{1, N}.$$

The comparison system $\frac{Z(t)}{dt} = CZ(t)$ tends asymptotically to zero if $\left|\lambda_{k,l}\right| < 1$. That is

$$\max_{k,l=\overline{1,N}} \left| \beta - \frac{4D}{\Delta^2} - \frac{2D}{\Delta^2} \left(\cos \frac{\pi k}{N+1} + \cos \frac{\pi l}{N+1} \right) \right| < 1.$$

2.4 *Numerical Simulation of Square* 4 × 4 *Pixels Array*

First of all, we calculate the basic reproductive numbers $\mathscr{R}_{0,i,j}$, $i, j = \overline{1, 4}$ due to (9) (See Table 1). We see that the conditions (11) hold. Thus, equilibrium without antibodies $\mathscr{E}^0_{i,j}$, $i, j = \overline{1, 4}$ is unstable and there exists endemic equilibrium $\mathscr{E}^{\star}_{i,j}$, $i, j = \overline{1, 4}$.

The numerical simulations were implemented at different values of $n \in (0, 1]$. Here we can see that when changing the value of τ we have changes in the qualitative behavior of pixels and the entire immunosensor. We considered the parameter value set given above and computed the long-time behavior of the system (3) for $\tau = 0.05, 0.22, 0.23, 0.2865$, and 0.28725. The phase diagrams of the antibody versus antigen populations for the pixel (1, 1) are shown in Table 2.

For example, at $\tau \in [0, 0.22]$, we can see trajectories corresponding to the stable node for all pixels.

For $\tau = 0.23$, the phase diagrams show that the solution is a limit cycle with two local extrema (one local maximum and one local minimum) per cycle. Then for $\tau = 0.2825$, the solution is a limit cycle with four local extrema per cycle, and,

Table 1 The values of $R_{0,i,j}$, $i, j = \overline{1, 4}$

$R^{\star}_{0,i,j}$	1	2	3	4
1	3.218727	3.425273	3.474323	3.224824
2	3.171270	3.235043	3.236289	3.126438
3	3.092287	3.107824	3.096617	3.040443
4	2.997269	3.020902	3.012915	2.971442

Table 2 The phase plane plots of the system (3) for antibody populations $F_{i,j}$ versus antigen populations $V_{i,j}, i, j = \overline{1, 4}$. Numerical simulation of the system (3) at $n = 0.9$, $\tau = 0.28725$. Here • indicates identical steady state, • indicates nonidentical steady state. Trajectories are constructed for $t \in [550, 800]$. The solution behavior looks chaotic

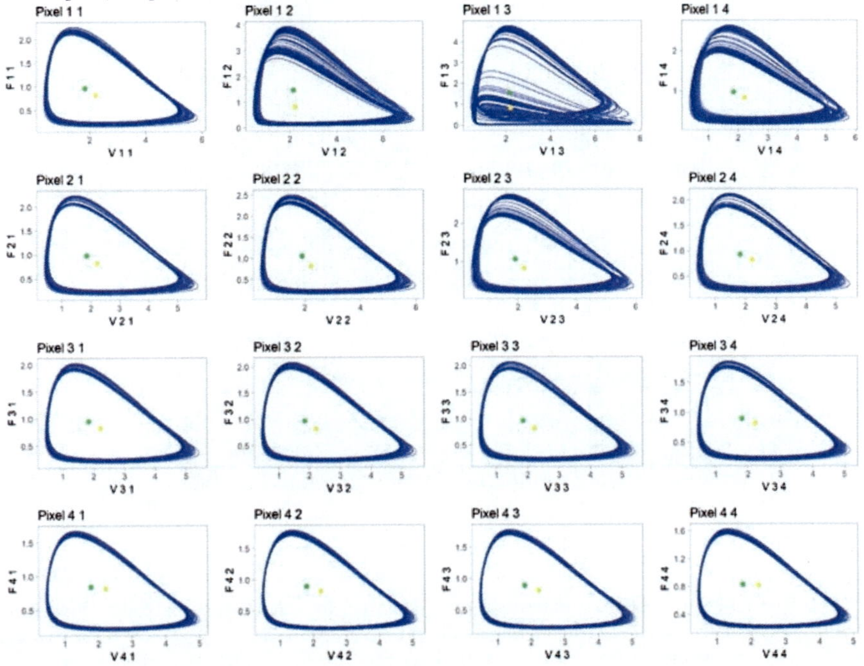

for $\tau = 0.2868, 0.2869, 0.28695$ the solutions are limit cycles with 8, 16, and 32 local extrema per cycle, respectively. Finally, for $\tau = 0.28725$, the behavior shown in Table 2 is obtained which looks like chaotic behavior. In this paper, we have regarded behavior as chaotic if no periodic behavior could be found in the long-time behavior of the solutions.

As a check that the solution is chaotic for $\tau = 0.28725$, we perturbed the initial conditions to test the sensitivity of the system. Figure 2 shows a comparison of the solutions for the antigen population $V_{1,3}$ with initial conditions $V_{1,3}(t) = 1$ and $V_{1,3}(t) = 1.001$, $t \in [-\tau, 0]$, and identical all the rest ones. Near the initial time, the two solutions appear to be the same, but as time increases, there is a marked difference between the solutions supporting the conclusion that the system behavior is chaotic at $\tau = 0.28725$.

We have also checked numerically that the solutions for the limit cycles are periodic and computed the periods for each of the local maxima and minima in the cycles. In the chaotic solution region, the numerical calculations (not shown in this paper) confirmed that no periodic behavior could be found.

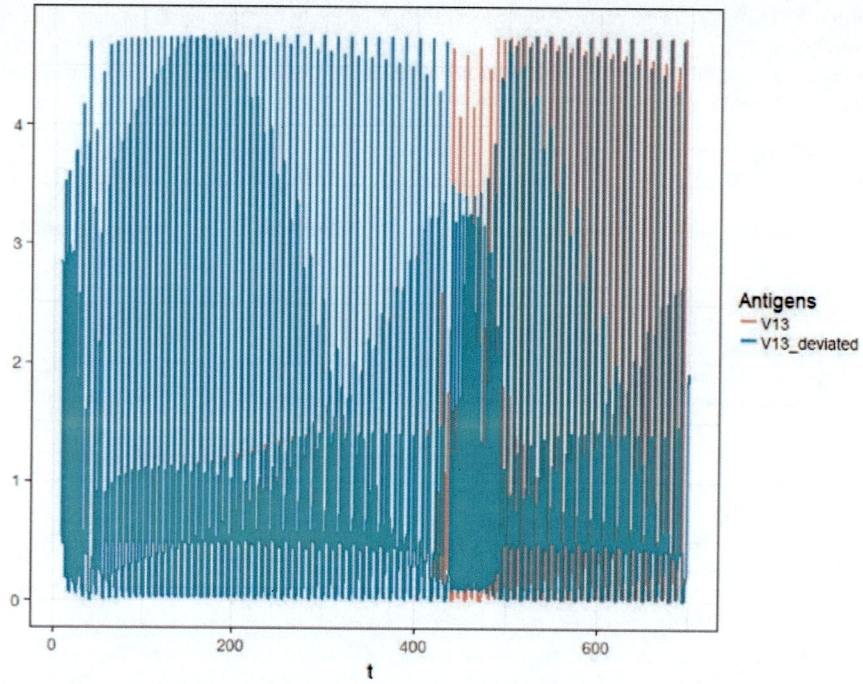

Fig. 2 The time series of the solutions to the system (3) for the antigen population $V_{1,3}$ from $t = 0$ to 700 with $\tau = 0.28725$ for initial conditions $V_{1,3}(t) = 1$ and $V_{1,3}(t) = 1.001$ (deviated), $t \in [-\tau, 0]$, and identical all the rest ones. At the beginning, the two solutions appear to be the same, but as time increases there is a marked difference between the solutions supporting the conclusion that the system behavior is chaotic

A bifurcation diagram showing the maximum and minimum points for the limit cycles for the antigen population $V_{1,3}$ as a function of time delay is given in Fig. 3. The Hopf bifurcation from the stable equilibrium point to a simple limit cycle and the sharp transitions at critical values of the time delay between limit cycles with increasing numbers of maximum and minimum points per cycle can be clearly seen.

3 Three-Dimensional Biopixels Array

When modeling three-dimensional pixels array, it is natural way to apply the model based on the hexagonal lattice. Such model may use the following assumption. Namely, antigens are assumed to diffuse from six neighboring pixels, $(i+1, j, k-1)$, $(i+1, j-1, k)$, $(i, j-1, k+1)$, $(i-1, j, k+1)$, $(i-1, j+1, k)$, $(i, j+1, k-1)$ (see Fig. 1), with diffusion rate $D\Delta^{-2}$, where $D > 0$ and $\Delta > 0$ is distance between pixels.

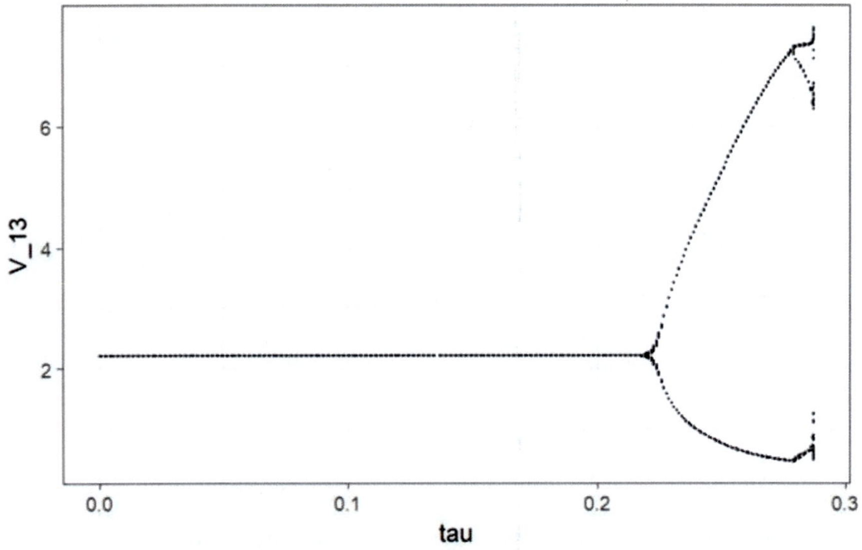

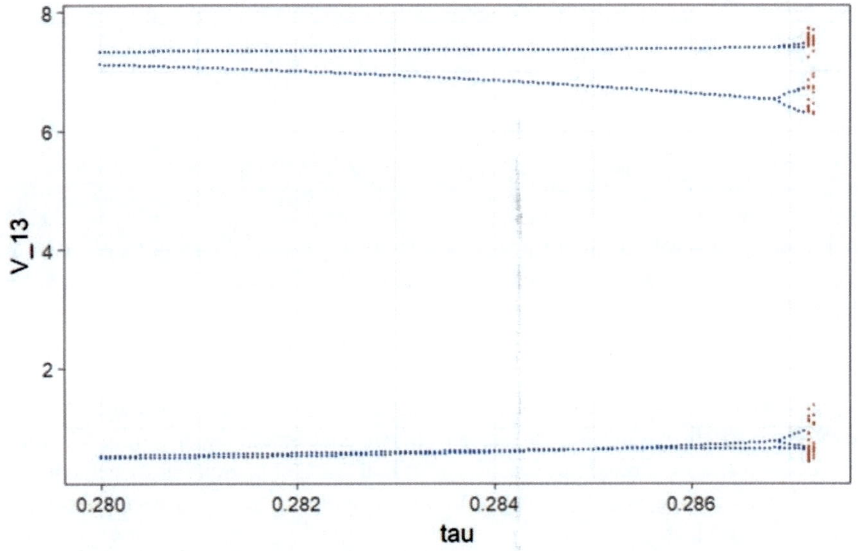

Fig. 3 A bifurcation diagram showing the "bifurcation path to chaos" as the time delay is increased. The points show the local extreme points per cycle for the $V_{1,3}$ population. Chaotic-type solutions occur at $\tau \approx 0.28725$ and are indicated in red in the figure with value 0 for the number of extreme points

Taking into account prerequisites mentioned above, we get a simplified antibody–antigen competition model with delay for a hexagonal array of biopixels, which uses Marchuk model of the immune response [20–23] and using spatial operator $\hat{S}$ which is constructed similarly to [24] (Supplementary information, p. 10)

$$\begin{aligned}\frac{dV_{i,j,k}(t)}{dt} &= (\beta - \gamma F_{i,j,k}(t-\tau) - \delta_v V_{i,j,k}(t-\tau))V_{i,j,k}(t) + \hat{S}\{V_{i,j,k}\},\\ \frac{dF_{i,j,k}(t)}{dt} &= \left(-\mu_f + \eta\gamma V_{i,j,k}(t-\tau) - \delta_f F_{i,j,k}(t)\right)F_{i,j,k}(t)\end{aligned} \tag{18}$$

with given initial functions

$$\begin{aligned}&V_{i,j,k}(t) = V^0_{i,j,k}(t) \geq 0, \quad F_{i,j,k}(t) = F^0_{i,j,k}(t) \geq 0, \quad t \in [-\tau, 0),\\ &V_{i,j,k}(0),\ F_{i,j,k}(0) > 0.\end{aligned} \tag{19}$$

We use the following spatial operator of discrete diffusion for a hexagonal array of pixels[4]

$$\begin{aligned}\hat{S}\{V_{i,j,k}\} = D\Delta^{-2}\Big[&V_{i+1,j,k-1} + V_{i+1,j-1,k} + V_{i,j-1,k+1} + V_{i-1,j,k+1} + V_{i-1,j+1,k}\\ &+ V_{i,j+1,k-1} - 6nV_{i,j,k}\Big]\\ &i,j,k \in \overline{-N+1, N-1}, \quad i+j+k=0.\end{aligned} \tag{20}$$

Each pixel is affected by the antigens flowing out six neighboring pixels, two in each of three directions of the hexagonal array. The adjoint pixels are separated by the distance Δ (Fig. 4).

Boundary conditions $V_{i,j,k} = 0$ for the edges of the hexagonal array, i.e., if $i \vee j \vee k \in \{-N-1, N+1\}$, are used.

We can present analytical results with respect to the model (18) in the form of restrictions for the parameters, enabling us persistence and global asymptotic stability. Moreover, we executed numerical research of the system qualitative behavior in dependence of changes of the time of immune response τ (delay of time), diffusion rate $D\Delta^{-2}$ and factor n.

3.1 Persistence and Extinction of Solutions

Concerning persistence, for the hexagonal lattice the similar result can be obtained as for square one (Theorem 1), just adding the third index.

[4] Without loss of generality we consider spatial operator for internal pixels only.

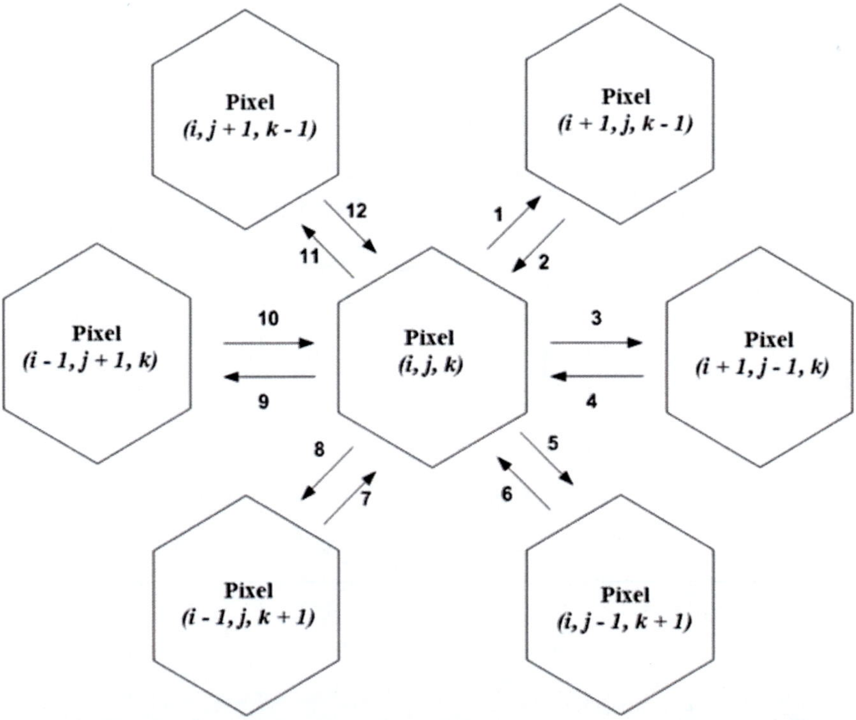

Fig. 4 Diffusion of antigens for the hexagonal lattice model. Antigens from six neighboring pixels interact, $n > 0$ is the constant of disbalance. Here '1', '3', '5', '8', '9', '11' have to be replaced with $D\Delta^{-2}V_{i,j,k}(t)$, '2' with $D\Delta^{-2}V_{i+1,j,k-1}(t)$, '4' with $D\Delta^{-2}V_{i+1,j-1,k}(t)$, '6' with $D\Delta^{-2}V_{i,j-1,k+1}(t)$, '7' with $D\Delta^{-2}V_{i-1,j,k+1}(t)$, '10' with $D\Delta^{-2}V_{i-1,j+1,k}(t)$, '12' with $D\Delta^{-2}V_{i,j+1,k-1}(t)$

Unfortunately, we didn't manage to present such a clear condition of extinction as in Theorem 2. We can check it only numerically in an experimental way.

3.2 Numerical Study

For numerical simulation, we consider model (18) of hexagonal pixels array at $N = 4$, $\beta = 2\ \text{min}^{-1}$, $\gamma = 2\frac{mL}{min\cdot\mu g}$, $\mu_f = 1\ \text{min}^{-1}$, $\eta = 0.8/\gamma$, $\delta_v = 0.5\frac{mL}{min\cdot\mu g}$, $\delta_f = 0.5\frac{mL}{min\cdot\mu g}$, $D = 0.2\frac{nm^2}{min}$, $\Delta = 0.3nm$. Numerical modeling was implemented at different values of $n \in (0, 1]$. For this purpose, we used RStudio environment.

Using local bifurcation plot, dynamics of the system (18) was analyzed for different values of $n \in (0, 1]$. We have concluded that oscillatory and then chaotic behavior starts for smaller values of τ at smaller values of n. Further, increasing the values of n, we can observe asymptotically stable steady solutions for a wider range of τ.

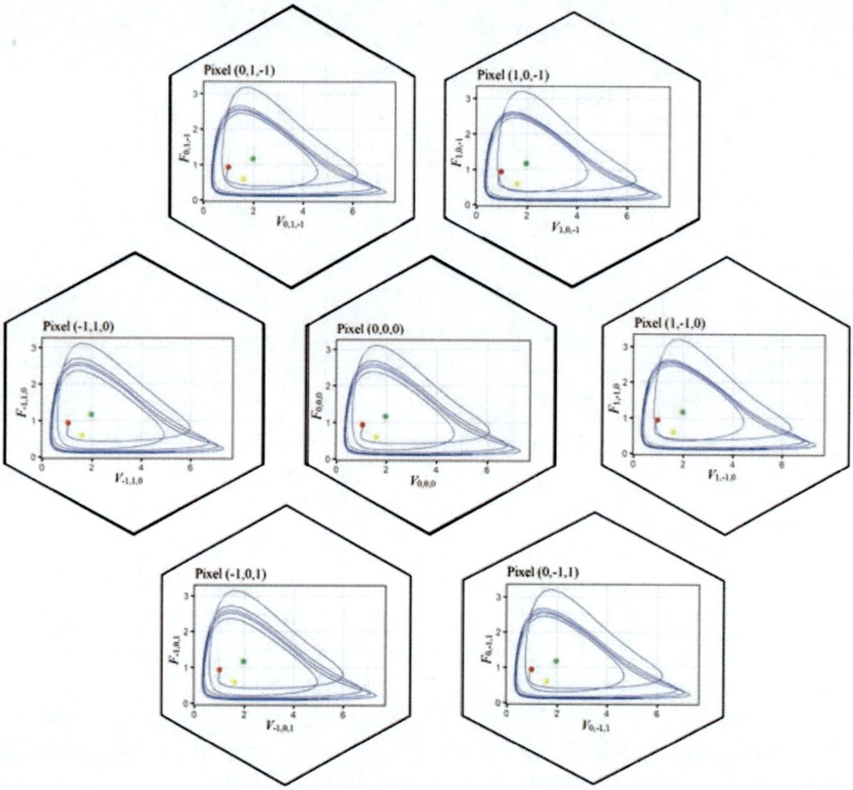

Fig. 5 Phase plots of the system (18) at $\tau = 0.287$. Here • indicates initial state, • indicates pixel-independent endemic state, • indicates pixel-dependent endemic state. The solution tends to a stable limit cycle with six local extrema per cycle

Numerical integration of the system has shown the influence of time delay τ. Namely, as it is agreed with the analytical results, we observe the stable focuses at pixel-dependent endemic states for small delays $\tau \in [0, 0.18)$. At $\tau \approx 0.18$ min the stable focus is transformed into a stable limit cycle of tiny radius, which corresponds to Hopf bifurcation. A deeper study of this phenomenon requires obtaining the condition of the appearance of the pair of purely imaginary roots of the characteristic quasipolynomial of the linearized system. The limit cycles of ellipsoidal form are observed till $\tau \approx 0.285$ min. Pay attention that when increasing τ, near $\tau = 0.285$, we get period doubling (see Fig. 5).[5]

The qualitative behavior of immunosensor model can be analyzed with help of hexagonal tiling plots also. For this purpose, we can use both plots for antigens (Fig. 6), antibodies (Fig. 7), and probabilities of binding antigens by antibodies (Fig. 8).

[5] It can be approximately seen from local bifurcation plot also.

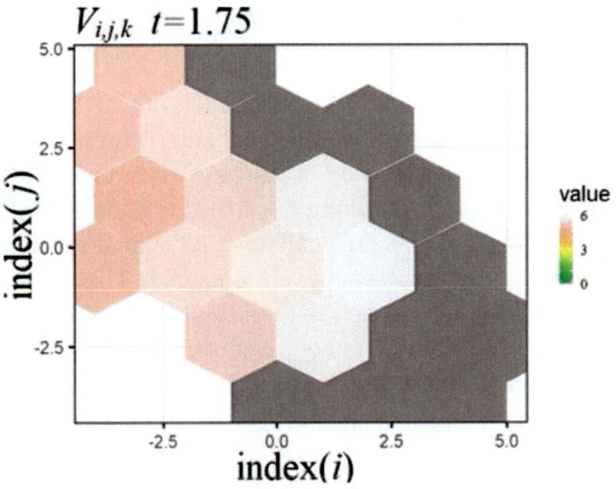

Fig. 6 Example of hexagonal tiling plot for V

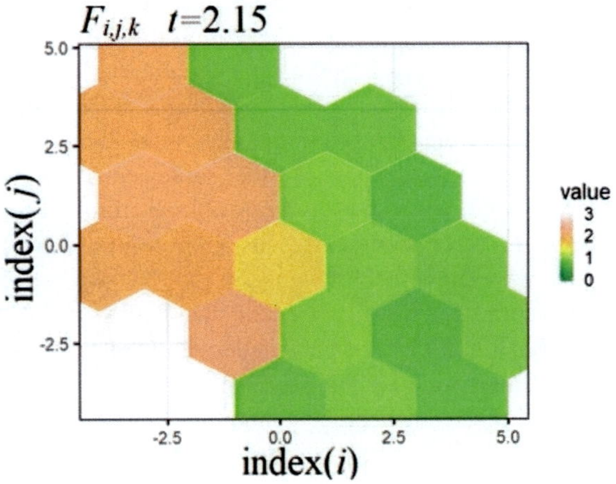

Fig. 7 Example of hexagonal tiling plot for F

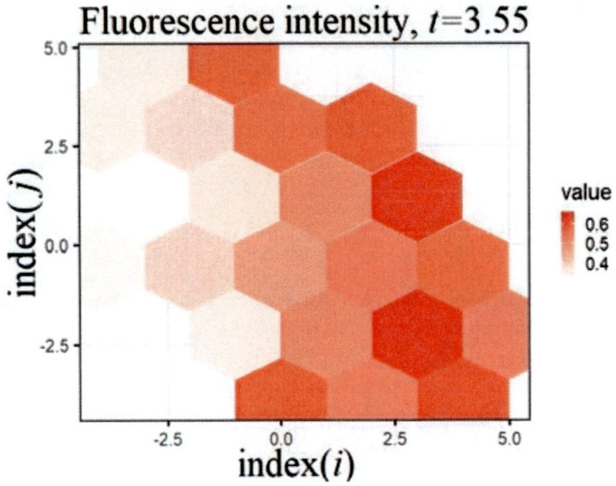

Fig. 8 Example of hexagonal tiling plot for probabilities of binding antigens by antibodies, i.e., $V \times F$. In case of optical immunosensor, it is fluorescence intensity

4 Conclusions

In this work, reaction-diffusion models of two- and three-dimensional immunopixels array were considered. Mathematically, it is described by the system of lattice delayed differential equations on rectangular or hexagonal grids. The systems include the spatial operator describing the diffusion of antigens between pixels.

The main results are dealing with the qualitative investigation of the model. The conditions of persistence were obtained. Also, we have managed to get the result dealing with the extinction of the solutions. Namely, it can be seen that the amount of pixels determines their non-vanishing. In a two-dimensional case, this dependence can be presented in a clear form.

The conditions of local or global asymptotic stability can be obtained using the construction of the Lyapunov functional. Because of the cumbersome evidence, we didn't include it here. They result in inequality including the system parameters and delay. So, estimation of the delay enabling us local or global asymptotic stability can be obtained.

Numerical analysis of the model qualitative behavior is performed with the help of the bifurcation diagram, phase trajectories, and rectangular or hexagonal tile portraits. It has shown the changes in qualitative behavior with respect to the growth of time delay. Namely, starting from the stable focus at small delay values, then through Hopf bifurcation to limit cycles, and finally through period doublings to deterministic chaos. It is agreed with the results on space-time chaos for reaction-diffusion systems, which were previously obtained in [1–3].

As compared with the rectangular lattice model, for the hexagonal model, we observe Hopf bifurcation at smaller values of τ. That is hexagonal lattice accelerates changes in qualitative behavior.

Note, that model can be applied for an arbitrary amount of pixels determined by natural $N \geq 1$. However, it can be numerically seen that qualitative behavior of the entire immunosensor is determined by 5 or 7 pixels array for square and hexagonal lattices, respectively.

Acknowledgements We are very thankful to the reviewers for valuable remarks and comments which allowed us to improve the work.

References

1. Rössler, O.E.: Chemical turbulence: Chaos in a simple reaction-diffusion system. Zeitschrift für Naturforschung A **31**(10) (1976). https://doi.org/10.1515/zna-1976-1006
2. Hildebrand, M., Bar, M., Eiswirth, M.: Statistics of topological defects and spatiotemporal chaos in a reaction-diffusion system. Phys. Rev. Lett. **75**(8), 1503–1506 (1995). https://doi.org/10.1103/physrevlett.75.1503
3. Zaitseva, M.F., Magnitskii, N.A.: Space-time chaos in a system of reaction-diffusion equations. Differ. Equ. **53**(11), 1519–1523 (2017)

4. Cahn, J.W., Chow, S., Van Vleck, E.S.: Spatially discrete nonlinear diffusion equations. Rocky Mount. J. Math., to appear (1995)
5. Chow, S.-N., Mallet-Paret, J., Van Vleck, E.S.: Dynamics of lattice differential equations. Int. J. Bifurc. Chaos **6**(09), 1605–1621 (1996)
6. Pan, S.: Propagation of delayed lattice differential equations without local quasimonotonicity (2014). arXiv:1405.1126
7. Huang, J., Lu, G., Zou, X.: Existence of traveling wave fronts of delayed lattice differential equations. J. Math. Anal. Appl. **298**(2), 538–558 (2004)
8. Niu, H.: Spreading speeds in a lattice differential equation with distributed delay. Turkish J. Math. **39**(2), 235–250 (2015)
9. Hoffman, A., Hupkes, H., Van Vleck, E.: Entire Solutions for Bistable Lattice Differential Equations with Obstacles. American Mathematical Society, Providence (2017)
10. Wu, F.: Asymptotic speed of spreading in a delay lattice differential equation without quasi-monotonicity. Electr. J. Differ. Equ. **2014**(213), 1–10 (2014)
11. Zhang, G.-B.: Global stability of traveling wave fronts for non-local delayed lattice differential equations. Nonlinear Anal.: Real World Appl. **13**(4), 1790–1801 (2012)
12. Luczak, E., Rosenfeld, A.: Distance on a hexagonal grid. IEEE Trans. Comput. **25**(5), 532–533 (1976). https://doi.org/10.1109/TC.1976.1674642
13. Hexagonal coordinate systems. https://homepages.inf.ed.ac.uk/rbf/CVonline/LOCAL_COPIES/AV0405/MARTIN/Hex.pdf. Accessed 12 May 2019
14. Middleton, L., Sivaswamy, J.: Edge detection in a hexagonal-image processing framework. Image Vis. Comput. **19**(14), 1071–1081 (2001)
15. Fayas, A., Nisar, H., Sultan, A.: Study on hexagonal grid in image processing. In: The 4th International Conference on Digital Image Processing, pp. 7–8 (2012)
16. Cruz, H.J., Rosa, C.C., Oliva, A.G.: Immunosensors for diagnostic applications. Parasitol. Res. **88**, S4–S7 (2002)
17. Paek, S.-H., Schramm, W.: Modeling of immunosensors under nonequilibrium conditions: I. Mathematic modeling of performance characteristics. Anal. Biochem. **196**(2), 319–325 (1991)
18. Bloomfield, V., Prager, S.: Diffusion-controlled reactions on spherical surfaces. application to bacteriophage tail fiber attachment. Biophys. J. **27**(3), 447–453 (1979)
19. Berg, O.: Orientation constraints in diffusion-limited macromolecular association. the role of surface diffusion as a rate-enhancing mechanism. Biophys. J. **47**(1), 1–14 (1985)
20. Marchuk, G., Petrov, R., Romanyukha, A., Bocharov, G.: Mathematical model of antiviral immune response. i. Data analysis, generalized picture construction and parameters evaluation for hepatitis b. J. Theor. Biol. **151**(1), 1–40 (1991), cited By 38. https://doi.org/10.1016/S0022-5193(05)80142-0. https://www.scopus.com/inward/record.uri?eid=2-s2.0-0025819779&doi=10.1016
21. Fory's, U.: Marchuk's model of immune system dynamics with application to tumour growth. J. Theor. Med. **4**(1), 85–93 (2002). https://doi.org/10.1080/10273660290052151. http://www.tandfonline.com/doi/pdf/10.1080/10273660290052151. http://www.tandfonline.com/doi/abs/10.1080/10273660290052151
22. Nakonechny, A., Marzeniuk, V.: Uncertainties in medical processes control. Lecture Notes in Economics and Mathematical Systems, vol. 581, pp. 185–192 (2006), cited By 2. https://doi.org/10.1007/3-540-35262-7_11. https://www.scopus.com/inward/record.uri?eid=2-s2.0-53749093113&doi=10.1007
23. Marzeniuk, V.: Taking into account delay in the problem of immune protection of organism. Nonlinear Anal.: Real World Appl. **2**(4), 483–496 (2001), cited By 2. https://doi.org/10.1016/S1468-1218(01)00005-0. https://www.scopus.com/inward/record.uri?eid=2-s2.0-0041331752&doi=10.1016
24. Prindle, A., Samayoa, P., Razinkov, I., Danino, T., Tsimring, L.S., Hasty, J.: A sensing array of radically coupled genetic 'biopixels'. Nature **481**(7379), 39–44 (2011). https://doi.org/10.1038/nature10722
25. Jones, J.H.: Notes on R0. Department of Anthropological Sciences, California (2007)

26. Yang, J., Wang, X., Zhang, F.: A differential equation model of hiv infection of cd t-cells with delay. Discrete Dynamics in Nature and Society, vol. 2008 (2008)
27. Kuang, Y.: Delay Differential Equations with Applications in Population Dynamics. Academic, New York (1993)
28. He, X.-z.: Stability and delays in a predator-prey system. J. Math. Anal. Appl. **198**(2), 355–370 (1996). https://doi.org/10.1006/jmaa.1996.0087
29. Wendi, W., Zhien, M.: Harmless delays for uniform persistence. J. Math. Anal. Appl. **158**(1), 256fffdfffdfffd268 (1991). https://doi.org/10.1016/0022-247x(91)90281-4
30. Lancaster, P., Tismenetsky, M.: The Theory of Matrices: With Applications. Elsevier, Amsterdam (1985)

Oscillation Criteria for Higher Order Linear Differential Equations

Roman Koplatadze

Abstract Oscillation criteria generalizing a series of earlier results are established for nth order linear differential equations.

Keywords Oscillation · Property A · Property B

1 Introduction

Consider the differential equation

$$u^{(n)} + p(t)\, u = 0, \tag{1.1}$$

where $p \in L_{\mathrm{loc}}(\mathbb{R}_+; \mathbb{R})$, $n \geq 2$.

For Eq. (1.1), A. Kneser posed the problem: What conditions must be satisfied for the function p, so that Eq. (1.1) has similar solutions such equations

$$u^{(n)} + u = 0, \quad \text{or} \quad u^{(n)} - u = 0.$$

Definition 1.1 We say that Eq. (1.1) has Property **A** if any of its non-trivial solutions is oscillatory, when n is even and either is oscillatory or satisfies

$$\left|u^{(i)}(t)\right| \downarrow 0 \quad \text{as} \quad t \uparrow +\infty \quad (k = 0, \dots, n-1), \tag{1.2}$$

when n is odd.

Definition 1.2 We say that Eq. (1.1) has Property **B** if any of its non-trivial solutions either is oscillatory or satisfies either (1.2) or

R. Koplatadze (✉)
Department of Mathematics and I.Vekua Institute of Applied Mathematics, I. Javakhishvili Tbilisi State University, 2 University Str., 0186 Tbilisi, Georgia
e-mail: roman.koplatadze@tsu.ge

A. Domoshnitsky et al. (eds.), *Functional Differential Equations and Applications*, Springer Proceedings in Mathematics & Statistics 379,
https://doi.org/10.1007/978-981-16-6297-3_15

$$\left|u^{(t)}\right| \uparrow +\infty \quad \text{as} \quad t \uparrow +\infty \quad (i = 0, \dots, n-1), \tag{1.3}$$

when n is even, and either is oscillatory or satisfies (1.3) when n is odd.

There is a lot of work dedicated to this problem. We will give some essential results on this problem and some new results about it.

2 Results of First Type

Theorem 2.1 (Kneser [1]) *If satisfying the condition*

$$\liminf_{t \to +\infty} t^{n/2} p(t) > 0,$$

then Eq. (1.1) has Property **A**.

This theorem was essentially generalized by Kondrat'ev.

His method was based on a comparison theorem which enables one to obtain optimal results for establishing oscillatory properties as solutions of Eq. (1.1).

Theorem 2.2 (Kondrat'ev [2]) *Let* $p(t) \geq 0$ $\big(p(t) \leq 0\big)$ *for* $t \in \mathbb{R}_+$ *and*

$$\liminf_{t \to +\infty} t^n \left|p(t)\right| > M_n^* \ \left(M_{*n}\right). \tag{2.1}$$

Then Eq. (1.1) has Property **A** *(Property* **B**), *where*

$$\begin{aligned} M_n^* &= \max\big\{-\lambda(\lambda-1)\cdots(\lambda-n+1) : \lambda \in [0, n-1]\big\}, \\ M_{*n} &= \max\big\{\lambda(\lambda-1)\cdots(\lambda-n+1) : \lambda \in [0, n-1]\big\}. \end{aligned} \tag{2.2}$$

Later, Chanturia proved integral comparison theorems which are integral generalized of the above-mentioned comparison theorems. Using these theorems, he succeeded in improving condition (2.1).

Theorem 2.3 (Chanturia [3]) *Let* $p(t) \geq 0$ $(p(t) \leq 0)$ *and the inequality*

$$\liminf_{t \to +\infty} t \int_t^{+\infty} s^{n-2} \left|p(s)\right| ds > M_n^* \ \left(M_{*n}\right)$$

be fulfilled. Then Eq. (1.1) has Property **A** (**B**), *where* M_n^* $\left(M_{*n}\right)$ *is given by (2.2).*

This results for an almost linear differential equation

$$u^{(n)}(t) + \sum_{i=1}^{m} p_i(t) \left|u(\sigma_i(t))\right|^{\mu_i(t)} \operatorname{sign} u(\sigma_i(t)) = 0, \tag{2.3}$$

where

$$\lim_{t\to+\infty}\sigma_i(t)=+\infty,\quad \lim_{t\to+\infty}\mu_i(t)=1,\quad p_i\in L_{\mathrm{loc}}(\mathbb{R}_+;\mathbb{R})$$

was generalized by Koplatadze.

For Eq. (2.3), give only a simple example:

$$u^{(n)}(t)+p(t)\,\big|u(t)\big|^{1+\frac{d}{\ln t}}\operatorname{sign}u(t)=0, \tag{2.4}$$

where $p\in L_{\mathrm{loc}}(\mathbb{R}_+;\mathbb{R})$, $d\in\mathbb{R}$.

Theorem 2.4 (Koplatadze [4, 5]) *Let* $p(t)\ge 0$ $\big(p(t)\le 0\big)$ *and*

$$\liminf_{t\to+\infty} t\int_t^{+\infty}s^{n-2}\big|p(s)\big|ds>M_n^*(d)\ \ \big(M_{*n}(d)\big). \tag{2.5}$$

Then Eq. (2.4) has Property **A** (**B**), *where*

$$\begin{aligned}&M_n^*(d)=\max\big\{-\lambda(\lambda-1)\cdots(\lambda-n+1)e^{-\lambda d}:\lambda\in[0,n-1]\big\},\\&\Big(M_{*n}(d)=\max\big\{\lambda(\lambda-1)\cdots(\lambda-n+1)e^{-\lambda d}:\lambda\in[0,n-1]\big\}\Big).\end{aligned} \tag{2.6}$$

Theorem 2.5 (Koplatadze [6]) *Let* $p(t)\ge 0$ $\big(p(t)\le 0\big)$ *and*

$$\liminf_{t\to+\infty}\frac{1}{t}\int_0^t s^n\big|p(s)\big|ds>M_n^*(d)\ \ \big(M_{*n}(d)\big). \tag{2.7}$$

Then Eq. (2.4) has Property **A** (**B**), *where* $M_n^*(d)$ $\big(M_{*n}(d)\big)$ *are given by (2.6).*

Remark 2.1 For any $d\in\mathbb{R}$, inequalities (2.5) and (2.7) cannot replace nonstrict ones. For $d=0$, Theorem 2.5 gives a theorem of Chanturia, which is an integral generalization of Kondrat'ev's theorem.

3 Results of Second Type

Theorem 3.1 (Kiguradze [7]) *Let* $p(t)\ge 0$ $(p(t)\le 0)$ *and there exists non-decreasing function* $\omega:\mathbb{R}_+\to(0,+\infty)$ *such that*

$$\int_1^{+\infty}\frac{dt}{t\,\omega(t)}<+\infty,\quad \int_0^{+\infty}\frac{t^{n-1}}{\omega(t)}\big|p(t)\big|dt=+\infty.$$

Then Eq. (1.1) has Property **A** (**B**).

Theorem 3.2 (Chanturia [8]) *Let $p(t) \geq 0$ and one of two conditions*

$$\limsup_{t\to+\infty} \frac{1}{\ln t}\int_1^t s^{n-1}p(s)ds > (n-1)!$$

or

$$\limsup_{t\to+\infty} t\int_t^{+\infty} s^{n-2}p(s)ds > (n-1)!$$

holds. Then Eq. (1.1) has Property **A**.

Theorem 3.3 (Chanturia [8]) *Let $p(t) \leq 0$ and for odd n, one of two conditions*

$$\limsup_{t\to+\infty} \frac{1}{\ln t}\int_0^t s^{n-1}\big|p(s)\big|ds > (n-1)!$$

or

$$\limsup_{t\to+\infty} t\int_t^{+\infty} s^{n-2}\big|p(s)\big|ds > (n-1)!$$

be fulfilled and for even n one of two conditions

$$\limsup_{t\to+\infty} \frac{1}{\ln t}\int_0^t s^{n-1}\big|p(s)\big|ds > 2(n-2)!$$

or

$$\limsup_{t\to+\infty} t\int_t^{+\infty} s^{n-2}\big|p(s)\big|ds > 2(n-2)!$$

be fulfilled. Then Eq. (1.1) has Property **B**.

Remark 3.1 It is interesting that if the conditions of the Kiguradze theorem are satisfied, then

$$\limsup_{t\to+\infty} t\int_t^{+\infty} s^{n-2}\big|p(s)\big|ds = +\infty.$$

That is, conditions of Chanturia theorems are fulfilled.

4 New Results

Theorem 4.1 *Let $p(t) \geq 0$ and*

$$\limsup_{t\to+\infty} \frac{1}{\ln t}\int_0^t s^{n-1}p(s)ds > \frac{(n-1)!}{4}.$$

Then Eq. (1.1) *has Property* **A**.

Theorem 4.2 *Let $p(t) \le 0$ and for odd n*

$$\limsup_{t\to+\infty} \frac{1}{\ln t}\int_0^t s^{n-1}\big|p(s)\big|ds > \frac{(n-1)!}{4}$$

and for even n

$$\limsup_{t\to+\infty} \frac{1}{\ln t}\int_0^t s^{n-1}\big|p(s)\big|ds > \frac{2\sqrt{3}}{9}\,(n-2)!$$

then Eq. (1.1) *has Property* **B**.

Denote

$$\rho(t,\lambda) = t^{-\lambda}\int_1^t\int_s^{+\infty} \xi^{n-2+\lambda}\big|p(\xi)\big|d\xi\,ds$$

and

$$\rho^*(\lambda) = \limsup_{t\to+\infty} \rho(t,\lambda).$$

Theorem 4.3 *Let $p(t) \ge 0$ for $t \in \mathbb{R}_+$ and*

$$\lim_{\lambda\to 1-} (1-\lambda)\,\rho^*(\lambda) > \frac{(n-1)!}{4}. \tag{4.1}$$

Then Eq. (1.1) *has Property* **A**.

Theorem 4.4 *Let $p(t) \le 0$, for odd n* (4.1) *be fulfilled and for even n*

$$\lim_{t\to 1-} (1-\lambda)\,\rho^*(\lambda) > \frac{2\sqrt{3}}{9}(n-2)!$$

Then Eq. (1.1) *has Property* **B**.

Remark 4.1 Here, we assume that

$$\int_0^{+\infty} s^{n-2+\lambda}\big|p(s)\big|ds < +\infty \quad \text{for} \quad \lambda \in [0,1).$$

Otherwise Eq. (1.1) has Property **A** (**B**).

Remark 4.2 These results are also valid for the equation

$$u^{(n)}(t) + \sum_{i=1}^{m} p_i(t)\,u\big(\tau_i(t)\big) = 0,$$

where $p_i \in L_{\text{loc}}(\mathbb{R}_+;\mathbb{R})$, $\tau_i(t) \le t$ for $t \in \mathbb{R}_+$, $\lim\limits_{t\to+\infty} \tau_i(t) = +\infty$ $(i = 1,\dots,m)$.

References

1. Kneser, A.: Untersuchugen uber die reelen nullstellen der integral linearer differentailgleichunder. Math. Ann. **42**, 409–435 (1893)
2. Kondratev, V.A.: Oscillatory properties of solutions of the equation $y^{(n)} + p(x)y = 0$. (Russian) Trudy Moskov. Mat. Obshch. **10**, 419–436 (1961)
3. Chanturia, T.A.: Integral criteria of oscillation of solutions of higher order linear differential equations I, II. (Russian) Differential'nye Uravneniya **16**(8), 470–482 (1980); **16**(4), 635–644 (1980)
4. Koplatadze, R.: Quasi-linear functional differential equations with Property A. J. Math. Anal. Appl. **330**(1), 483–510 (2007)
5. Koplatadze, R., Litsyn, E.: Oscillation criteria for higher order "almost linear" functional differential equations. Funct. Differ. Equ. **16**(3), 387–434 (2009)
6. Koplatadze, R.: Almost linear functional differential equations with properties A and B. Trans. A. Razmadze Math. Inst. **170**(2), 215–242 (2016)
7. Kiguradze, I.: On the oscillation of solutions of some ordinary differential equations. (Russian) Dokl. Akad. Nauk SSSR **144**(1), 33–36 (1962)
8. Chanturia, T.A.: On the oscillation of solutions of linear differential equations of higher order. (Russian) Report of Enlanged Session of the Seminar of I. Vekua Inst. Appl. Math. **16**, 3–72 (1982)

On Necessary Conditions of Optimality to the Extremum Problem for Parabolic Equations

Irina Astashova, Alexey Filinovskiy, and Dmitriy Lashin

Abstract We consider a control problem associated with a mathematical model of temperature control in industrial greenhouses. It is based on a one-dimensional non-self-adjoint parabolic equation with variable coefficients. Defining the optimal control as the minimizer quadratic functional with point observation, we prove the existence of the minimizer and obtain necessary conditions.

Keywords Parabolic equation · Extremum problem · Quadratic functional · Minimizer · Necessary condition

Mathematics Subject Classification (2010). 35Q93 · 49J20 · 49K20

1 Introduction

The mixed boundary value problem for the parabolic equation with the convection term is considered:

The reported study of I. Astashova and A. Filinovskiy was partially supported by RSF, project 20-11-20272.

I. Astashova (✉)
Lomonosov Moscow State University, Plekhanov Russian University of Economics, Moscow, Russia
e-mail: ast.diffiety@gmail.com

A. Filinovskiy
Bauman Moscow State Technical University, Lomonosov Moscow State University, Moscow, Russia
e-mail: flnv@yandex.ru

D. Lashin
SPC FITO, Moscow, Russia

A. Domoshnitsky et al. (eds.), *Functional Differential Equations and Applications*, Springer Proceedings in Mathematics & Statistics 379,
https://doi.org/10.1007/978-981-16-6297-3_16

$$u_t = (a(x,t)u_x)_x + b(x,t)u_x, \quad (x,t) \in Q_T = (0,1) \times (0,T), \quad T > 0, \tag{1.1}$$
$$u(0,t) = \varphi(t), \quad u_x(1,t) = \psi(t), \quad t > 0, \tag{1.2}$$
$$u(x,0) = \xi(x), \quad 0 < x < 1. \tag{1.3}$$

The functions a and b are smooth functions in $\overline{Q}_T$ satisfying inequalities $0 < a_0 \leq a(x,t) \leq a_1 < \infty$, $| b(x,t) | \leq b_1 < \infty$ for some a_0, a_1, b_1, and $\varphi, \psi \in W_2^1(0,T)$, $\xi \in L_2(0,1)$.

We considered this problem (1.1)–(1.3) for the first time in connection with the temperature control problem in an industrial hothouse with flow heating and ceiling ventilation (see [3–5])). We suppose that the functions ξ and ψ are fixed, and the function φ is a control function to be found. Our task is to find a control function $\varphi = \varphi_0$ making the temperature $u(x,t)$ at some fixed point $x = c \in (0,1)$ maximally close to a given one, $z(t)$, during the whole time interval $(0,T)$. We estimate the control quality by the quadratic cost functional. It is so-called point-wise observation (see [17]):

$$J[z,\varphi] := \int_0^T (u_\varphi(c,t) - z(t))^2 dt. \tag{1.4}$$

The function u_φ is a solution (as $x = c$) to problem (1.1)–(1.3) with a control function φ.

Authors' previous results are obtained in [3–8].

Now, we consider a more general problem (equation with variable coefficient $a = a(x,t)$, convective term $b(x,t)u_x$), and prove new results on necessary conditions to minimizer. We prove these results by methods of qualitative theory of differential equations and, in particular, by some methods described in [1, 2].

Various extremum problems for partial differential equations with integral functionals were studied earlier in [9, 11, 12, 17]. In particular, the problem of minimization of a functional with final observation and the problem of the optimal time control were studied in [9–12, 24]. A review of results can be found in [3, 10, 13, 20, 24]. A more detailed review of known results is given at the end of this paper.

2 Basic Notations, Definitions and Results

Some of the following notations were introduced in [7, 8].

Definition 2.1 By $V_2^{1,0}(Q_T)$, we denote the Banach space of all functions $u \in W_2^{1,0}(Q_T)$ with the finite norm

$$\|u\|_{V_2^{1,0}(Q_T)} := \sup_{0 \leq t \leq T} \|u(x,t)\|_{L_2(0,1)} + \|u_x\|_{L_2(Q_T)} \tag{2.1}$$

such that $t \mapsto u(\cdot,t)$ is a continuous mapping $[0,T] \to L_2(0,1)$.

Properties of $V_2^{1,0}(Q_T)$ can be found in [15, 16], where this space is introduced.

Definition 2.2 By $\widetilde{W}_2^1(Q_T)$, we denote the space of all functions $\eta \in W_2^1(Q_T)$ such that $\eta(x, T) = 0$ for all $x \in (0, 1)$ and $\eta(0, t) = 0$ for all $t \in (0, T)$.

Definition 2.3 We say that a function $u \in V_2^{1,0}(Q_T)$ is a weak solution to problem (1.1)–(1.3) if it satisfies the boundary condition $u(0, t) = \varphi(t)$ and the integral identity

$$\int_{Q_T} (a(x,t)u_x\eta_x - b(x,t)u_x\eta - u\eta_t)\,dx\,dt = \int_0^1 \xi(x)\eta(x,0)\,dx + \int_0^T a(1,t)\psi(t)\,\eta(1,t)\,dt \tag{2.2}$$

for any function $\eta \in \widetilde{W}_2^1(Q_T)$.

Theorem 2.4 ([5]) *Problem (1.1)–(1.3) has a unique weak solution. This solution satisfies the estimate*

$$\|u\|_{V_2^{1,0}(Q_T)} \le C_1\big(\|\varphi\|_{W_2^1(0,T)} + \|\psi\|_{W_2^1(0,T)} + \|\xi\|_{L_2(0,1)}\big), \tag{2.3}$$

where the constant C_1 is independent of φ, ψ, and ξ.

Let us further denote by u_φ the solution of problem (1.1)–(1.3) with $\varphi, \psi \in W_2^1(0, T)$, $\xi \in L_2(0, 1)$.

Suppose $z \in L_2(0, T)$. Let $\Phi \subset W_2^1(0, T)$ be a bounded closed convex set of control functions. For some $c \in (0, 1)$ consider the functional $J[z, \varphi]$ defined by (1.4) and put

$$m[z, \Phi] := \inf_{\varphi\in\Phi} J[z, \varphi]. \tag{2.4}$$

Definition 2.5 We say that minimization problem (1.1)–(1.3), (2.4), $z \in L_2(0, T)$, $\varphi \in \Phi$, has a solution if there exists a function $\varphi_0 \in \Phi$ such that

$$m[z, \Phi] = J[z, \varphi_0]. \tag{2.5}$$

The function $\varphi_0 \in \Phi$ we call a minimizer for the problem (1.1)–(1.3), (2.4).

For a necessary condition of optimality, we also consider the adjoint to (1.1)–(1.3), (2.4) mixed problem for the inverse parabolic equation

$$p_t + (a(x,t)p_x)_x - (b(x,t)p)_x = \delta(x-c)\otimes\big(u_\varphi(c,t) - z(t)\big), \tag{2.6}$$
$$(x,t) \in Q_T,$$
$$p(0,t) = 0, \quad a(1,t)p_x(1,t) - b(1,t)p(1,t) = 0, \quad 0 < t < T, \tag{2.7}$$
$$p(x,T) = 0, \quad 0 < x < 1, \tag{2.8}$$

where u_φ is a solution of problem (1.1)–(1.3).

Definition 2.6 We say that a function $p \in V_2^{1,0}(Q_T)$ is a weak solution to problem (2.6)–(2.8) if it satisfies the boundary condition $p(0,t) = 0$, and the integral identity

$$\int_{Q_T} ((a(x,t)p_x - b(x,t)p)\eta_x + p\eta_t)\, dx\, dt$$
$$= -\int_0^T \big(u_\varphi(c,t) - z(t)\big)\, \eta(c,t)\, dt \tag{2.9}$$

holds for any function $\eta \in W_2^1(Q_T)$ such that $\eta(0,t) = 0$, and $\eta(x,0) = 0$.

Theorem 2.7 *For any $z \in L_2(0,T)$ and any bounded closed convex set $\Phi \subset W_2^1(0,T)$ problem (1.1)–(1.3), (2.4) has unique solution $\varphi_0 \in \Phi$.*

Theorem 2.8 *Let $\varphi_0 \in \Phi$ be a minimizer. Then for any $\varphi \in \Phi$ the following inequality holds:*

$$\int_0^T \big(u_{\varphi_0}(c,t) - z(t)\big)\big(u_\varphi(c,t) - u_{\varphi_0}(c,t)\big)\, dt \ge 0.$$

Theorem 2.9 *There exists a unique weak solution $p \in V_2^{1,0}(Q_T)$ to problem (2.6)–(2.8), and this solution satisfies the following inequality*

$$\|p\|_{V_2^{1,0}(Q_T)} \le C_2\big(\|\varphi\|_{W_2^1(0,T)} + \|\psi\|_{W_2^1(0,T)} + \|\xi\|_{L_2(0,1)} + \|z\|_{L_2(0,T)}\big), \tag{2.10}$$

where the constant C_2 is independent of φ, ψ, ξ and z.

Theorem 2.10 *Let $\varphi_0 \in \Phi$ be a minimizer. Then for any $\varphi \in \Phi$ the following inequality holds:*

$$\int_0^T a(0,t)p_x(0,t)\,(\varphi(t) - \varphi_0(t))\, dt \le 0, \tag{2.11}$$

where p is a weak solution of the problem (2.6)–(2.8) with $\varphi = \varphi_0$.

Theorems 2.8 and 2.10 give necessary conditions to minimizer.

3 Proofs

At first we prove Theorem 2.7.

Proof Let us define the set

$$B =: \{y = u_\varphi(c,\cdot) : \varphi \in \Phi\} \subset L_2(0,T).$$

Since Φ is a convex set, B is a convex set, too. Therefore, by (2.3), we have the inequality

$$\|u_\varphi\|_{V_2^{1,0}(Q_T)} \le C_1\big(\|\xi\|_{L_2(0,1)} + \|\varphi\|_{W_2^1(0,T)} + \|\psi\|_{W_2^1(0,T)}\big). \tag{3.1}$$

Note that the constant C_1 does not depend on ξ, φ and ψ.

The set Φ is bounded and closed in $W_2^1(0,T)$ and, by estimate (3.1), we obtain that B is a bounded set in $L_2(0,T)$.

Now we prove that B is a closed subset of $L_2(0,T)$. Consider a fundamental sequence $\{y_k\}_{k=1}^\infty \subset B$ in $L_2(0,T)$ and its limit $y \in L_2(0,T)$. Therefore, $\|y - y_k\|_{L_2(0,T)} \to 0$, $k \to \infty$. The corresponding sequence $\{\varphi_k\} \subset \Phi$, by the boundedness of Φ, is a weakly pre-compact set in $W_2^1(0,T)$. Denote by $\varphi \in \Phi$ the weak limit of any its subsequence. So, by theorem [22], ch. 2, par. 38, (see also [21]), there exists a new subsequence $\{\varphi_{k_j}\}$ such that

$$\|\tilde{\varphi}_l - \varphi\|_{W_2^1(0,T)} \to 0, \quad l \to \infty, \quad \tilde{\varphi}_l = \frac{1}{l}\sum_{j=1}^{l} \varphi_{k_j}. \tag{3.2}$$

Consequently, for the solutions

$$u_{\tilde{\varphi}_l} = \frac{1}{l}\sum_{j=1}^{l} u_{\varphi_{k_j}}$$

we get

$$\|u_{\tilde{\varphi}_l} - u_{\tilde{\varphi}_m}\|_{V_2^{1,0}(Q_T)} \le C_1\|\tilde{\varphi}_l - \tilde{\varphi}_m\|_{W_2^1(0,T)} \to 0, \quad l, m \to \infty. \tag{3.3}$$

This means that $u_{\tilde{\varphi}_l}(0,t) = \tilde{\varphi}_l(t)$, and the equality

$$\begin{aligned} &\int_{Q_T} (a(x,t)(u_{\tilde{\varphi}_l})_x\eta_x - b(x,t)\eta_x - u_{\tilde{\varphi}_l}\eta_t)\,dx\,dt \\ &= \int_0^1 \xi(x)\eta(x,0)\,dx + \int_0^T a(1,t)\psi(t)\,\eta(1,t)\,dt \end{aligned} \tag{3.4}$$

fulfils for any $\eta \in \widetilde{W}_2^1(Q_T)$.

The limit function $u \in V_2^{1,0}(Q_T)$ exists in accordance with (3.2), (3.3) and (3.4). So, u is a weak solution of (1.1)–(1.3) with the boundary function φ satisfying the inequality

$$\|u - u_{\tilde{\varphi}_l}\|_{V_2^{1,0}(Q_T)} \le C_1\|\varphi - \tilde{\varphi}_l\|_{W_2^1(0,T)}.$$

So, by ((6.15), [16], ch. 1) we get

$$\|u(c,\cdot) - u_{\tilde{\varphi}_l}(c,\cdot)\|_{L_2(0,T)} \le C_3\|u - u_{\tilde{\varphi}_l}\|_{V_2^{1,0}(Q_T)}$$
$$\le C_1C_3\|\varphi - \tilde{\varphi}_l\|_{W_2^1(0,T)},$$

whence $y = u(c,\cdot) \in B$, and B is a closed subset in $L_2(0,T)$.

To continue the proof, we formulate the following lemma (see, for example, [3]):

Lemma 3.1 *Let A be a closed convex set in a Hilbert space H. Then for any $x \in H$ there exists a unique element $y \in A$ such that*

$$\|x - y\| = \inf_{z\in A} \|x - z\|. \tag{3.5}$$

Consequently, by Lemma 3.1, there exists a unique function $y = u(c,\cdot)$, where $u \in V_2^{1,0}(Q_T)$ is a solution to problem (1.1)–(1.3) with some $\varphi_0 \in \Phi$, such that

$$m[z,\Phi] = J[z,\varphi_0].$$

Now, we prove the uniqueness of such function $\varphi_0 \in \Phi$. If it is not true, consider a pair of such functions φ_1, φ_2. The function $\tilde{u} = u_{\varphi_1} - u_{\varphi_2}$ is a solution to the problem

$$\tilde{u}_t = (a(x,t)\tilde{u}_x)_x + b(x,t)\tilde{u}_x, \quad 0 < t < T, \quad 0 < x < 1, \tag{3.6}$$
$$\tilde{u}(0,t) = \tilde{\varphi}(t), \quad 0 < t < T, \quad \tilde{\varphi}(t) = \varphi_1(t) - \varphi_2(t), \tag{3.7}$$
$$\tilde{u}(c,t) = 0, \quad \tilde{u}_x(1,t) = 0, \quad 0 < t < T, \tag{3.8}$$
$$\tilde{u}(x,0) = 0, \quad 0 < x < 1. \tag{3.9}$$

Using integral identity (2.2) with the function η equal to 0 on $[0,c] \times [0,T]$, we obtain that the function $\tilde{u}$ on the rectangle $Q_T^{(c)} = (c,1) \times (0,T)$ is equal to the solution of the problem

$$\hat{u}_t = (a(x,t)\hat{u}_x)_x + b(x,t)\hat{u}_x, \quad 0 < t < T, \quad c < x < 1, \tag{3.10}$$
$$\hat{u}(c,t) = 0, \quad \hat{u}_x(1,t) = 0, \quad 0 < t < T, \tag{3.11}$$
$$\hat{u}(x,0) = 0, \quad c < x < 1. \tag{3.12}$$

But the solution of problem (3.10)–(3.12) vanishes on $[c,1] \times [0,T]$, so we have

$$\tilde{u}(x,t) = 0, \quad c < x < 1, \quad 0 < t < T. \tag{3.13}$$

Let us prove that

$$\tilde{u}(x,t) = 0, \quad 0 < x < 1, \quad 0 < t < T. \tag{3.14}$$

We apply the unique continuation theorem for parabolic equations (th. 1.1, [23]) to the solution $\tilde{u}$ of problem (3.6)–(3.9) and obtain that (3.14) follows from (3.13). So, $\tilde{u}(x,t) = 0$ for all $x \in (0,1)$ and $t \in (0,T)$. Therefore, $\tilde{\varphi}(t) = \tilde{u}(0,t) = 0$. □

Now we prove Theorem 2.8.

Proof Denote by $\varphi_0 \in \Phi$ a minimizer for the problem (1.1)–(1.3), (2.4) (minimizer exists by virtue of theorem 2.7). Now, for an arbitrary $\varphi \in \Phi$, by the convexity of Φ we obtain that $\varphi_0 + \gamma(\varphi - \varphi_0) \in \Phi$ for $\gamma \in [0,1]$. Then

$$\begin{aligned}
0 &\leq \frac{d}{d\gamma} J[z, \varphi_0 + \gamma(\varphi - \varphi_0)]|_{\gamma=0} \\
&= \frac{d}{d\gamma} \int_0^T (u_{\varphi_0+\gamma(\varphi-\varphi_0)}(c,t) - z(t))^2 dt|_{\gamma=0} \\
&= 2\int_0^T (u_{\varphi_0+\gamma(\varphi-\varphi_0)}(c,t) - z(t))(u_\varphi(c,t) - u_{\varphi_0}(c,t))dt|_{\gamma=0} \\
&= 2\int_0^T (u_{\varphi_0}(c,t) - z(t))(u_\varphi(c,t) - u_{\varphi_0}(c,t))dt, \qquad (3.15)
\end{aligned}$$

and Theorem 2.8 is proved. □

Let us prove the existence and uniqueness of the solution to (2.6)–(2.8) and also the estimate (2.10).

Proof We can prove the uniqueness of solution by traditional methods (see [16], ch. 3). To prove the existence of solutions, we can use Galerkin's method. The scheme of such type of proof is presented in ([16], ch. 3, par. 3). The main problem is to obtain estimate (2.10) from the energy balance equation, which has the following form:

$$\begin{aligned}
&\|p(x,t_1)\|^2_{L_2(0,1)} + 2\int_{Q_{t_1,T}} a(x,t)p_x^2 dxdt \qquad (3.16) \\
&= \|p(x,T)\|^2_{L_2(0,1)} + 2\int_{Q_{t_1,T}} b(x,t)pp_x dxdt \\
&- \int_{t_1}^T p(c,t)(u_\varphi(c,t) - z(t))dt, \\
&Q_{t_1,T} = (t_1,T) \times (0,1), \quad 0 < t_1 < T.
\end{aligned}$$

Now we apply the inequality

$$\int_{t_1}^T p^2(c,t)dt \leq \int_0^1 \int_{t_1}^T p_x^2(x,t)dxdt, \qquad (3.17)$$

which is valid for functions $p \in V_2^{1,0}(Q_T)$, satisfying the condition $p(0,t) = 0$. From (3.16) and (3.17), we obtain

$$\begin{aligned} &\|p(x,t_1)\|^2_{L_2(0,1)} + 2a_0\|p_x\|^2_{L_2(Q_{t_1,T})} \\ &\le \|p(x,T)\|^2_{L_2(0,1)} + \left(\nu_1 + \frac{\nu_2}{2}\right)\|p_x\|^2_{L_2(Q_{t_1,T})} \\ &+ \frac{b_1^2}{\nu_1}(T-t_1)\sup_{t_1\le t\le T}\|p(x,t)\|^2_{L_2(0,1)} + \frac{1}{2\nu_2}\|u_\varphi(c,t)-z(t)\|^2_{L_2(t_1,T)}, \\ &\nu_1,\nu_2 > 0. \end{aligned} \tag{3.18}$$

Denote $y(t) = \sup\limits_{t\le\tau\le T}\|p(x,\tau)\|_{L_2(0,1)}$. Therefore,

$$\begin{aligned} &y^2(t_1) + 2a_0\|p_x\|^2_{L_2(Q_{t_1,T})} \\ &\le y^2(T) + \left(\nu_1 + \frac{\nu_2}{2}\right)\|p_x\|^2_{L_2(Q_{t_1,T})} \\ &+ \frac{b_1^2}{\nu_1}(T-t_1)y^2(t_1) + \frac{1}{2\nu_2}\|u_\varphi(c,t)-z(t)\|^2_{L_2(t_1,T)}. \end{aligned} \tag{3.19}$$

Then, by choosing $\nu_1 + \frac{\nu_2}{2} < a_0$ and $t_1 > T - \frac{\nu_1}{2b_1^2}$, we obtain

$$y^2(t_1) + \|p_x\|^2_{L_2(Q_{t_1,T})} \le C_4\left(y^2(T) + \|u_\varphi(c,t)-z(t)\|^2_{L_2(t_1,T)}\right). \tag{3.20}$$

By (3.20) we have the following estimate:

$$\|p\|_{V_2^{1,0}(Q_{t_1,T})} \le C_5\left(\|p(x,T)\|_{L_2(0,1)} + \|u_\varphi(c,t)-z(t)\|_{L_2(t_1,T)}\right). \tag{3.21}$$

We divide the segment $[0,T]$ from right to left to the segments $[T,t_1]$, $[t_1,t_2]$, ..., $[t_n,0]$, with a length less than $\frac{\nu_1}{2b_1^2}$. For each of them estimate (3.21) holds. We derive from these estimates, then, taking into account (2.3) and the equality $p(x,T)=0$, the required inequality

$$\begin{aligned} &\|p\|_{V_2^{1,0}(Q_T)} \le C_6\left(\|u_\varphi(c,t)-z(t)\|_{L_2(t_1,T)}\right) \\ &\le C_2\left(\|\varphi\|_{W_2^1(0,T)} + \|\psi\|_{W_2^1(0,T)} + \|\xi\|_{L_2(0,1)} + \|z\|_{L_2(0,T)}\right) \end{aligned}$$

fulfill. Further, the proof is completed according to the scheme presented in ([16], ch. 3, par. 3). □

Now we prove the necessary condition for the extremum in terms of the adjoint problem (Theorem 2.10).

Proof Let $\varphi_0 \in \Phi$ be a minimizer, and $\varphi \in \Phi$ be an arbitrary control function. Put $w = u_\varphi - u_{\varphi_0} \in V_2^{1,0}(Q_T)$. It follows from (2.2) that w satisfies the integral identity

$$\int_{Q_T}(a(x,t)w_x\eta_x - b(x,t)w_x\eta - w\eta_t)\,dx\,dt = 0. \tag{3.22}$$

Consider a sequence of weak solutions (from $V_2^{1,0}(Q_T)$) to problems

$$(p_k)_t + (a(x,t)(p_k)_x)_x - (b(x,t)p_k)_x = \theta_k(t)\zeta_k(x), \quad (x,t) \in Q_T, \tag{3.23}$$
$$p_k(0,t) = 0, \quad a(1,t)(p_k)_x(1,t) - b(1,t)p_k(1,t) = 0, \quad 0 < t < T, \tag{3.24}$$
$$p_k(x,T) = 0, \quad 0 < x < 1, \quad k = 1, 2, \ldots, \tag{3.25}$$

where

$$\theta_k(t) \in D(0,T), \quad \|\theta_k(t) - u_{\varphi_0}(c,t) + z(t)\|_{L_2(0,T)} \to 0, \quad k \to \infty, \tag{3.26}$$
$$\zeta_k(x) \in D(0,1), \quad \zeta_k(x) \to \delta(x-c) \quad \text{in} \quad H^{-1}(0,1), \quad k \to \infty. \tag{3.27}$$

Note, that p_k are smooth functions in $\overline{Q}_T$. By definition (2.9) we have

$$\|p_k - p\|_{V_2^{1,0}(Q_T)} \to 0, \quad k \to \infty. \tag{3.28}$$

Now we can put $\eta = p_k$ in (3.22). Then, we obtain the equality

$$\int_{Q_T} (a(x,t)w_x(p_k)_x - b(x,t)w_x p_k - w(p_k)_t)\, dx\, dt = 0. \tag{3.29}$$

From (3.23)–(3.25), (3.29) and the theorem on the regularity of solutions to parabolic boundary value problems (see [15], ch. 3, par. 12) for $k \to \infty$ we get the equalities

$$\begin{aligned} 0 &= \int_{Q_T} ((a(x,t)p_x)_x - (b(x,t)p)_x + p_t)\, w\, dx\, dt \\ &+ \int_0^T a(0,t)p_x(0,t)(\varphi(t) - \varphi_0(t))dt \\ &= \int_0^T \left(u_{\varphi_0}(c,t) - z(t)\right)\left(u_{\varphi}(c,t) - u_{\varphi_0}(c,t)\right) dt \\ &+ \int_0^T a(0,t)p_x(0,t)(\varphi(t) - \varphi_0(t))dt. \end{aligned} \tag{3.30}$$

Applying (3.15) and (3.30), we obtain the required inequality (2.11).

Theorem 2.10 is proved. □

4 Conclusions. Notes and Comments

Let us note the main difference of considered problem from the previous ones. It consists in the type of observation. In previous articles of different authors, control problems with final and distributed observation are studied (see, for example, [11,

14, 17]), whereas, in our articles, we consider the point-wise observation for different parabolic control problems.

One of the first studies is described in [12], where the heat equation with the third-type boundary condition containing the control function is considered. In [12], for the extremum problem with the final observation functional, the existence of minimizer is proved in some class of measurable control functions. The author also proves the existence and uniqueness of minimizer in the case of a functional with an additional quadratic term. Some of the later results deal with a non-homogeneous equation having a distribution in Q_T control function at the right-hand side and with the distributed or boundary observation (see [17–19]). Other problems of minimization with final observation and the problem of the optimal time control are considered in [9–11, 24]. Note that our formulation of the extremum problem with pointwise observation is different from those formulated in the papers listed. Besides, the case of the equation with a free convection term was not previously considered. Closely related to our formulation of problem is the problem with distributed control and point-wise observation but for different cost functional. For such a problem related to the parabolic equation with symmetric elliptic operator, the existence and uniqueness of minimizer are proved in [17].

In our articles [3–8] we prove the existence and uniqueness of the control function $\varphi_0(t)$ from a convex set (the minimizer) giving the minimum to this functional [4] and study the structure of the set of accessible temperature functions [4, 8]. We also prove the "dense controllability" of the problem for some set of control functions [6] and obtain qualitative properties of minimizers [7]. In comparison with our previous results, in the present paper we consider the parabolic equation with non-self-adjoint elliptic operator and an arbitrary convex closed bounded set of control functions.

References

1. Astashova, I.V., Filinovskiy, A.V., Kondratiev, V.A., Muravei, L.A.: Some problems in the qualitative theory of differential equations. J. Nat. Geom. **23**(1–2), 1–126 (2003)
2. Astashova, I.V. (ed.): Qualitative Properties of Solutions to Differential Equations and Related Topics of Spectral Analysis. UNITY-DANA Publ, Moscow (2012).[in Russian]
3. Astashova, I.V., Filinovskiy, A.V., Lashin, D.A.: On maintaining optimal temperatures in greenhouses. WSEAS Trans. Circuits Syst. **15**, 198–204 (2016)
4. Astashova, I., Filinovskiy, A., Lashin, D.: On a model of maintaining the optimal temperature in greenhouse. Funct. Differ. Equ. **23**(3–4), 97–108 (2016)
5. Astashova, I., Filinovskiy, A., Lashin, D.: On optimal temperature control in hothouses, In: Th. Simos, Ch. Tsitouras (eds.) Proceedings of 14th International Conference on Numerical Analysis and Applied Mathematics (ICNAAM 2016) (19–25 September 2016, Rhodes, Greece), AIP Conf. Proc., 1863, pp. 4–8 (2017)
6. Astashova, I.V., Filinovskiy, A.V.: On the dense controllability for the parabolic problem with time-distributed functional. Tatr. Mt. Math. Publ. **71**, 9–25 (2018)
7. Astashova, I.V., Filinovskiy, A.V.: On properties of minimizers of a control problem with time-distributed functional related to parabolic equations. Opusc. Math. **39**(5), 595–609 (2019)

8. Astashova, Irina, Filinovskiy, Alexey: On the controllability problem with pointwise observation for the parabolic equation with free convection term. WSEAS Trans. Syst. Control **14**, 224–231 (2019)
9. Butkovsky, A.G.: Optimal control in the systems with distributed parameters. Avtomat. i Telemekh. **22**(1), 17–26 (1961). [in Russian]
10. Butkovsky, A.G., Egorov, A.I., Lurie, K.A.: Optimal control of distributed systems. SIAM J. Control **6**(3), 437–476 (1968)
11. Egorov, A.I.: Optimal control by heat and diffusion processes. Nauka, Moscow (1978). [in Russian]
12. Egorov, Yu.V.: Some problems of theory of optimal control. Zh. Vychisl. Mat. Mat. Fiz. **3**(5), 887–904 (1963). [in Russian]
13. Farag, M.H., Talaat, T.A., Kamal, E.M.: Existence and uniqueness solution of a class of quasilinear parabolic boundary control problem. Cubo **15**(2), 111–119 (2013)
14. Friedman, A.: Optimal control for parabolic equations. J. Math. Anal. Appl. **18**(3), 479–491 (1968)
15. Ladyzhenskaya, O.A., Solonnikov, V.A., Ural'seva, N.N.: Linear and Quasi-linear Equations of Parabolic Type. Translations of Mathematical Monographs, vol. 23. American Mathematical Society, Providence, RI (1968)
16. Ladyzhenskaya, O.A.: Boundary Value Problems of Mathematical Physics. Fizmatlit, Moscow (1973).[in Russian]
17. Lions, J.L.: Optimal Control of Systems Governed by Partial Differential Equations. Springer, Berlin (1971)
18. Lions, J.L.: The optimal control of distributed systems. Russ. Math. Surv. **28**(4), 13–46 (1973)
19. Lions, J.L.: Contrôle des systémes distribués singuliers. Gauthier-Villars, Paris (1983)
20. Lurie, K.A.: Applied Optimal Control Theory of Distributed Systems. Springer, Berlin (2013)
21. Mazur, S.: Über convexe Mengen in linearen normierte Raumen. Studia Math. **4**(1), 70–84 (1933)
22. Riesz, F., Szökefalvi-Nagy, B.: Functional Analysis. Dover Books on Advanced Mathematics, Dover Publ, New-York (1990)
23. Saut, J.-C., Scheurer, B.: Unique continuation for some evolution equations. J. Diff. Equ. **66**(1), 118–139 (1987)
24. Troltzsch, F.: Optimal Control of Partial Differential Equations. Theory, Methods and Applications, Providence, AMS, Graduate Studies in Mathematics, 112 (2010)

Homogenization of a Parabolic Equation for p-Laplace Operator in a Domain Perforated Along $(n-1)$-Dimensional Manifold with Dynamical Boundary Condition Specified on Perforations Boundary: Critical Case

A. V. Podolskiy and T. A. Shaposhnikova

Abstract The present paper focuses on the study of a homogenized limit of a parabolic equation for the p-Laplace operator with a nonlinear dynamical boundary condition set in a perforated domain that is obtained by removing "tiny" balls from a fixed domain. We investigate a "critical" case that is characterized by a relation between the size of holes, period of the structure, and coefficient in the boundary condition. The main result of the paper is a theorem that states weak convergence of an original problem solution to a solution of the limit problem containing transmission condition with a nonlocal "strange" term.

Keywords 35B27 · 35J95 · Homogenization · Perforated media · p-Laplacian · Nonlinear parabolic equation · Dynamical boundary conditions · Strange term · Nonlocal term

1 Introduction

The present paper investigates asymptotic behavior as $\varepsilon \to 0$ of a solution u_ε to an initial boundary value problem for the equation $\partial_t u_\varepsilon - \Delta_p u_\varepsilon = f(x,t)$, where $\Delta_p u_\varepsilon \equiv \mathrm{div}(|\nabla u_\varepsilon|^{p-2}\nabla u_\varepsilon)$, $2 \le p < n$ in a perforated domain Ω_ε that is obtained by removing balls G_ε with a diameter of order $O(\varepsilon^\alpha)$, $\alpha > 1$, from a fixed domain $\Omega \subset \mathbb{R}^n$. It is supposed that balls are ε-periodically distributed along $(n-1)$-dimensional manifold γ. On the perforations boundary, nonlinear dynamical condition $\varepsilon^{-k}\partial_t u_\varepsilon + |\nabla u_\varepsilon|^{p-2}\partial_\nu u_\varepsilon + \varepsilon^{-k}\sigma(u_\varepsilon) = \varepsilon^{-k} g(x,t), k \in \mathbb{R}$, is specified. Dynamical boundary conditions arise while modeling different physical processes; we refer

A. V. Podolskiy · T. A. Shaposhnikova (✉)
Moscow State University, Moscow, Russia
e-mail: shaposh.tan@mail.ru

A. V. Podolskiy
e-mail: AVPodolskiy@yandex.ru

A. Domoshnitsky et al. (eds.), *Functional Differential Equations and Applications*, Springer Proceedings in Mathematics & Statistics 379,
https://doi.org/10.1007/978-981-16-6297-3_17

to the following works [1, 2] which give an overview of this type of problems. The combination of a dynamical boundary condition and a parabolic equation was considered, for example, in [3, 4].

Papers [5–7] are devoted to homogenization of initial boundary value problems with dynamical boundary condition in the case $\alpha = 1$. In the present paper, we consider the case $\alpha > 1$, so-called "tiny" cavities. Such a structure of the perforated domain is noteworthy because the limit problem may contain "strange" term. That behavior for different problems was studied, for example, in works [8–15]. Papers [8, 9] investigate asymptotic behavior of the solution to the diffusion equation with the Laplace operator with a linear dynamical boundary condition in the critical case. They show that homogenization leads to a rise of a nonlocal memory "strange" term that is a solution to an ODE.

The purpose of this paper is to construct a homogenized model and prove weak convergence as $\varepsilon \to 0$ of the original problem solution to a solution of the homogenized problem in the critical case: $\alpha = (n-1)/(n-p)$, $k = \alpha(p-1)$. We prove that the limit problem contains a transmission condition with a new nonlinear nonlocal term that could be obtained by solving an ODE. Therefore, this paper extends results of [8] to the case of the nonlinear parabolic equation for the p-Laplace operator and a nonlinear dynamical boundary condition.

2 Statement of Results

2.1 Problem Statement

Let Ω be a bounded domain in $\mathbb{R}^n$, $n \geq 3$, with a smooth boundary $\partial\Omega$; $\gamma = \Omega \bigcap \{x_1 = 0\} \neq \emptyset$ is a domain on the plane $x_1 = 0$, $Y = (-1/2, 1/2)^n$, $G_0 = \{x : |x| < 1\}$. Define $\delta B = \{x : \delta^{-1}x \in B\}$, $\delta > 0$. Consider $\tilde{G}_\varepsilon = \bigcup\limits_{j\in\mathbb{Z}'} (a_\varepsilon G_0 + \varepsilon z) = \bigcup\limits_{j\in\mathbb{Z}'} G_\varepsilon^j$, where $\mathbb{Z}'$ is a set of vectors $(0, z_2, \ldots, z_n)$, with whole coefficients z_i, $i = 2, \ldots, n$, $\varepsilon > 0$, $a_\varepsilon = C_0\varepsilon^\alpha$, $\alpha = \frac{n-1}{n-p}$, $C_0 = const > 0$.

We define a set $G_\varepsilon = \bigcup\limits_{j\in\Upsilon_\varepsilon} G_\varepsilon^j$, where $\Upsilon_\varepsilon = \{j \in \mathbb{Z}' : \rho(\partial\Omega, \overline{G_\varepsilon^j}) \geq 2\varepsilon)\}$, and $S_\varepsilon^j = \partial G_\varepsilon^j$. Note that $|\Upsilon_\varepsilon| = d\varepsilon^{n-1}$, $d = const > 0$. By T_r^j, we denote a ball of radius r with the center in P_ε^j, where P_ε^j is the center of a cube $Y_\varepsilon^j = \varepsilon Y + j\varepsilon$, $j \in \Upsilon_\varepsilon$.

Let us introduce sets: $\Omega_\varepsilon = \Omega \setminus \overline{G_\varepsilon}$, $S_\varepsilon = \partial G_\varepsilon$, $\partial\Omega_\varepsilon = S_\varepsilon \bigcup \partial\Omega$.

In $Q_\varepsilon^T = \Omega_\varepsilon \times (0, T)$, where $0 < T < \infty$, we consider the following initial boundary value problem:

$$\begin{cases} \partial_t u_\varepsilon - \Delta_p u_\varepsilon = f(x,t), & (x,t) \in Q_\varepsilon^T, \\ \varepsilon^{-k}\partial_t u_\varepsilon + \partial_{\nu_p} u_\varepsilon + \varepsilon^{-k}\sigma(u_\varepsilon) = \varepsilon^{-k} g(x,t), & (x,t) \in S_\varepsilon^T = S_\varepsilon \times (0,T), \\ u_\varepsilon = 0, & (x,t) \in \Gamma^T = \partial\Omega \times (0,T), \\ u_\varepsilon(x,0) = 0, & x \in \Omega_\varepsilon, \\ u_\varepsilon(x,0) = 0, & x \in S_\varepsilon, \end{cases} \quad (1)$$

where $\Delta_p u \equiv \operatorname{div}(|\nabla u|^{p-2}\nabla u)$, $2 \le p < n$, $f \in L^2(Q^T)$, $Q^T = \Omega \times (0,T)$, $\partial_{\nu_p} u \equiv |\nabla u|^{p-2}(\nabla u, \nu)$, ν is the unit outward normal vector to S_ε^T, and $k = \frac{(n-1)(p-1)}{n-p}$, $n \ge 3$. Also it is considered that $g(x,t) \in C^1(\overline{Q^T})$, $\sigma(u) \in C^1(\mathbb{R})$, $\sigma(0) = 0$ and there exist such positive constants k_1, k_2 that

$$\begin{aligned} (\sigma(u) - \sigma(v))(u - v) &\ge k_1 |u - v|^p, \\ |\sigma(u)| &\le k_2 |u|^{p-1}. \end{aligned} \quad (2)$$

By $W^{1,p}(\Omega_\varepsilon, \partial\Omega)$, we denote a space that is obtained as a closure in $W^{1,p}(\Omega_\varepsilon)$ of the set of infinitely differentiable in $\overline{\Omega}_\varepsilon$ functions that vanish near the boundary $\partial\Omega$. Define $W^{-1,q}(\Omega_\varepsilon, \partial\Omega)$, $q = \frac{p}{p-1}$, as an adjoint to $W^{1,p}(\Omega_\varepsilon, \partial\Omega)$, and by $< .,. >_{\Omega_\varepsilon}$ we denote duality relation.

The trace theorem applied to functions from the space $W^{1,p}(\Omega_\varepsilon, \partial\Omega)$ implies that their traces belong to $W^{1-1/p,p}(S_\varepsilon)$. By $W^{-1/q,q}(S_\varepsilon)$, we denote a space adjoint to $W^{1-1/p,p}(S_\varepsilon)$, and by $\langle .,. \rangle_{S_\varepsilon}$ we express duality relation.

Next, we define several spaces that naturally arise in considering the problem (1). First, we introduce a space $\mathbb{L}_\varepsilon^s = L^s(\Omega_\varepsilon) \times L^s(S_\varepsilon) = \{U = (u,z) : u \in L^s(\Omega_\varepsilon),\ z \in L^s(S_\varepsilon)\}$, $s \in (1,\infty)$ with a norm

$$\|U\|_{\mathbb{L}_\varepsilon^s} = \left(\int_{\Omega_\varepsilon} |u|^s dx + \varepsilon^{-k} \int_{S_\varepsilon} |z|^s ds \right)^{\frac{1}{s}}.$$

The space $\mathbb{H}_\varepsilon = \mathbb{L}_\varepsilon^2$ is a Hilbert space with an inner product

$$(U,V)_{\mathbb{H}_\varepsilon} = (u,v)_{L^2(\Omega_\varepsilon)} + \varepsilon^{-k}(\tilde{u},\tilde{v})_{L_2(S_\varepsilon)},$$

where $U = (u,\tilde{u})$, $V = (v,\tilde{v}) \in \mathbb{H}_\varepsilon$. Then, for $p > 1$ we consider $\mathbb{V}_\varepsilon^p = \{U = (u, u|_{S_\varepsilon}) : u \in W^{1,p}(\Omega_\varepsilon, \partial\Omega)\}$ (by $u|_{S_\varepsilon}$, we denote the trace of function u on S_ε), and equip it with the following norm:

$$\|(u, u|_{S_\varepsilon})\|_{\mathbb{V}_\varepsilon^p}^p = \|u\|_{W^{1,p}(\Omega_\varepsilon,\partial\Omega)}^p + \varepsilon^{-k}\|u|_{S_\varepsilon}\|_{W^{1-1/p,p}(S_\varepsilon)}^p.$$

Note that $\mathbb{V}_\varepsilon^p$ can be identified with the closed subspace in the $W^{1,p}(\Omega_\varepsilon, \partial\Omega) \times W^{1-1/p,p}(S_\varepsilon)$ regarding this norm. According to the embedding $W^{1,p}(\Omega_\varepsilon, \partial\Omega) \hookrightarrow W^{1-1/p,p}(S_\varepsilon)$, we conclude that norms in spaces $\mathbb{V}_\varepsilon^p$ and $W^{1,p}(\Omega_\varepsilon, \partial\Omega)$ are equivalent. For the sake of convenience, we would identify, when useful, $(u, u|_{S_\varepsilon})$

with u. Also note that $\mathbb{V}_\varepsilon^p$ is compactly embedded into $\mathbb{H}_\varepsilon$ for $p > \frac{2n}{n+2}$. It is clear that $(\mathbb{V}_\varepsilon^p)^*$ can be identified with a closed subspace in $W^{-1,q}(\Omega_\varepsilon, \partial\Omega) \times W^{-1/q,q}(S_\varepsilon)$, also we have $\mathbb{V}_\varepsilon^p \hookrightarrow \mathbb{L}_\varepsilon^2 = (\mathbb{L}_\varepsilon^2)^* \hookrightarrow (\mathbb{V}_\varepsilon^p)^*$.

We say that a pair of functions $U_\varepsilon = (u_\varepsilon, \tilde{u}_\varepsilon)$ is a weak solution of (1) if

$$u_\varepsilon \in L^p(0,T; W^{1,p}(\Omega_\varepsilon, \partial\Omega)),\ \partial_t u_\varepsilon \in L^q(0,T; W^{-1,q}(\Omega_\varepsilon, \partial\Omega)),$$
$$\tilde{u}_\varepsilon \in L^p(0,T; W^{1-1/p,p}(S_\varepsilon, \partial\Omega)),\ \partial_t \tilde{u}_\varepsilon \in L^q(0,T; W^{-1/q,q}(S_\varepsilon, \partial\Omega)),$$

$\tilde{u}_\varepsilon(t) = u_\varepsilon|_{S_\varepsilon}(t)$ for a.e. $t \in (0,T)$, and if it satisfies integral identity

$$\begin{aligned} &\int_0^T \langle \partial_t u_\varepsilon, v\rangle_{\Omega_\varepsilon} dt + \varepsilon^{-k} \int_0^T \langle \partial_t \tilde{u}_\varepsilon, v\rangle_{S_\varepsilon} dt + \\ &+ \int_{Q_\varepsilon^T} |\nabla u_\varepsilon|^{p-2} \nabla u_\varepsilon \nabla v dx dt + \varepsilon^{-k} \int_{S_\varepsilon^T} \sigma(\tilde{u}_\varepsilon) v ds dt = \\ &= \varepsilon^{-k} \int_{S_\varepsilon^T} g v ds dt + \int_{Q_\varepsilon^T} f v dx dt, \end{aligned} \tag{3}$$

for an arbitrary pair $\Theta = (v, v|_{S_\varepsilon})$, where v is from $L^p(0,T; W^{1,p}(\Omega_\varepsilon, \partial\Omega))$, i.e. $\Theta \in L^p(0,T; \mathbb{V}_\varepsilon^p))$. Moreover, $U_\varepsilon|_{t=0} = (0,0)$ in $\mathbb{H}_\varepsilon$.

2.2 Existence, Uniqueness, and Extension

Using Galerkin's method, the existence and uniqueness theorem for problem (1) is proved.

Theorem 1 *There exists a unique weak solution u_ε of the problem* (1) *and the following estimations are valid:*

$$\|U_\varepsilon\|_{L^\infty(0,T;\mathbb{V}_\varepsilon^p)} + \|U_\varepsilon\|_{L^p(0,T;\mathbb{V}_\varepsilon^p)} + \|\partial_t U_\varepsilon\|_{L^2(0,T;\mathbb{H}_\varepsilon)} \le K; \tag{4}$$

here and below, constant K is independent of ε.

By using extension results from [18], we have the following:

Theorem 2 *Let Ω_ε be a perforated domain defined above. Then there exist extension operator $R_\varepsilon : L^p(0,T; W^{1,p}(\Omega_\varepsilon, \partial\Omega)) \to L^p(0,T; W_0^{1,p}(\Omega))$ and $R_\varepsilon : L^p(Q_\varepsilon^T) \to L^p(Q^T)$, such that $\forall u \in L^p(0,T; W^{1,p}(\Omega_\varepsilon))$ we have*

$$R_\varepsilon u = u, \text{ in } Q_\varepsilon^T,$$
$$\|R_\varepsilon u\|_{L^p(Q^T)} \le K\|u\|_{L^p(Q_\varepsilon^T)},\quad \|\nabla R_\varepsilon u\|_{L^p(Q^T)} \le K\|\nabla u\|_{L^p(Q_\varepsilon^T)}.$$

Then, we use the well-known result that there exists linear extension operator $P_\varepsilon : H^1(Q_\varepsilon^T) \to H^1(Q^T)$ such that

$$\|\partial_t(P_\varepsilon u_\varepsilon)\|^2_{L^2(Q^T)} + \|\nabla_x(P_\varepsilon u_\varepsilon)\|^2_{L^2(Q^T)} \le K(\|\partial_t u_\varepsilon\|^2_{L^2(Q_\varepsilon^T)} + \|\nabla_x u_\varepsilon\|^2_{L^2(Q_\varepsilon^T)}).$$

Hence, estimations (4) imply that a subsequence exists (we preserve the notation of the original one) such that as $\varepsilon \to 0$, we have

$$\begin{aligned} R_\varepsilon u_\varepsilon &\rightharpoonup u_0 \text{ weakly in } L^p(0, T; W_0^{1,p}(\Omega)), \\ R_\varepsilon u_\varepsilon &\to u_0 \text{ strongly in } L^p(Q^T), \\ \partial_t(P_\varepsilon u_\varepsilon) &\rightharpoonup \partial_t u_0 \text{ weakly in } L^2(Q^T), \end{aligned} \tag{5}$$

where u_ε is a solution to (1).

2.3 Homogenization Theorem

The next theorem gives a description of the limit function u_0 from (5).

Theorem 3 *Let $n \ge 3$, $\alpha = \frac{n-1}{n-p}$, $k = \frac{n-1}{n-p}(p-1)$, u_ε be a weak solution of (1). Then the function u_0, defined in (5), is a weak solution of the following problem:*

$$\begin{cases} \partial_t u_0 - \Delta_p u_0 = f(x,t), & (x,t) \in Q^T, \\ \left[u_0\right]\Big|_\gamma = 0, & t \in (0,T), \\ \left[\partial_{\nu_p} u_0\right]\Big|_\gamma = \mathcal{A}_{n,p}|u_0 - H_{u_0}|^{p-2}(u_0 - H_{u_0}), & t \in (0,T), \\ \partial_t H_{u_0} + \sigma(H_{u_0}) = & \\ \mathcal{B}_{n,p}|u_0 - H_{u_0}|^{p-2}(u_0 - H_{u_0}) + g(x,t), & (x,t) \in \gamma^T \\ u_0 = 0, & (x,t) \in \Gamma^T, \\ u_0(x,0) = 0, & x \in \Omega, \\ H_{u_0}(x,0) = 0, & x \in \gamma, \end{cases} \tag{6}$$

where $\partial_{\nu_p} u_0 = |\nabla u_0|^{p-2}\partial_{x_1} u_0$, $\gamma^T = \gamma \times (0,T)$, $\mathcal{A}_{n,p} = \left(\frac{n-p}{p-1}\right)^{p-1} C_0^{n-p}\omega_n$, $\mathcal{B}_{n,p} = \left(\frac{n-p}{p-1}\right)^{p-1} C_0^{1-p}$, and by $\left[\zeta\right]\Big|_\gamma$ we denote a jump of a function ζ in points $x \in \gamma$.

Remark 1 Denote by H_φ, a solution to the following Cauchy problem:

$$\begin{cases} \partial_t H_\varphi + \sigma(H_\varphi) = \mathcal{B}_{n,p}|\varphi - H_\varphi|^{p-2}(\varphi - H_\varphi) + g(x,t), \\ H_\varphi(x,0) = 0, \end{cases} \tag{7}$$

where $\varphi \in L^p(\gamma^T)$. It's well known that there exists a solution to this problem. Using H_φ as a test function in the corresponding integral identity and monotonicity of the function σ, we get

$$\|H_\varphi(x,t)\|^2_{L^2(\gamma)} + k_1 \int_0^t \|H_\varphi\|^p_{L^p(\gamma)} d\tau + C_1 \int_0^t \|\varphi - H_\varphi\|^p_{L^p(\gamma)} d\tau \le$$

$$\le \mathscr{B}_{n,p} \int_0^t \int_\gamma |\varphi - H_\varphi|^{p-1} |\varphi| dx' d\tau + \int_0^t \int_\gamma |g||H_\varphi| dx' d\tau.$$

Then we use Young's inequality and obtain

$$\|H_\varphi(x,t)\|^2_{L^2(\gamma)} + C_2 \int_0^t \|H_\varphi\|^p_{L^p(\gamma)} d\tau + C_3 \int_0^t \|\varphi - H_\varphi\|^p_{L^p(\gamma)} d\tau \le$$

$$\le K \left(\int_0^t \|\varphi\|^p_{L^p(\gamma)} d\tau + \int_0^t \|H_\varphi\|^2_{L^2(\gamma)} d\tau + \max_{\gamma^T} |g|^2 \right).$$

Using Gronwall's lemma, we finally prove an estimation

$$\|H_\varphi\|^2_{L^\infty(0,T;L^2(\gamma))} + \|H_\varphi\|^p_{L^p(\gamma^T)} \le K \left(\|\varphi\|^p_{L^p(\gamma^T)} + \max_{\gamma^T} |g|^2 \right).$$

Next, for two solutions of the Cauchy problem related to functions φ and ζ, we take $H_\varphi - H_\zeta$ as a test function in integral identities and subtract one from another. As a result, we get

$$\|(H_\varphi - H_\zeta)(x,t)\|^2_{L^2(\gamma)} + K_1 \int_0^t \|H_\varphi - H_\zeta\|^p_{L^p(\gamma)} d\tau +$$

$$K_2 \int_0^t \|\varphi - H_\varphi - \zeta + H_\zeta\|^p_{L^p(\gamma)} d\tau \le$$

$$\mathscr{B}_{n,p} \int_0^t \int_\gamma |(|\varphi - H_\varphi|^{p-2}(\varphi - H_\varphi) - |\zeta - H_\zeta|^{p-2}(\zeta - H_\zeta))||\varphi - \zeta| dx' d\tau \le$$

$$K_3 \int_0^t \int_\gamma (|\varphi - H_\varphi| + |\zeta - H_\zeta|)^{p-2} |\varphi - \zeta||\varphi - H_\varphi - \zeta + H_\zeta| dx' d\tau \le$$

$$K_3 \int_0^t \int_\gamma (|\varphi - H_\varphi| + |\zeta - H_\zeta|)^{p-2} (|\varphi - \zeta|^2 + |\varphi - \zeta||H_\varphi - H_\zeta|) dx' d\tau.$$

Using Holder's inequality, we derive

$$\int_0^t \int_\gamma (|\varphi - H_\varphi| + |\zeta - H_\zeta|)^{p-2} |\varphi - \zeta|^2 dx' d\tau \leq$$

$$(\|\varphi - H_\varphi\|^p_{L^p(\gamma^T)} + \|\zeta - H_\zeta\|^p_{L^p(\gamma^T)})^{\frac{p-2}{p}} \|\varphi - \zeta\|^2_{L^p(\gamma^T)} \leq$$

$$K(\|\varphi\|^{p-2}_{L^p(\gamma^T)} + \|\zeta\|^{p-2}_{L^p(\gamma^T)} + \max_{\gamma^T} |g|^{\frac{2(p-2)}{p}})\|\varphi - \zeta\|^2_{L^p(\gamma^T)}$$

and

$$\int_0^t \int_\gamma (|\varphi - H_\varphi| + |\zeta - H_\zeta|)^{p-2} |\varphi - \zeta||H_\varphi - H_\zeta| dx' d\tau \leq$$

$$(\|\varphi - H_\varphi\|^p_{L^p(\gamma^T)} + \|\zeta - H_\zeta\|^p_{L^p(\gamma^T)})^{\frac{p-2}{p}} \|H_\varphi - H_\zeta\|_{L^p(\gamma^T)} \|\varphi - \zeta\|_{L^p(\gamma^T)} \leq$$

$$K(\|\varphi\|^{p-2}_{L^p(\gamma^T)} + \|\zeta\|^{p-2}_{L^p(\gamma^T)} + \max_{\gamma^T} |g|^{\frac{2(p-2)}{p}})^2 \|\varphi - \zeta\|_{L^p(\gamma^T)}.$$

Using these estimations, we obtain

$$\|H_\varphi - H_\zeta\|^p_{L^p(\gamma^T)} \leq K(1 + (\|\varphi\|^{p-2}_{L^p(\gamma^T)} + \|\zeta\|^{p-2}_{L^p(\gamma^T)} + \max_{\gamma^T} |g|^{\frac{2(p-2)}{p}})^2) \times \\ (1 + \|\varphi\|_{L^p(\gamma^T)} + \|\zeta\|_{L^p(\gamma^T)}) \|\varphi - \zeta\|_{L^p(\gamma^T)}. \tag{8}$$

3 Proof of the Homogenization Theorem

3.1 Integral Inequality Construction

Using monotonicity of the p-Laplace operator and functions $|\lambda|^{p-2}\lambda$, $\sigma(\lambda)$ we derive that u_ε satisfies the following integral inequality:

$$\int_0^T \langle \partial_t \zeta, \zeta - u_\varepsilon \rangle_{\Omega_\varepsilon} dt + \varepsilon^{-k} \int_0^T \langle \partial_t \zeta, \zeta - u_\varepsilon \rangle_{S_\varepsilon} dt + \\ \int_{Q^T_\varepsilon} |\nabla \zeta|^{p-2} \nabla \zeta \nabla (\zeta - u_\varepsilon) dx dt + \varepsilon^{-k} \int_{S^T_\varepsilon} \sigma(\zeta)(\zeta - u_\varepsilon) ds dt \geq \\ \varepsilon^{-k} \int_{S^T_\varepsilon} g(\zeta - u_\varepsilon) ds dt + \int_{Q^T_\varepsilon} f(\zeta - u_\varepsilon) dx dt - \\ \frac{1}{2} \|\zeta(x, 0)\|^2_{L_2(\Omega_\varepsilon)} - \frac{1}{2} \varepsilon^{-k} \|\zeta(x, 0)\|^2_{L_2(S_\varepsilon)}, \tag{9}$$

where ζ is an arbitrary function from the space $L^p(0,T;\mathbb{V}^p_\varepsilon)$ such that $\partial_t\zeta \in L^q(0,T;(\mathbb{V}^p_\varepsilon)^*)$.

3.2 *Test Function Selection*

3.2.1 Time Component

We consider $\varphi = \eta(t)\psi(x)$, $\eta(t) \in C^1([0,T])$, $\psi(x) \in C_0^\infty(\Omega)$ and define functions $H_\varphi^{\varepsilon,j}$ as a solution to ODEs:

$$\begin{cases} \partial_t H_\varphi^{\varepsilon,j} + \sigma(H_\varphi^{\varepsilon,j}) = \\ \mathscr{B}_{n,p}|\varphi(P_\varepsilon^j,t) - H_\varphi^{\varepsilon,j}|^{p-2}(\varphi(P_\varepsilon^j,t) - H_\varphi^{\varepsilon,j}) + g(P_\varepsilon^j,t), \\ H_\varphi^{\varepsilon,j} = 0. \end{cases} \tag{10}$$

Note that $H_\varphi^{\varepsilon,j}(t) = H_\varphi(P_\varepsilon^j,t)$ where H_φ is a solution to the problem (7).

3.2.2 Spatial Component

Then we consider auxiliary functions w_ε^j, $j \in \Upsilon_\varepsilon$, as a solution to boundary value problems:

$$\begin{cases} \Delta_p w_\varepsilon^j = 0, \ x \in T_{\varepsilon/4}^j \setminus \overline{G_\varepsilon^j}, \\ w_\varepsilon^j = 1, \quad x \in \partial G_\varepsilon^j, \\ w_\varepsilon^j = 0, \quad x \in \partial T_{\varepsilon/4}^j. \end{cases} \tag{11}$$

We can find explicit form of a solution to (11):

$$w_\varepsilon^j = \frac{|x - P_\varepsilon^j|^{\frac{p-n}{p-1}} - (\varepsilon/4)^{\frac{p-n}{p-1}}}{a_\varepsilon^{\frac{p-n}{p-1}} - (\varepsilon/4)^{\frac{p-n}{p-1}}}. \tag{12}$$

3.2.3 Test Function

Introduce an auxiliary function

$$W_{\varphi,\varepsilon} = \begin{cases} w_\varepsilon^j(x)(\varphi(P_\varepsilon^j,t) - H_\varphi^{\varepsilon,j}(t)), \ x \in T_{\varepsilon/4}^j \setminus \overline{G_\varepsilon^j}, \ j \in \Upsilon_\varepsilon \ t \in (0,T), \\ 0, \ x \in \Omega \setminus \bigcup\limits_{j\in\Upsilon_\varepsilon} T_{\varepsilon/4}^j, \qquad t \in (0,T). \end{cases} \tag{13}$$

Note that $W_{\varphi,\varepsilon} \in H^1(Q_\varepsilon^T) \bigcap L^p(0,T;W^{1,p}(\Omega_\varepsilon,\partial\Omega))$. Theorem 2 implies that

$$R_\varepsilon W_{\varphi,\varepsilon} \rightharpoonup 0 \text{ in } L^p(0,T;W_0^{1,p}(\Omega)),\ R_\varepsilon W_{\varphi,\varepsilon} \to 0 \text{ in } L^2(0,T;L^p(\Omega)),$$
$$\partial_t(P_\varepsilon W_{\varphi,\varepsilon}) \rightharpoonup 0 \text{ weakly in } L^2(Q^T). \tag{14}$$

3.3 Integral Inequality Transformation

Taking $\zeta = \varphi - W_{\varphi,\varepsilon}$ as a test function in the integral inequality (9), we get

$$\begin{gathered}
\int\limits_{Q_\varepsilon^T} (\partial_t\varphi - \partial_t W_{\varphi,\varepsilon})(\varphi - W_{\varphi,\varepsilon} - u_\varepsilon)dxdt+ \\
\varepsilon^{-k}\sum_{j\in\Upsilon_\varepsilon}\int\limits_0^T\int\limits_{S_\varepsilon^j}(\partial_t(\varphi(x,t)-\varphi(P_\varepsilon^j,t))+\partial_t H_\varphi^{\varepsilon,j})\times \\
(\varphi(x,t)-\varphi(P_\varepsilon^j,t)+H_\varphi^{\varepsilon,j}-u_\varepsilon)dsdt+ \\
\varepsilon^{-k}\sum_{j\in\Upsilon_\varepsilon}\int\limits_0^T\int\limits_{S_\varepsilon^j}\sigma(\varphi(x,t)-\varphi(P_\varepsilon^j,t)+H_\varphi^{\varepsilon,j})\times \\
(\varphi(x,t)-\varphi(P_\varepsilon^j,t)+H_\varphi^{\varepsilon,j}-u_\varepsilon)dsdt+ \\
\int\limits_{Q_\varepsilon^T}|\nabla(\varphi - W_{\varphi,\varepsilon})|^{p-2}\nabla(\varphi - W_{\varphi,\varepsilon})\nabla(\varphi - W_{\varphi,\varepsilon} - u_\varepsilon)dxdt \geq \\
\int\limits_{Q_\varepsilon^T} f(\varphi - W_{\varphi,\varepsilon} - u_\varepsilon)dxdt + \varepsilon^{-k}\int\limits_{S_\varepsilon^T} g(\varphi - W_{\varphi,\varepsilon} - u_\varepsilon)dsdt- \\
\frac{1}{2}\|\varphi(x,0) - W_{\varphi,\varepsilon}|_{t=0}\|^2_{L_2(\Omega_\varepsilon)} - \frac{1}{2}\varepsilon^{-k}\|\varphi(x,0) - W_{\varphi,\varepsilon}|_{t=0}\|^2_{L_2(S_\varepsilon)}.
\end{gathered} \tag{15}$$

Due to the convergences (5) and (14) as $\varepsilon \to 0$, we conclude

$$\begin{gathered}
\int\limits_{Q_\varepsilon^T} (\partial_t\varphi - \partial_t W_{\varphi,\varepsilon})(\varphi - W_{\varphi,\varepsilon} - u_\varepsilon)dxdt \to \int\limits_{Q^T}\partial_t\varphi(\varphi - u_0)dxdt, \\
\int\limits_{Q_\varepsilon^T} f(x,t)(\varphi - W_{\varphi,\varepsilon} - u_\varepsilon)dxdt \to \int\limits_{Q^T} f(x,t)(\varphi - u_0)dxdt \\
\frac{1}{2}\|\varphi(x,0) - W_{\varphi,\varepsilon}|_{t=0}\|^2_{L_2(\Omega_\varepsilon)} \to \frac{1}{2}\|\varphi(x,0)\|^2_{L_2(\Omega)}.
\end{gathered} \tag{16}$$

It is easy to see that

$$\lim_{\varepsilon\to 0}\varepsilon^{-k}\sum_{j\in\Upsilon_\varepsilon}\int_0^T\int_{S_\varepsilon^j}\partial_t(\varphi(x,t)-\varphi(P_\varepsilon^j,t))\times$$
$$(\varphi(x,t)-\varphi(P_\varepsilon^j,t)+H_\varphi^{\varepsilon,j}-u_\varepsilon)dsdt=0,$$
$$\lim_{\varepsilon\to 0}\varepsilon^{-k}\sum_{j\in\Upsilon_\varepsilon}\int_0^T\int_{S_\varepsilon^j}(\sigma(\varphi(x,t)-\varphi(P_\varepsilon^j,t)+H_\varphi^{\varepsilon,j})-\sigma(H_\varphi^{\varepsilon,j}))\times \tag{17}$$
$$(\varphi(x,t)-\varphi(P_\varepsilon^j,t)+H_\varphi^{\varepsilon,j}-u_\varepsilon)dsdt=0.$$

The following lemma has been proven in work [13].

Lemma 1 *Let* $v\in W^{1,\infty}(\Omega)$, $\varphi\in W_0^{1,p}(\Omega)$ *and a sequence of functions* $\eta_\varepsilon\in W_0^{1,p}(\Omega)$ *be such that* $\|\eta_\varepsilon\|_{L_m(\Omega)}\to 0$ *as* $\varepsilon\to 0$ *and* $m\in[1,p)$. *Then*

$$\lim_{\varepsilon\to 0}\int_{\Omega_\varepsilon}\left(|\nabla(v+\eta_\varepsilon)|^{p-2}\nabla(v+\eta_\varepsilon)-|\nabla v|^{p-2}\nabla v\right)\nabla\varphi dx=$$
$$=\lim_{\varepsilon\to 0}\int_{\Omega_\varepsilon}|\nabla\eta_\varepsilon|^{p-2}\nabla\eta_\varepsilon\nabla\varphi dx. \tag{18}$$

The relation (18) *is also valid when* φ *depends on* ε, *but* $\|\varphi\|_{L^p(\Omega)}\le K$, *where* K *is independent of* ε.

Equipped with this lemma we conclude

$$\int_{Q_\varepsilon^T}|\nabla(\varphi-W_{\varphi,\varepsilon})|^{p-2}\nabla(\varphi-W_{\varphi,\varepsilon})\nabla(\varphi-W_{\varphi,\varepsilon}-u_\varepsilon)dxdt=$$
$$\int_{Q_\varepsilon^T}|\nabla\varphi|^{p-2}\nabla\varphi\nabla(\varphi-W_{\varphi,\varepsilon}-u_\varepsilon)dxdt- \tag{19}$$
$$\int_{Q_\varepsilon^T}|\nabla W_{\varphi,\varepsilon}|^{p-2}\nabla W_{\varphi,\varepsilon}\nabla(\varphi-W_{\varphi,\varepsilon}-u_\varepsilon)dxdt+\mathscr{R}_\varepsilon,$$

where $\mathscr{R}_\varepsilon\to 0$ as $\varepsilon\to 0$. Using (5) and (14), we get for the first term in (19) that as $\varepsilon\to 0$

$$\int_{Q_\varepsilon^T}|\nabla\varphi|^{p-2}\nabla\varphi\nabla(\varphi-W_{\varphi,\varepsilon}-u_\varepsilon)dxdt\to\int_{Q^T}|\nabla\varphi|^{p-2}\nabla\varphi\nabla(\varphi-u_0)dxdt. \tag{20}$$

The second term in (19) is transformed with Green's formula in the following manner:

$$\int\limits_{Q_\varepsilon^T} |\nabla W_{\varphi,\varepsilon}|^{p-2}\nabla W_{\varphi,\varepsilon}\nabla(\varphi - W_{\varphi,\varepsilon} - u_\varepsilon)dxdt =$$

$$\sum_{j\in\Upsilon_\varepsilon}\int_0^T\int\limits_{\partial T^j_{\varepsilon/4}} \partial_{\nu_p} w^j_\varepsilon |\varphi(P^j_\varepsilon,t) - H^{\varepsilon,j}_\varphi|^{p-2}(\varphi(P^j_\varepsilon,t) - H^{\varepsilon,j}_\varphi)(\varphi - u_\varepsilon)dsdt+ \tag{21}$$

$$\sum_{j\in\Upsilon_\varepsilon}\int_0^T\int\limits_{S^j_\varepsilon} \partial_{\nu_p} w^j_\varepsilon |\varphi(P^j_\varepsilon,t) - H^{\varepsilon,j}_\varphi|^{p-2}(\varphi(P^j_\varepsilon,t) - H^{\varepsilon,j}_\varphi)\times$$

$$\times(\varphi(x,t) - \varphi(P^j_\varepsilon,t) + H^{\varepsilon,j}_\varphi - u_\varepsilon)dsdt + \kappa_\varepsilon = \mathcal{I}_{1,\varepsilon} + \mathcal{I}_{2,\varepsilon} + \kappa_\varepsilon,$$

where $\kappa_\varepsilon \to 0$ as $\varepsilon \to 0$. Using formula (13) for functions w^j_ε, we obtain

$$\begin{aligned}\partial_\nu w^j_\varepsilon|_{\partial T^j_{\varepsilon/4}} &= -\left(\frac{n-p}{p-1}\right)\frac{C_0^{\frac{n-p}{p-1}}2^{\frac{2n-2}{p-1}}}{1 - a_\varepsilon^{\frac{n-p}{p-1}}\varepsilon^{\frac{p-n}{p-1}}2^{\frac{2n-2p}{p-1}}}\\ \partial_\nu w^j_\varepsilon|_{S^j_\varepsilon} &= \left(\frac{n-p}{p-1}\right)\frac{\varepsilon^{-\frac{n-1}{n-p}}}{C_0(1 - a_\varepsilon^{\frac{n-p}{p-1}}\varepsilon^{\frac{p-n}{p-1}}2^{\frac{2n-2p}{p-1}})}.\end{aligned} \tag{22}$$

We substitute formulas (22) into expressions (21) and get,

$$\mathcal{I}_{1,\varepsilon} = -\left(\frac{n-p}{p-1}\right)^{p-1}\frac{C_0^{n-p}2^{2n-2}}{(1 - a_\varepsilon^{\frac{n-p}{p-1}}\varepsilon^{\frac{p-n}{p-1}}2^{\frac{2n-2p}{p-1}})^{p-1}}\times$$

$$\sum_{j\in\Upsilon}\int_0^T\int\limits_{\partial T^j_{\varepsilon/4}} |\varphi(P^j_\varepsilon,t) - H^{\varepsilon,j}_\varphi|^{p-2}(\varphi(P^j_\varepsilon,t) - H^{\varepsilon,j}_\varphi)(\varphi - u_\varepsilon)dsdt,$$

$$\mathcal{I}_{2,\varepsilon} = \left(\frac{n-p}{p-1}\right)^{p-1}\frac{\varepsilon^{-\frac{n-1}{n-p}(p-1)}C_0^{1-p}}{(1 - a_\varepsilon^{\frac{n-p}{p-1}}\varepsilon^{\frac{p-n}{p-1}}2^{\frac{2n-2p}{p-1}})^{p-1}}\times \tag{23}$$

$$\sum_{j\in\Upsilon_\varepsilon}\int_0^T\int\limits_{S^j_\varepsilon} |\varphi(P^j_\varepsilon,t) - H^{\varepsilon,j}_\varphi|^{p-2}(\varphi(P^j_\varepsilon,t) - H^{\varepsilon,j}_\varphi)\times$$

$$(\varphi(x,t) - \varphi(P^j_\varepsilon,t) + H^{\varepsilon,j}_\varphi - u_\varepsilon)dsdt.$$

3.4 Deduction of the Effective Term

To find the $\lim\limits_{\varepsilon\to 0}\mathcal{I}_{1,\varepsilon}$, we use lemma introduced in [8].

Lemma 2 *Let $h \in H_0^1(\Omega)$, then*

$$\Big|\sum_{j\in\Upsilon_\varepsilon}\int_{\partial T^j_{\varepsilon/4}} h ds - 2^{2-2n}\omega_n\int_\gamma h dx'\Big| \le K\sqrt{\varepsilon}\|h\|_{H^1(\Omega,\partial\Omega)},$$

where ω_n is an area of a unit sphere in $\mathbb{R}^n$.

Hence, we get

$$\lim_{\varepsilon\to 0}\mathscr{I}_{1,\varepsilon} = -\left(\tfrac{n-p}{p-1}\right)^{p-1} C_0^{n-p}\omega_n\int_0^T\int_\gamma |\varphi - H_\varphi|^{p-2}(\varphi - H_\varphi)(\varphi - u_0)dx'dt, \tag{24}$$

where $x' = (0, x_2, \dots, x_n)$.

3.5 Integrals Over The Boundary of Inclusions

Then we have

$$\mathscr{I}_{2,\varepsilon} = Q_\varepsilon + \left(\frac{n-p}{p-1}\right)^{p-1} C_0^{1-p}\varepsilon^{-k}\times \sum_{j\in\Upsilon_\varepsilon}\int_0^T\int_{S^j_\varepsilon}|\varphi(P^j_\varepsilon,t) - H^{\varepsilon,j}_\varphi|^{p-2}(\varphi(P^j_\varepsilon,t) - H^{\varepsilon,j}_\varphi)\times (\varphi(x,t) - \varphi(P^j_\varepsilon,t) + H^{\varepsilon,j}_\varphi - u_\varepsilon)dsdt, \tag{25}$$

where

$$Q_\varepsilon = \left(\frac{n-p}{p-1}\right)^{p-1}\frac{(1-(1-a_\varepsilon^{\frac{n-p}{p-1}}\varepsilon^{\frac{p-n}{p-1}}2^{\frac{2n-2p}{p-1}})^{p-1})C_0^{1-p}\varepsilon^{-k}}{(1-a_\varepsilon^{\frac{n-p}{p-1}}\varepsilon^{\frac{p-n}{p-1}}2^{\frac{2n-2p}{p-1}})^{p-1}}\times \times\sum_{j\in\Upsilon_\varepsilon}\int_0^T\int_{S^j_\varepsilon}|\varphi(P^j_\varepsilon,t) - H^{\varepsilon,j}_\varphi|^{p-2}(\varphi(P^j_\varepsilon,t) - H^{\varepsilon,j}_\varphi)\times (\varphi(x,t) - \varphi(P^j_\varepsilon,t) + H^{\varepsilon,j}_\varphi - u_\varepsilon)dsdt. \tag{26}$$

It is easy to see that $\lim_{\varepsilon\to 0} Q_\varepsilon = 0$.

Since functions $H^{\varepsilon,j}_\varphi$ satisfy the Cauchy problem (10), then expression

$$
\begin{aligned}
\varepsilon^{-k}\sum_{j\in\Upsilon_\varepsilon}\int_0^T\int_{S_\varepsilon^j}(\partial_t H_\varphi^{\varepsilon,j}+\sigma(H_\varphi^{\varepsilon,j})-\\
\mathcal{B}_{n,p}|\varphi(P_\varepsilon^j,t)-H_\varphi^{\varepsilon,j}|^{p-2}(\varphi(P_\varepsilon^j,t)-H_\varphi^{\varepsilon,j})-g)\times\\
(\varphi(x,t)-\varphi(P_\varepsilon^j,t)+H_\varphi^{\varepsilon,j}(t)-u_\varepsilon)dsdt
\end{aligned}
\tag{27}
$$

tends to zero as $\varepsilon\to 0$.

3.6 Homogenized Equation for u_0

Using (16)–(27), we conclude that u_0 satisfies the following integral inequality:

$$
\begin{aligned}
&\int_{Q^T}\partial_t\varphi(\varphi-u_0)dxdt+\int_{Q^T}|\nabla\varphi|^{p-2}\nabla\varphi\nabla(\varphi-u_0)dxdt+\\
&+\mathcal{A}_{n,p}\int_{\gamma^T}|\varphi-H_\varphi|^{p-2}(\varphi-H_\varphi)(\varphi-u_0)dx'dt\geq\\
&\geq\int_{Q^T}f(x,t)(\varphi-u_0)dxdt-\frac{1}{2}\|\psi(x)\eta(0)\|^2_{L_2(\Omega)}.
\end{aligned}
\tag{28}
$$

Taking into account that the linear span of functions

$$
\{\psi(x)\eta(t):\psi\in C_0^\infty(\Omega),\ \eta\in C^1([0,T])\}
$$

is dense in a space $\mathcal{W}=\{u\in L^p(0,T;W_0^{1,p}(\Omega)):\partial_t u\in L^q(0,T;W^{-1,q}(\Omega))\}$, we derive that the inequality (28) is valid for an arbitrary function $\varphi\in\mathcal{W}$. Then we take $\varphi=u_0\pm\lambda w$, where $\lambda\geq 0$, $w\in\mathcal{W}$ as a test function in the inequality the (28) and pass to the limit as $\lambda\to 0$. Hence, we conclude that u_0 satisfies an integral identity

$$
\begin{aligned}
&\int_0^T\langle\partial_t u_0,w\rangle_\Omega dt+\int_{Q^T}|\nabla u_0|^{p-2}\nabla u_0\nabla wdxdt+\\
&+\mathcal{A}_{n,p}\int_{\gamma^T}|u_0-H_{u_0}|^{p-2}(u_0-H_{u_0})wdx'dt=\int_{Q^T}f(x,t)wdxdt.
\end{aligned}
\tag{29}
$$

Therefore, u_0 is a weak solution to (6).

4 Existence and Uniqueness of a Solution to Homogenized Problem

4.1 Uniqueness

Let there be two different solutions (u_1, H_{u_1}) and (u_2, H_{u_2}) to problem (6). Taking $u_1 - u_2$ as a test function for the first equation and subtracting the one from the other, we get by integration on $(0, t)$

$$\|(u_1 - u_2)(x,t)\|^2_{L^2(\Omega)} + \int_0^t \|\nabla(u_1 - u_2)\|^p_{L^p(\Omega)} d\tau + \\ + \mathscr{A}_{n,p} \int_{\gamma^t} (|u_1 - H_{u_1}|^{p-2}(u_1 - H_{u_1}) - \\ |u_2 - H_{u_2}|^{p-2}(u_2 - H_{u_2}))(u_1 - u_2) dx' d\tau \le 0. \tag{30}$$

Then we take $H_{u_1} - H_{u_2}$ as a test function for the second equation and obtain

$$\|(H_{u_1} - H_{u_2})(x,t)\|^2_{L^2(\gamma)} + \int_0^t \|H_{u_1} - H_{u_2}\|^p_{L^p(\gamma)} d\tau - \\ - \mathscr{B}_{n,p} \int_{\gamma^t} (|u_1 - H_{u_1}|^{p-2}(u_1 - H_{u_1}) - \\ |u_2 - H_{u_2}|^{p-2}(u_2 - H_{u_2}))(H_{u_1} - H_{u_2}) dx' d\tau \le 0. \tag{31}$$

Summing the inequality (30) with the inequality (31) multiplied by $\mathscr{C}_{n,p} = \frac{\mathscr{A}_{n,p}}{\mathscr{B}_{n,p}}$, we derive that

$$\|(u_1 - u_2)(x,t)\|^2_{L^2(\Omega)} + \int_0^t \|\nabla(u_1 - u_2)\|^p_{L^p(\Omega)} d\tau + \\ + \mathscr{C}_{n,p} \|(H_{u_1} - H_{u_2})(x,t)\|^2_{L^2(\gamma)} + \mathscr{C}_{n,p} k_1 \int_0^t \|H_{u_1} - H_{u_2}\|^p_{L^p(\gamma)} d\tau \le 0. \tag{32}$$

Hence, we conclude that $u_1 = u_2$ and $H_{u_1} = H_{u_2}$.

4.2 Existence

We use the Galerkin method by searching for a solution u_m in a linear span of "basis" in $W_0^{1,p}(\Omega)$ functions $\{\psi_i\}_{i=1}^m$ that is orthonormal in the space $L^2(\Omega)$, i.e. we define $u_m(t) = \sum\limits_{i=1}^{m} c_{i,m}(t)\psi_i$. Let H_m be a solution to (7) with $\varphi = u_m$. Then, u_m must satisfy the following equation:

$$(\partial_t u_m, \psi_j)_{L^2(\Omega)} + (|\nabla u_m|^{p-2}\nabla u_m, \nabla\psi_j)_{L^2(\Omega)}+$$
$$+\mathscr{A}_{n,p}(|u_m - H_m|^{p-2}(u_m - H_m), \psi_j)_{L^2(\gamma)} = (f, \psi_j)_{L^2(\Omega)},$$

with zero initial conditions for $j = 1, \dots, m$. The acquired system is a Cauchy problem on $c_{i,m}(t)$. It's well known that such a system has a unique absolutely continuous solution defined on $[0, T_m]$ for some $T_m > 0$. Let us get the priory estimates. Multiplying the above equations by $c_{j,m}$ and summing over all $j = 1, \dots, m$ and integrating over $(0, t)$, we get

$$\frac{1}{2}\|u_m(x,t)\|^2_{L^2(\Omega)} + \int\limits_0^t \|\nabla u_m\|^p_{L^p(\Omega)} d\tau+$$
$$+\mathscr{A}_{n,p}\int\limits_0^t\int\limits_\gamma |u_m - H_m|^{p-2}(u_m - H_m)u_m dx'd\tau \le K_1(\|f\|^2_{L^2(Q^T)} + \int\limits_0^t \|u_m\|^2_{L^2(\Omega)} d\tau).$$

Next, we multiple the equation of system (7), with $\varphi = u_m$, by H_m and integrate over γ and $(0, t)$. As a result, we obtain

$$\frac{1}{2}\|H_m(x,t)\|^2_{L^2(\gamma)} + k_1\int\limits_0^t \|H_m\|^p_{L^p(\gamma)} \le$$
$$\frac{1}{2}\|H_m(x,t)\|^2_{L^2(\gamma)} + \int\limits_0^t\int\limits_\gamma \sigma(H_m)H_m dx'd\tau \le K_2(\max_{(x,t)\in Q^T}|g|^2 + \int\limits_0^t \|H_m\|^2_{L^2(\gamma)} d\tau)+$$
$$+\mathscr{B}_{n,p}\int\limits_0^t\int\limits_\gamma |u_m - H_m|^{p-2}(u_m - H_m)H_m dx'd\tau.$$

Multiplying this estimation by $\mathscr{C}_{n,p}$ and summing it with the previous estimation, we get

$$\|u_m(x,t)\|^2_{L^2(\Omega)} + \mathscr{C}_{n,p}\|H_m(x,t)\|^2_{L^2(\gamma)}+$$
$$+2\int\limits_0^t \|\nabla u_m\|^p_{L^p} d\tau + 2\mathscr{A}_{n,p}\int\limits_0^t \|u_m - H_m\|^p_{L^p(\gamma)} d\tau + K_3\int\limits_0^t \|H_m\|^p_{L^p(\gamma)} \le$$
$$\le K_4(\|f\|^2_{L^2(Q^T)} + \max_{(x,t)\in Q^T}|g|^2) + K_5(\int\limits_0^t \|u_m\|^2_{L^2(\Omega)} d\tau + \mathscr{C}_{n,p}\int\limits_0^t \|H_m\|^2_{L^2(\gamma)} d\tau).$$

Using Gronwall's inequality, we conclude

$$\|u_m(x,t)\|^2_{L^2(\Omega)} \le C, \quad \|H_m(x,t)\|^2_{L^2(\gamma)} \le C,$$

for all $t \in (0, T)$. Hence, solution $(u_m,\ H_m)$ is defined on $[0, T]$ and the following estimations are valid:

$$\|\nabla u_m\|^p_{L^p(0,T;L^p(\Omega))} + \|H_m\|^p_{L^p(0,T;L^p(\gamma))} + \|u_m - H_m\|^p_{L^p(0,T;L^p(\gamma))} \le C.$$

Analogously, multiplying equations by $c'_{i,m}$ and summing over all $j = 1, \dots, m$ and integrating over $(0, t)$, we obtain

$$\frac{1}{2}\int_0^t \|\partial_t u_m\|^2_{L^2(\Omega)} d\tau + \frac{1}{p}\|\nabla u_m(x,t)\|^p_{L^p(\Omega)} +$$

$$\mathscr{A}_{n,p}\int_0^t\int_\gamma |u_m - H_m|^{p-2}(u_m - H_m)\partial_t u_m dx' d\tau \le$$

$$K_6\|f\|^2_{L^2(Q^T)} + \frac{1}{4}\int_0^t \|\partial_t u_m\|^2_{L^2(\Omega)} d\tau.$$

Multiplying the equation for function H_m by $\partial_t H_m$ and integrating over γ and $(0, t)$, we get

$$\frac{1}{2}\int_0^t \|\partial_t H_m\|^2_{L^2(\gamma)} d\tau + K_7\|H_m(x,t)\|^p_{L^p(\gamma)} \le$$

$$\mathscr{B}_{n,p}\int_0^t\int_\gamma |u_m - H_m|^{p-2}(u_m - H_m)\partial_t H_m dx' d\tau +$$

$$+K_8 \max_{(x,t)\in Q^T} |g|^2 + \frac{1}{4}\int_0^t \|\partial_t H_m\|^2_{L^2(\gamma)} d\tau.$$

In a similar way as for proving previous estimations, we multiply the last estimation by $C_{n,p}$ and sum it with the former estimation. Thus, we obtain

$$\int_0^t \|\partial_t u_m\|^2_{L^2(\Omega)} d\tau + \tfrac{2}{p}\|\nabla u_m(x,t)\|^p_{L^p(\Omega)} +$$

$$\mathscr{C}_{n,p}\int_0^t \|\partial_t H_m\|^2_{L^2(\gamma)} d\tau + K_8\|H_m(x,t)\|^p_{L^p(\gamma)} + K_9\|(u_m - H_m)(x,t)\|^p_{L^p(\gamma)} \le$$

$$K_{10}(\|f\|^2_{L^2(Q^T)} + \max_{(x,t)\in Q^T} |g|^2) + \tfrac{1}{2}\int_0^t \|\partial_t u_m\|^2_{L^2(\Omega)} d\tau + \tfrac{1}{2}\int_0^t \|\partial_t H_m\|^2_{L^2(\gamma)} d\tau.$$

Therefore, we obtain

$$\|\partial_t u_m\|^2_{L^2(Q^T)} + \|\partial_t H_m\|^2_{L^2(\gamma^T)} + \max_{t\in[0,T]} \|\nabla u_m(x,t)\|^p_{L^p(\Omega)}+$$
$$+ \max_{t\in[0,T]} \|H_m(x,t)\|^p_{L^p(\gamma)} + \max_{t\in[0,T]} \|(u_m - H_m)\|^p_{L^p(\gamma)} \leq C.$$

Hence, there exists a subsequence (u_μ, H_μ) such that

$$u_\mu \rightharpoonup^* u \text{ weakly star in } L^\infty(0,T; W^{1,p}(\Omega)),$$
$$u_\mu \rightharpoonup u \text{ weakly in } L^p(0,T; W^{1,p}(\Omega)),$$
$$u_\mu \to u,\ H_\mu \to H_u \text{ strongly in } L^p(\gamma^T),$$
$$\partial_t u_\mu \to \partial_t u \text{ strongly in } L^2(Q^T),\ \partial_t H_\mu \to \partial_t H_u \text{ strongly in } L^2(\gamma^T).$$

Also, due to the monotonicity method (see, for example, [16]), we have

$$|\nabla u_\mu|^{p-2}\nabla u_\mu \rightharpoonup^* |\nabla u|^{p-2}\nabla u \text{ weakly-* in } L^q(0,T; W^{-1,q}(\Omega)).$$

Then we have estimation

$$\||u_\mu - H_\mu|^{p-2}(u_\mu - H_\mu)\|_{L^\infty(0,T;L^q(\gamma))} \leq C$$

which implies that $|u_\mu - H_\mu|^{p-2}(u_\mu - H_\mu) \rightharpoonup^* \omega$ weakly star in $L^\infty(0,T; L^q(\gamma))$. Using Lemma 1.3 from [16], we conclude $w = |u - H_u|^{p-2}(u - H_u)$. Analogously, one can show that $\sigma(H_\mu) \to \sigma(H_u)$ weakly in $L^q(\gamma^T)$. Therefore, the pair of functions (u, H_u) is a solution to system (6).

References

1. Bandle, C., von Below, J., Reichel, W.: Parabolic problems with dynamical boundary conditions: eigenvalue expansions and blow up, Atti della Accademia Nazionale dei Lincei, Classe di Scienze Fisiche, Matematiche e Naturali, Rendiconti Lincei Matematica E Applicazioni, pp. 35–67 (2006)
2. Bejenaru, I., Diaz, J.I., Vrabie, I.I.: An abstract approximate controllability result and applications to elliptic and parabolic systems with dynamical boundary conditions. Electron. J. Differ. Equ. **50**, 1–19 (2001)
3. Arrieta, J.M., Quittner, P., Rodriguez-Bernal, A.: Parabolic problems with nonlinear dynamical boundary conditions and singular initial data. Differ. Int. Equ. **14**(12), 1487–1510 (2011)
4. Escher, J.: Quasilinear parabolic systems with dynamical boundary conditions. Commun. Part. Differ. Equ. **18**, 1309–1364 (1993)
5. Angulano, M.: Existence, uniqueness and homogenization of nonlinear parabolic problems with dynamical boundary conditions in perforated media (2017). arXiv: 1712.01183
6. Timofte, C.: Parabolic problems with dynamical boundary conditions in perforated media. Math. Model. Anal. **8**, 337–350 (2003)
7. Wang, W., Duan, J.: Homogenized dynamics of stochastic partial differential equations with dynamical boundary conditions. Commun. Math. Phys. **275**(1), 163–186 (2007)

8. Zubova, M.N., Shaposhnikova, T.A.: Homgenization limit for the diffusion equation in a domain, perforated along (n–1)-dimensional manifold with dynamic conditions on the boundary of the perforations: critical case. Dokl. Math. **99**(3), 39–45 (2019)
9. Diaz, J.I., Gomez-Castro, D., Shaposhnikova, T.A., Zubova, M.N.: A nonlocal memory strange term arising in the critical scale homogenisation of a diffusion equation with a dynamic boundary condition. Electron. J. Differ. Equ. **77**, 1–13 (2019)
10. Zubova, M.N., Shaposhnikova, T.A.: Homogenization of boundary value problems in perforated domains with third boundary conditions and the resulting change in the character of the nonlinearity in the problem. Differ. Equ. **47**(1), 78–90 (2011)
11. Zubova, M.N., Shaposhnikova, T.A.: Homogenization of a boundary-value problem in a domain perforated by cavities of arbitrary shape with a general nonlinear boundary conditions on their boundaries:the case of critical values of the parameters. J. Math. Sci. Plenum Publishers (United States) **244**(2), pp. 235–253 (2020)
12. Gomez, D., Perez, M.E., Podolskii, A.V., Shaposhnikova, T.A.: Homogenization for the p-Laplace operator and nonlinear Robin boundary conditions in perforated media along $(n-1)$-dimensional manifolds. Dokl. Math. **89**(1), 11–15 (2014)
13. Gomez, D., Perez, M.E., Podolskii, A.V., Shaposhnikova, T.A.: Homogenization of variational inequalities for the p-Laplace operator in perforated media along manifolds. Appl. Math. Optim. **475**, 1–19 (2017)
14. Diaz, J.I., Gomez-Castro, D., Shaposhnikova, T.A., Zubova, M.N.: Change of homogenized absorption term in diffusion processes with reaction on the boundary of periodically distributed asymmetric particles of critical size. Electron. J. Differ. Equ. **2017**(178), 1–25 (2017)
15. Diaz, J.I., Gomez-Castro, D., Podolskiy, A.V., Shaoshnikova, T.A.: Homogenization of a net of periodic critically scaled boundary obstacles related to reverse osmosis "nano-composite" membranes. Adv. Nonlinear Anal. ISSN (Online) 2191-950X, ISSN (Print) 2191-9496. https://doi.org/10.1515/anona-2018-0158
16. Lions, J.-L.: Quelques methodes de resolution des problemes aux limites non lineaires (Dunod, Paris, 1969). Editorial URSS, Moscow (2010)
17. Gal, C.G.: Well posedness and the global attractor of some quasi-linear parabolic equations with nonlinear dynamic boundary conditions. Differ. Integr. Equ. **23**, 327–358 (2010)
18. Podol'skii, A.V.: Solution continuation and homogenization of a boundary value problem for the p-Laplacian in a perforated domain with a nonlinear third boundary condition on the boundary of holes. Dokl. Math. **91**(1), 30–34 (2015)

Poisson Problem for a Functional–Differential Equation. Positivity of a Quadratic Functional. Jacobi Condition

Sergey Labovskiy and Manuel Alves

Abstract For the Poisson problem

$$\mathcal{L}u := -\Delta u + p(x)u - \int_{\Omega} u(s)\, r(x, \mathrm{d}s) = \rho f, \quad u\big|_{\Gamma(\Omega)} = 0$$

equivalence of positivity of the quadratic functional

$$\int_{\Omega} u'_x u'_x \,\mathrm{d}x + \int_{\Omega} p(x)u(x)^2 \,\mathrm{d}x - \int_{\Omega\times\Omega} u(x)u(s)\,\xi(\mathrm{d}x \times \mathrm{d}s),$$

($\mathrm{d}x := \mathrm{d}x_1 \cdots \mathrm{d}x_n$), the corresponding Jacobi condition, and positivity of the Green function are showed.

Keywords Positive solutions · Quadratic functional · Jacobi condition

1 The Poisson Problem

In book [3], the theory of functional differential equations is developed, oriented to the generality of problems from the point of view of functional analysis. This book is devoted to generalizations of ordinary differential equations. Boundary value problems for functional differential equations generalizing partial differential equations have been poorly studied.

S. Labovskiy (✉)
Plekhanov Russian University of Economics, Moscow, Russia
e-mail: labovski@gmail.com

M. Alves
Eduardo Mondlane University, Maputo, Mozambique
e-mail: mjalves.moz@gmail.com

A. Domoshnitsky et al. (eds.), *Functional Differential Equations and Applications*, Springer Proceedings in Mathematics & Statistics 379,
https://doi.org/10.1007/978-981-16-6297-3_18

Let Ω be a bounded connected open set in $\mathbb{R}^n$, $\Gamma(\Omega)$ be the boundary of the Ω, and $X = \overline{\Omega}$ be the closure of Ω. For a function $u = u(x)$, $x \in \Omega$, $\Delta u := u''_{x_1 x_1} + \cdots + u''_{x_n x_n}$,[1] where $x := (x_1, \ldots, x_n)$. In the Poisson problem

$$-\Delta u + p(x)u - \int\limits_{\Omega} u(s)\, r(x, \mathrm{d}s) = \rho(x) f(x), \tag{1}$$

$$u\big|_{\Gamma(\Omega)} = 0 \tag{2}$$

for almost all $x \in \Omega$, the function $r(x, \cdot)$ is assumed to be a measure. The function ρ is a positive weight.

In [6, 7], the Fredholm property of the boundary value problem (1), (2) is showed under assumptions below. In the case of unique solvability, its solution can be represented by means of Green's function

$$u(x) = \int_{\Omega} G(x, s) f(s) \rho(x)\, \mathrm{d}s.$$

Green's function is symmetric: $G(x, s) = G(s, x)$, $x, s \in \Omega$.

This article establishes an analogue of the Jacobi condition for the positive definiteness of the quadratic functional corresponding to Problem (1), (2). In [5], an analogue of the Jacobi condition for a second-order functional differential equation is considered. Note that this condition is an analogue of non-oscillation for an ordinary differential equation of the second order. In this regard, it is worth noting the book [2] devoted to non-oscillation.

2 Assumptions and Notation

For a real function $u = u(x)$ defined on Ω and having derivative of the first order, $u'_x := \left(u'_{x_1}, \ldots, u'_{x_n}\right)$, where $x = (x_1, \ldots, x_n)$. For two such functions $u(x)$ and $v(x)$, $u'_x v'_x := u'_{x_1} v'_{x_1} + \cdots + u'_{x_n} v'_{x_n}$.

Let $\mathcal{M}$ be the set of all Lebesgue measurable subsets in the closure $X = \overline{\Omega}$. Assume that the function $r_0 \colon X \times \mathcal{M} \to R$ satisfy two conditions: for almost all $x \in X$, the function $r_0(x, \cdot)$ is a measure on $\mathcal{M}$, and for any $e \in \mathcal{M}$, $r_0(\cdot, e)$ is measurable on X. Assume that $p(x) = r_0(x, \Omega)$. For any set $E \subset X \times X$, $E_x := \{y \colon (x, y) \in E\}$. The set functiom ξ_0 defined by

$$\xi_0(E) = \int_X r_0(x, E_x)\, \mathrm{d}x \tag{3}$$

[1] the sign := means 'is equal by definition'.

is a measure. Assume that ξ_0 is symmetric, that is,

$$\xi_0(e_1 \times e_2) = \xi_0(e_2 \times e_1), \ \forall e_1, e_2 \in \mathcal{M}.$$

The measure η has the same properties and is defined by

$$\eta(E) = \int_X q(x, E_x)\,\mathrm{d}x, \tag{4}$$

where q has properties analogous to r_0. The measures ξ and $r(x, \cdot)$ are defined by

$$\xi := \xi_0 + \eta, \ r := r_0 + q. \tag{5}$$

Assume that

- $\rho(x)$, $x \in X$ is a positive measurable function and $\mu(E) := \int_E \rho(x)\,\mathrm{d}x$, and $\int_\Omega \rho(x)\,\mathrm{d}x < \infty$.
- $[u, v]$, $\langle u, v\rangle$, (f, g), and $Q(u, v)$ are bilinear forms defined by

$$[u, v] := \int\limits_\Omega u'_x v'_x\,\mathrm{d}x + \int\limits_\Omega p(x)uv\,\mathrm{d}x - \int\limits_{\Omega\times\Omega} v(x)u(s)\,\xi_0(\mathrm{d}x \times \mathrm{d}s),$$

$$\langle u, v\rangle := [u, v] - Q(u, v),$$

$$(f, g) := \int\limits_\Omega f(x)g(x)\rho(x)\,\mathrm{d}x,$$

$$Q(u, v) := \int\limits_{\Omega\times\Omega} v(x)u(s)\eta(\mathrm{d}x \times \mathrm{d}s).$$

- $L_2(\Omega, \mu)$ be the space of all μ-measurable functions on Ω with finite integral $\int\limits_\Omega f(x)^2\rho(x)\,\mathrm{d}x$ and scalar product (f, g).
- Let $q(x) := q(x, \Omega)$. Assume that

$$\frac{q}{\rho} \in L_2(\Omega, \mu). \tag{6}$$

- We use the Sobolev spaces $W_0^{1,2}(\Omega)$ and $W_0^{2,2}(\Omega)$ [1].
- Let W be the vector subspace of all elements from $W_0^{1,2}(\Omega)$ with finite value $[u, u] < \infty$. The bilinear form $[u, v]$ serves as inner product in the Hilbert space W.
- Define the operator $T\colon W \to L_2(\Omega, \mu)$ by the equality $Tu(x) = u(x)$, $x \in \Omega$. The operator T is continuous.
- T^* is the adjoint to T operator.

As we will see below, the following expressions representing two linear operators correspond to the forms $[u, v]$ and $\langle u, v\rangle$:

$$\mathcal{L}_0 u(x) := \frac{1}{\rho}\left(-\Delta u + p(x)u - \int_{\Omega} u(s) r_0(x, \mathrm{d}s)\right), \tag{7}$$

$$\mathcal{L} u(x) := \frac{1}{\rho}\left(-\Delta u + p(x)u - \int_{\Omega} u(s) r(x, \mathrm{d}s)\right). \tag{8}$$

3 Variational Method

We give a brief outline of the form of variational method from [5, 7].

3.1 The Scheme

The equation with relation to u in variational form

$$\int\limits_{\Omega} u'_x v'_x \,\mathrm{d}x + \int\limits_{\Omega} p(x)u(x)v(x)\,\mathrm{d}x - \int\limits_{\Omega\times\Omega} v(x)u(s)\,\xi_0(\mathrm{d}x \times \mathrm{d}s)$$
$$= \int_{\Omega} f(x)v(x)\rho(x)\,\mathrm{d}x, \quad \forall v \in W,$$

can be written in the short form

$$[u, v] = (f, Tv), \quad (\forall v \in W). \tag{9}$$

Equation (9) has the unique solution $u = T^* f$ for any $f \in L_2(\Omega, \mu)$. The image $T(W)$ of the operator T is dense in $L_2(\Omega, \mu)$ and $\dim\ker T = 0$. Thus, (9) is equivalent to $\mathcal{L}_0 u = f$ where $\mathcal{L}_0 = (T^*)^{-1}$.

Theorem 1 *If the operator T is compact, then the spectrum of $\mathcal{L}_0$ is discrete and positive:* $0 < \lambda_0 \le \lambda_1 \le \ldots, \lambda_n \to \infty$.

Remark 1 The spectrum of $\mathcal{L}_0$ is the spectrum of the problem $\mathcal{L}_0 u = \lambda T u$.

3.2 Euler Equation

Lemma 1 $W_0^{2,2}(\Omega) \subset D(\mathcal{L}_0)$*, operator* $\mathcal{L}_0$ *has representation* (7) *in* $W_0^{2,2}(\Omega)$.

The equation

$$\int_{\Omega} u'_x v'_x \, \mathrm{d}x = -\int_{\Omega} \Delta u \cdot v \, \mathrm{d}x$$

can be used as *definition of operator* Δ on the space $D(\mathcal{L}_0)$ in a weak sense. Using this definition of operator Δ, operator $\mathcal{L}_0$ can be represented as (7).

3.3 Eigenvalue Problem and Spectrum

Theorem 2 *Let* Ω *satisfy the* cone condition *[1, Paragraph 4.6]. The spectral problem*

$$-\Delta u + pu - \int_{\Omega} u(s)\, r_0(x, \mathrm{d}s) = \lambda \rho u, \quad u\big|_{\Gamma(\Omega)} = 0 \tag{10}$$

has in W a system of nontrivial solutions $u_n(x)$ *corresponding to positive eigenvalues* λ_n $(\lambda_0 \le \lambda_1 \le \ldots)$*. This system forms an orthogonal basis in the space* $W_0^{1,2}(\Omega)$.

Note that the minimal eigenvalue λ_0 of the problem (10) can be estimated by the relation

$$\lambda_0 = \inf_{[u,u] \le 1} \frac{[u, u]}{(Tu, Tu)}.$$

3.4 Positivity of Solutions

Theorem 3 *Suppose* $f \geq\not\equiv 0$ *and* $u(x)$ *is the solution of the problem*

$$-\Delta u + pu - \int_{\Omega} u(s)\, r_0(x, \mathrm{d}s) = \rho f, \tag{11}$$

$$u\big|_{\Gamma(\Omega)} = 0, \tag{12}$$

then $u(x) > 0$ *in* Ω.

Corollary 1 *The minimal eigenvalue* λ_0 *of the problem* (10) *is positive. It is associated with a positive in* Ω *eigenfunction.*

This follows from Theorem 3 and the following theorem.

Theorem 4 (M. Krein, M. Rutman [4]) *If spectrum of A contains points different from zero, then its spectral radius r is eigenvalue of both the A and its adjoint A^*, this eigenvalue is associated with an eigenvector $v_0 \in K$: $Av_0 = rv_0$.*

4 General Case

Here, we consider the form $\langle u, v\rangle$ and the operator (8). The equation in variational form

$$\langle u, v\rangle = (f, Tv), \quad \forall v \in W, \tag{13}$$

is equivalent to the boundary value problem

$$\mathcal{L}u := \mathcal{L}_0 u - Qu = f, \ u\big|_{\Gamma(\Omega)} = 0, \tag{14}$$

where the operator $Q\colon W \to L_2(\Omega, \mu)$ has the representation

$$Qu(x) = (1/\rho)\int_\Omega u(s)q(x, ds). \tag{15}$$

Under condition (6), this operator acts from W to $L_2(\Omega, \mu)$ and is continuous.

Theorem 5 *Let Ω satisfy the* cone condition *[1, Paragraph 4.6]. The eigenvalue problem*

$$-\Delta u + pu - \int_\Omega u(s)\, r(x, ds) = \lambda\rho u, \quad u\big|_{\Gamma(\Omega)} = 0 \tag{16}$$

has in W (see below) a system of nontrivial solutions $u_n(x)$ corresponding to eigenvalues $\lambda_0 \le \lambda_1 \le \dots$ This system forms an orthogonal basis in the space $L_2(\Omega, \mu)$ (μ is defined below).

Theorem 6 *The following affirmations are equivalent:*

1. *the quadratic functional $\langle u, u\rangle$ defined by $\langle u, v\rangle$ is positive definite,*
2. *the problem (14) is uniquely resolvable, and its Green function is positive in $\Omega \times \Omega$,*
3. *the inequality $-\Delta v + pv - \int_\Omega v(s)\, r(x, ds) \ge\not\equiv 0$ has positive in Ω solution,*
4. *the minimal eigenvalue of the problem (16) is positive, and*
5. *spectral radius r of the operator QT^* is less than unit.*

The minimum eigenvalue λ_0 satisfies the relation

$$\lambda_0 = \inf_{[u,u]\le 1} \frac{\langle u, u\rangle}{(Tu, Tu)}. \tag{17}$$

5 An Analogue of the Jacobi Condition for the Positivity of the Quadratic Functional

In the one-dimensional case, the Jacobi condition is a criterion for the positive definiteness of a quadratic functional. Jacobi's condition is the absence of two zeros of the corresponding homogeneous Euler equation. In the case of R^n, such a condition is the absence of a nonzero solution of the homogeneous Poisson problem that vanishes at the boundary of some subdomain.

Let Ω' be an open connected subset of Ω. For the functional differential equation (8), to formulate the corresponding condition, it is necessary to consider the so-called truncated problem

$$-\Delta u + p(x)u - \int_{\Omega'} u(s)\, r(x, \mathrm{d}s) = 0, x \in \Omega', \tag{18}$$

$$u\big|_{\Gamma(\Omega')} = 0. \tag{19}$$

All the constructions made above are also valid for the set Ω'. For this set, Theorem 6 holds. Let W' be the corresponding working space (instead of W), and λ_0' be the corresponding minimal eigenvalue of the problem

$$-\Delta u + p(x)u - \int_{\Omega'} u(s)\, r(x, \mathrm{d}s) = \lambda u, x \in \Omega', \tag{20}$$

$$u\big|_{\Gamma(\Omega')} = 0. \tag{21}$$

In this case, we have the following analogue of the Jacobi criterion.

Theorem 7 *The quadratic functional $\langle u, u\rangle$ is positive definite iff for any subdomain Ω', the problem* (18) *has no nonzero solutions.*

Proof According to Theorem 6, the positivity of a quadratic functional is equivalent to the positivity of the first eigenvalue λ_0. The space W' corresponding to the set Ω' can be considered as a subspace of W. Elements of W' are elements of W that are equal to zero outside of Ω'.

Therefore, by virtue of (17), $\lambda_0' \geq \lambda_0$. Thus, the condition $\lambda_0' > 0$ is necessary for positiveness of the quadratic functional $\langle u, u\rangle$.

Sufficiency. Suppose $\lambda_0 \leq 0$. For an region of small dimension, $\lambda_0' > 0$. Using the continuous dependence (Lemma 2) of the λ_0' on the region Ω', we can find the region Ω' for which $\lambda_0' = 0$, which contradicts the unique solvability of the homogeneous problem.

Lemma 2 *The minimum eigenvalue λ_0' of the problem* (20) *continuously depends on region Ω'.*

Continuity is understood as follows. If $\Omega' \subset \Omega''$, then the distance from Ω' to Ω'' is the supremum of the distances of points of Ω'' to the set Ω'.

The proof of the lemma does not contain significant difficulties, but is cumbersome. Therefore, we omit it.

References

1. Adams, R.A., Fournier, J.: Sobolev Spaces. Elsevier (2003)
2. Agarwal, R.P., Berezansky, L., Braverman, E., Domoshnitsky, A.: Nonoscillation Theory of Functional Differential Equations with Applications. Springer, Berlin (2012). Zbl 1253.34002
3. Azbelev, N.V., Maksimov, V.P., Rakhmatullina, L.F.: Introduction to the theory of functional differential equations. Methods and applications. New York, NY: Hindawi Publishing Corporation (2007). Zbl 1202.34002
4. Kreĭn, M., Rutman, M.: Linear operators leaving invariant a cone in a Banach space. Usp. Mat. Nauk **3**(1(23)), 3–95 (1948). Zbl 0030.12902
5. Labovskiy, S.: On spectral problem and positive solutions of a linear singular functional-differential equation. Funct. Differ. Equ. **20**(3–4), 179–200 (2013). Zbl 1318.34088
6. Labovskiy, S.: On positivity of the Green function for Poisson problem for a linear functional differential equation. Vestnik Tambovskogo universiteta. Seriya Estestvennye i tekhnicheskie nauki—Tambov University Reports. Ser.: Nat. Tech. Sci. **22**(6), 1229–1234 (2017). https://doi.org/10.20310/1810-0198-2017-22-6-1229-1234
7. Labovskiy, S., Alves, M.: On positive solutions for Poisson problem for a functional differential equation. Funct. Differ. Equ. **26**(1–2), 61–74 (2019). https://doi.org/10.26351/FDE/26/1-2/4

On Asymptotic Behavior of the First Derivatives of Bounded Solutions to Second-Order Differential Equations with General Power-Law Nonlinearity

Tatiana Korchemkina

Abstract We consider second-order differential equations with general power-law nonlinearity. All nontrivial solutions to these equations are monotonous and can be bounded or unbounded depending on the nonlinearity exponents. We study the asymptotic behavior of the first derivatives of bounded solutions.

Keywords Power-law nonlinearity · Asymptotic behavior · Second order · Asymptotic of first derivatives · Bounded solutions

1 Introduction

Consider the second-order nonlinear differential equation with general power-law nonlinearity

$$y'' = p\left(x, y, y'\right) |y|^{k_0} |y'|^{k_1} \operatorname{sgn}(yy'), \tag{1}$$

with $k_0, k_1 > 0$ and continuous in x and Lipschitz continuous in u, v function $p(x, u, v)$ satisfying the inequalities $0 < m \leq p(x, u, v) \leq M < \infty$.

For $p = p(x)$, Evtukhov [1] obtained sufficient conditions for the existence of solutions with prescribed asymptotic behavior. The results on the qualitative properties of solutions to Eq. (1) depending on the values of k_0, k_1 can be found in [8]. Using the methods represented by Astashova in [2–4], asymptotic behavior of unbounded solutions and their first derivatives was obtained in the case $p = p\left(x, y, y'\right)$ in [9]. For bounded solutions, however, asymptotic behavior of first derivatives was an open question. In this paper, we give the answer to this question.

T. Korchemkina (✉)
Lomonosov Moscow State University, Moscow, Russian Federation
e-mail: krtaalex@gmail.com

A. Domoshnitsky et al. (eds.), *Functional Differential Equations and Applications*, Springer Proceedings in Mathematics & Statistics 379,
https://doi.org/10.1007/978-981-16-6297-3_19

2 Preliminary Results

Because of the fact that for $k_1 < 1$ any constant positive (negative) solution can be extended to the right (to the left) not in a unique way, μ-solutions to the equation are considered as in [3, 5].

Definition 1 [5] A solution $y\colon (a,b) \to \mathbb{R}$, $-\infty \le a < b \le +\infty$, to an ordinary differential equation is called a *μ-solution* if

(1) the equation has no other solutions equal to y on some subinterval $(a,\ b)$ and not equal to y at some point in $(a,\ b)$;

(2) the equation either has no solution equal to y on $(a,\ b)$ and is defined on another interval containing $(a,\ b)$ or has at least two such solutions which differ from each other at points arbitrary close to the boundary of the interval $(a,\ b)$.

Theorem 1 *[8] All nontrivial μ-solutions to Eq. (1) are strictly monotonous.*

In [8], it is proved that in the case $k_1 > 2$, all bounded μ-solutions to Eq. (1) are increasing, and in the case $0 < k_1 < 2$, all bounded μ-solutions to Eq. (1) are decreasing. In particular, for $k_1 > 2$, increasing solutions are *black hole* solutions.

Definition 2 [6] A solution satisfying at some finite point x^* the conditions $\lim\limits_{x\to x^*} |y'(x)| = \infty$ and $\lim\limits_{x\to x^*} |y(x)| < \infty$ is called a *black hole* solution.

It is shown in [8] that for $0 < k_1 < 1$, all decreasing μ-solutions are so-called *white hole* solutions near domain's left and right boundaries.

Definition 3 [7] A μ-solution satisfying at a finite point (its domain's boundary) x_+ the conditions $\lim\limits_{x\to x_+} y'(x) = 0$ and $\lim\limits_{x\to x_+} y(x) \ne 0$ is called a *white hole* solution.

Finally, it is proved in [8] that in the case $1 \le k_1 < 2$, every decreasing solution is defined on the whole axis and has horizontal asymptotes $y = y_- > 0$ as $x \to -\infty$, and $y = y_+ < 0$ as $x \to +\infty$.

3 Asymptotic Behavior of Bounded Solutions

Consider now the asymptotic behavior of bounded μ-solutions to (1) and of their first derivatives.

Since the substitution $y(x) \mapsto -y(-x)$ does not change the type of Eq. (1), it is sufficient to study the behavior of these solutions only near the domain's right boundaries. The behavior of a solution near its domain's left boundary is similar, but with the opposite sign.

Denote

$$C_0(s,t) = (s\,|1-k_1|)^{\frac{1}{1-k_1}}\,|t|^{\frac{k_0}{1-k_1}}\,.$$

First, let us consider increasing bounded solutions. As it is stated in [8], those are black hole solutions appearing in the case $k_1 > 2$.

Theorem 2 *Suppose $k_1 > 2$. Let $y(x)$ be an increasing solution to Eq. (1), let $x^* < +\infty$ be its domain's right boundary. Denote $y^* = \lim\limits_{x \to x^*-0} y(x)$ and let $p(x, u, v) \to p^*$ as $x \to x^*$, $u \to y^*$, $v \to +\infty$. Then*

$$y'(x) = C_0(p^*, y^*)\left(x^* - x\right)^{-\frac{1}{k_1-1}} (1 + o(1)),\ x \to x^* - 0.$$

Proof Since $y' \to +\infty$ as $x \to x^* - 0$, we derive from Eq. (1) that

$$y''(x) = p^*(y^*)^{k_0}(y'(x))^{k_1} (1 + o(1)), \quad x \to x^* - 0,$$

$$y''(x)(y'(x))^{-k_1} = p^*(y^*)^{k_0} (1 + o(1)), \quad x \to x^* - 0,$$

$$\left.\frac{(y'(x))^{1-k_1}}{1-k_1}\right|_x^{x^*} = p^*(y^*)^{k_0}(x^* - x)(1 + o(1)), \quad x \to x^* - 0.$$

Due to the fact that $k_1 > 2$ and $1 - k_1 < 0$, we obtain

$$(y'(x))^{1-k_1} = p^*(k_1 - 1)(y^*)^{k_0}(x^* - x)(1 + o(1)), \quad x \to x^* - 0,$$

whence

$$y'(x) = C_0(p^*, y^*)\left(x^* - x\right)^{-\frac{1}{k_1-1}} (1 + o(1)),\ x \to x^* - 0,$$

and the theorem is proved. □

Now consider decreasing μ-solutions.

It is shown in [8] that in this case every decreasing μ-solution has a finite right boundary of domain, which is further denoted by x_+, and a finite limit $y_+ = \lim\limits_{x \to x_+} y(x)$, $-\infty < y_+ < 0$.

Let us start with the case $0 < k_1 < 1$.

Theorem 3 *Suppose $0 < k_1 < 1$ and $p(x, u, v) \to p_+$ as $x \to x_+ - 0$, $u \to y_+$, $v \to 0$. Let $y(x)$ be a decreasing μ-solution to Eq. (1), $y(x_0) \le 0$, $x_0 \in \mathbb{R}$. Then*

$$y'(x) = -C_0(p_+, y_+)\,(x_+ - x)^{\frac{1}{1-k_1}} (1 + o(1)), \quad x \to x_+ - 0.$$

Proof For a decreasing solution, we derive from Eq. (1) that

$$y'' = p_+|y_+|^{k_0}|y'|^{k_1} (1 + o(1)), \quad x \to x_+ - 0,$$

$$y''(-y'(x))^{-k_1} = p_+|y_+|^{k_0} (1 + o(1)), \quad x \to x_+ - 0,$$

and, since $y' \to -0$ as $x \to x_+ - 0$,

$$\frac{(-y'(x))^{1-k_1}}{1-k_1}\Bigg|_x^{x_+} = -p_+|y_+|^{k_0}(x_+ - x)\,(1+o(1)), \quad x \to x_+ - 0.$$

Due to the fact that $0 < k_1 < 1$, $1 - k_1 > 0$, we obtain

$$(-y'(x))^{1-k_1} = p_+(1-k_1)|y_+|^{k_0}(x_+ - x)\,(1+o(1)), \quad x \to x_+ - 0,$$

whence

$$y'(x) = -C_0(p_+, y_+)\,(x_+ - x)^{\frac{1}{1-k_1}}\,(1+o(1)), \quad x \to x_+ - 0,$$

and Theorem 3 is proved. □

It is shown in [8] that in the case $1 \le k_1 < 2$, every decreasing μ-solution is defined on the whole real axis and has a finite limit $\tilde{y} = \lim\limits_{x\to+\infty} y(x)$, $-\infty < \tilde{y} < 0$.

Theorem 4 *Suppose* $k_1 = 1$ *and* $p(x, u, v) \to p_+$ *as* $x \to +\infty$, $u \to \tilde{y}$, $v \to 0$. *Let* $y(x)$ *be a decreasing* μ*-solution to Eq.* (1), $y(x_0) \le 0$, $x_0 \in \mathbb{R}$. *Then*

$$y'(x) = -|y'(x_0)|\,e^{-p_+|\tilde{y}|^{k_0}\,(x-x_0)}(1+o(1)), \quad x \to +\infty.$$

Proof Analogously to the proof of the previous theorem, we obtain that

$$(-y'(x))^{-1}y'' = p_+|\tilde{y}|^{k_0}\,(1+o(1)), \quad x \to +\infty,$$

$$\ln(-y')\Bigg|_{x_0}^{x} = -p_+|\tilde{y}|^{k_0}(x - x_0)\,(1+o(1)), \quad x \to +\infty,$$

and

$$\ln(-y'(x)) - \ln(-y'(x_0)) = -p_+|\tilde{y}|^{k_0}(x - x_0)\,(1+o(1)), \quad x \to +\infty,$$

whence

$$y'(x) = y'(x_0)\,e^{-p_+|\tilde{y}|^{k_0}(x-x_0)\,(1+o(1))}, \quad x \to +\infty,$$

$$y'(x) = -|y'(x_0)|\,e^{-p_+|\tilde{y}|^{k_0}\,(x-x_0)}\,(1+o(1)), \quad x \to +\infty,$$

and Theorem 4 is proved. □

Theorem 5 *Suppose* $1 < k_1 < 2$ *and* $p(x, u, v) \to p_+$ *as* $x \to +\infty$, $u \to \tilde{y}$, $v \to 0$. *Let* $y(x)$ *be a decreasing* μ*-solution to Eq.* (1), $y(x_0) \le 0$, $x_0 \in \mathbb{R}$. *Then*

$$y'(x) = -C_0(p_+, \tilde{y})\,(x - x_0)^{\frac{1}{1-k_1}}\,(1+o(1)), \quad x \to +\infty.$$

Proof Again, from Eq. (1), we derive that

$$(-y'(x))^{-k_1} y'' = p_+ |\tilde{y}|^{k_0} \, (1 + o(1)), \quad x \to +\infty,$$

and due to the fact that $y' \to -0$ as $x \to x_+ - 0$,

$$\left. \frac{(-y')^{1-k_1}}{1-k_1} \right|_{x_0}^{x} = -p_+ |\tilde{y}|^{k_0} (x - x_0) \, (1 + o(1)), \quad x \to +\infty.$$

Since $1 < k_1 < 2$, $1 - k_1 < 0$, then

$$(-y'(x))^{1-k_1} - (-y'(x_0))^{1-k_1} = p_+ (k_1 - 1) |\tilde{y}|^{k_0} (x - x_0) \, (1 + o(1)), \quad x \to +\infty,$$

whence

$$(-y'(x))^{1-k_1} = p_+ (k_1 - 1) |\tilde{y}|^{k_0} (x - x_0) \, (1 + o(1)), \quad x \to +\infty,$$

and

$$y'(x) = -C_0(p_+, \tilde{y}) \; (x - x_0)^{\frac{1}{1-k_1}} \; (1 + o(1)), \quad x \to +\infty,$$

and Theorem 5 is proved. □

Acknowledgements The reported study was funded by RFBR, project number 19-31-90168.

References

1. Evtukhov, V.M.: On asymptotic behavior of monotonous solutions to nonlinear differential equations. Differ. Equ. **28**(6), 1076–1078 (1992)
2. Astashova, I.V.: Qualitative properties of solutions to quasilinear ordinary differential equations. In: Astashova, I.V. (ed.) Qualitative Properties of Solutions to Differential Equations and Related Topics of Spectral Analysis, scientific, pp. 22–290. UNITY-DANA Publ, Moscow (2012) [in Russian]
3. Astashova, I.V.: On asymptotic classification of solutions to fourth-order differential equations with singular power nonlinearity. Math. Model. Anal. **21**(4), 502–521 (2016)
4. Astashova, I.V.: Asymptotic behavior of singular solutions of Emden-Fowler type equations. Differ. Equ. **55**(5), 581–590 (2019)
5. Astashova, I.V.: On asymptotic classification of solutions to nonlinear regular and singular third- and fourth-order differential equations with power nonlinearity. In: Differential and Difference Equations with Applications, vol. 164 of Springer Proceedings in Mathematics & Statistics, pp. 191–204. Springer International Publishing (2016)
6. Jaroš, J., Kusano, T.: On black hole solutions of second order differential equations with a singular nonlinearity in the differential operator. Funkcialaj Ekvacioj **43**(5), 491–509 (2000)
7. Jaroš, J., Kusano, T.: On white hole solutions of a class of nonlinear ordinary differential equations of the second order. Funkcialaj Ekvacioj **45**(3), 319–339 (2002)
8. Korchemkina, T.: On the behavior of solutions to second-order differential equation with general power-law nonlinearity. Mem. Differ. Equ. Math. Phys. **73**, 101–111 (2018)

9. Korchemkina, T.A.: Asymptotic behavior of unbounded solutions of second-order differential equations with general nonlinearities. J. Math. Sci. **244**(2), 267–277 (2019)

Periodic Solutions for a Class of Impulsive Delay Differential Equations

Dan Gamliel

Abstract We study two coupled linear delay differential equations (DDEs) with additive impulses at regular time intervals. The equations are transformed to a DDE coupled to an ODE. Conditions are found for positive periodic solutions, and some examples are given for periodic solutions and for non-periodic solutions.

Keywords Delay differential equations · Impulses · Periodic solutions

1 Introduction

Periodic solutions to delay differential equations (DDE) have been studied by analogy to Floquet theory of ODE [1], by lower and upper solutions [2], by Lyapunov's second method and the contraction mapping principle [3], or by fixed point arguments [4–6]. In this work, we use the results of [4] to investigate the conditions for periodic solutions for the following linear DDE with impulses:

$$\frac{d}{dt}\mathbf{x}(t) + \mathbf{A}(t) \cdot \mathbf{x}(t) + \mathbf{B}(t) \cdot \mathbf{x}(t-r) = 0 \qquad (t \geq t_{in}) \tag{1}$$

$$\mathbf{x}\left({t_k}^{+}\right) - \mathbf{x}\left(t_k\right) = \mathbf{I}_{(k)} \qquad t_k = t_0 + k \cdot T \qquad (k \in N) \tag{2}$$

where the constant time delay satisfies: $r > 0$, and t_0 is related to the initial time value t_{in} by $t_0 - r \geq t_{in}$. The impulses are assumed to be additive, as in [6]. The arrays in (1) are defined as follows:

D. Gamliel (✉)
Physics Department, Ariel University, Ariel, Israel
e-mail: dang@ariel.ac.il

A. Domoshnitsky et al. (eds.), *Functional Differential Equations and Applications*, Springer Proceedings in Mathematics & Statistics 379,
https://doi.org/10.1007/978-981-16-6297-3_20

$$\mathbf{x}(t) = \begin{pmatrix} x_1(t) \\ x_2(t) \end{pmatrix} \tag{3}$$

$$\mathbf{A}(t) = \begin{pmatrix} a_1(t) & -a_1(t) \\ -a_2(t) & a_2(t) \end{pmatrix} \tag{4}$$

$$\mathbf{B(t)} = b_1 \cdot h(t) \cdot \begin{pmatrix} 1 & 1 \\ 1 & 1 \end{pmatrix} \tag{5}$$

Here, b_1 is a constant, and the time interval T is the common period of the functions h(t), $a_1(t)$, $a_2(t)$.

This is given together with the initial condition

$$\begin{aligned} x_1(t) &= \phi_1(t) \\ x_2(t) &= \phi_2(t) \end{aligned} \tag{6}$$

for $t_0 - r < t < t_0$, with $x_1(t_0) = m_1 \quad x_2(t_0) = m_2$.

2 Solutions for the DDE

In order to simplify the treatment of the coupled equations presented above, we define the transformation

$$\mathbf{y}(t) = \begin{pmatrix} y_1(t) \\ y_2(t) \end{pmatrix} = \frac{1}{2} \begin{pmatrix} 1 & 1 \\ 1 & -1 \end{pmatrix} \begin{pmatrix} x_1(t) \\ x_2(t) \end{pmatrix} \tag{7}$$

Then Eq. (1) leads to

$$\frac{d}{dt} y_1(t) + (a_1(t) - a_2(t)) \cdot y_2(t) + 2 \cdot b_1 \cdot h(t) \cdot y_1(t-r) = 0 \tag{8}$$

$$\frac{d}{dt} y_2(t) + (a_1(t) + a_2(t)) \cdot y_2(t) = 0 \tag{9}$$

There are still two coupled functions, but now only one function, $y_1(t)$, satisfies a DDE, whereas $y_2(t)$ satisfies an ODE. Impulses can be considered for each of these functions. The initial conditions for the two functions are

$$\begin{aligned} y_1(t) &= \tfrac{1}{2}(\phi_1(t) + \phi_2(t)) \\ y_2(t) &= \tfrac{1}{2}(\phi_1(t) - \phi_2(t)) \end{aligned} \tag{10}$$

for $t_0 - r < t < t_0$, with $y_1(t_0) = \frac{1}{2}(m_1 + m_2) \quad y_2(t_0) = \frac{1}{2}(m_1 - m_2)$

Using the notation

$$a(t) = (a_1(t) + a_2(t))$$

the function $y_2(t)$ is calculated (for $t_0 < t$) as

$$y_2\,(t) = exp\left\{-\int_{t_0}^{t} a(s)\cdot \mathrm{d}s\right\}\cdot y_2\,(t_0) \tag{11}$$

Note that if the initial conditions include $m_1 = m_2$, then $y_2(t_0) = 0$ so the function remains zero for all times. We shall assume here that $m_1 \neq m_2$ so that $y_2(t)$ is not trivial.

Proposition 1 *For the solution of Eq. (9), if*

$$\int_{t^+}^{t+T} a(s)\cdot \mathrm{d}s = 0$$

then the solution is periodic. Otherwise, if the solution is modified by adding for each t_k *(k = 1, 2, …) the impulse*

$$I_{2(k)} = \{1 - exp\{-\int_{t_k^+}^{t_k+T} a(s)\cdot \mathrm{d}s\}\cdot y_2\left(t_k^+\right)\},$$

the resulting modified solution is periodic. If the function a(t) is continuous in the interval $[t_k, t_k + T]$, *then the solution y(t) is bounded.*

The periodicity is checked by the evolution of the solution between t_k and $t_{k+1} = t_k + T$:

$$y_2\,(t_k + T) = exp\{-\int_{t_k^+}^{t_k+T} a(s)\cdot \mathrm{d}s\}\cdot y_2\left(t_k^+\right) \tag{12}$$

In the trivial case where

$$\int_{t^+}^{t+T} a(s)\cdot \mathrm{d}s = 0,$$

the solution for $y_2(t)$ is already periodic, without any need for impulses. If

$$\int_{t^+}^{t+T} a(s)\cdot \mathrm{d}s > 0$$

and no impulses are applied, then the solution tends to zero for $t \to \infty$, so that the zero solution is stable, but there is no periodicity. If

$$\int_{t^+}^{t+T} a(s)\cdot \mathrm{d}s < 0$$

and no impulses are applied, the solution diverges as $t \to \infty$. In the last two cases, if the additive impulse

$$I_{2(k)} = \{1 - exp\{-\int_{t_k^+}^{t_k+T} a(s)\cdot \mathrm{d}s\}\cdot y_2\left(t_k^+\right)\} \tag{13}$$

is applied at the times $t_k = k \cdot T$, i.e.,

$$y_2\left(t_k{}^+\right) - y_2\left(t_k\right) = y_2\left(t_k\right) + I_{2(k)}, \tag{14}$$

then $y_2(t)$ is periodic. If the function $a(t)$ is continuous in each interval $[t_k, t_k + T]$, then $y_2(t)$ is bounded.

Equation (8) for $y_1(t)$ will be re-written as

$$\frac{d}{dt} y_1(t) + b(t) \cdot y_1(t - r) = -\left(a_1(t) - a_2(t)\right) \cdot y_2(t) \tag{15}$$

where $b(t) \equiv 2 \cdot b_1 \cdot h(t)$. In [4], Schauder's fixed point theorem is used in order to prove that if there exists a continuous function w(t) such that

$$\int_t^{t+T} b(s) \cdot w(s) \cdot \mathrm{d}s = 0 \qquad (for \quad t - r \geq t_0) \tag{16}$$

and also

$$\int_{t-r}^{t} b(s) \cdot w(s) \cdot \mathrm{d}s = ln\left(w(t)\right), \tag{17}$$

then there is a positive periodic solution to the homogeneous part of Eq. (15). A way to construct the solution is given in [4]. If this periodic solution is denoted by $y_0(t)$, then the solution to full Eq. (15) is

$$y_1(t) = y_0(t) + \int_{t_0}^{t} X(t, s) \cdot \{-\left(a_1(s) - a_2(s)\right)\} \cdot y_2(s) \cdot \mathrm{d}s \tag{18}$$

where $X(t, s)$ is the fundamental solution to Eq. (15) [7]. This solution evolves over one period of $y_0(t)$ as

$$\begin{aligned} &y_1(t + T) - y_0(t + T) = \\ &\qquad y_1(t) - y_0(t) + \int_{t^+}^{t+T} X(t, s) \cdot \{-\left(a_1(s) - a_2(s)\right)\} \cdot y_2(s) \cdot \mathrm{d}s \end{aligned} \tag{19}$$

Proposition 2 *For the equation as Eq. (15) above, if*

$$\int_{t^+}^{t+T} X(t, s) \cdot \{-\left(a_1(s) - a_2(s)\right)\} \cdot y_2(s) \cdot \mathrm{d}s = 0,$$

then the solution to the equation is positive and periodic. If the functions b(t) and $a(t)$ are continuous in each interval $[t_k, t_k + T]$ (with at most a finite number of finite discontinuities), then the solution is bounded.

Note: If the integral in Eq. (19) is not zero, stability for Eq. (15) can hold if: (a) the integral tends to zero as $t \to \infty$ and (b) the equation for $y_0(t)$ is stable. The stability of $y_0(t)$ can be checked as in [8]. However, if the integral in Eq. (18) diverges for $t \to \infty$, then the equation for $y_1(t)$ is not stable, even if the equation for $y_0(t)$ is stable.

The original Equation (1) is solved (for $t_k < t \le t_k + T$) by

$$x_1(t) = exp\{-\int_{t_k}^{t} a(s) \cdot ds\} \cdot y_2(t_k) + \tag{20}$$

$$y_0(t) + \int_{t_k}^{t} X(t,s) \cdot \{-(a_1(s) - a_2(s))\} \cdot y_2(s) \cdot ds$$

$$x_2(t) = -exp\{-\int_{t_k}^{t} a(s) \cdot ds\} \cdot y_2(t_k) +$$

$$y_0(t) + \int_{t_k}^{t} X(t,s) \cdot \{-(a_1(s) - a_2(s))\} \cdot y_2(s) \cdot ds \tag{21}$$

where the properties of the individual terms (y_2 and y_1) determine the properties of the original variables $x_1(t), x_2(t)$.

3 Examples

3.1 Example 1

Consider a delay of $r = 6\pi$ and the following functions:

$h(t) = \cos(t)$

$a_1(t) = c_0 + c_1 \cdot \cos(t)$, $a_2(t) = c_2 \cdot \cos(t)$ where c_0, c_1, c_2 are constants. For initial conditions, let us choose $m_1 \ne m_2$, so that $y_2(t_0) \ne 0$ and take $t_0 = 0$. Then for $0 < t$,

$$y_2(t) = exp\{-c_0 \cdot (t - t_0) - (c_1 + c_2) \cdot (\sin(t) - \sin(t_0))\} \cdot y_2\left(0^+\right) \tag{22}$$

As for $y_1(t)$, the solution for the homogenous equation of Eq. (15) can be obtained by choosing $w(t) = 1$, and then Eqs. (16) and (17) become

$$\int_{t}^{t+2\cdot\pi} 2b_1 \cdot \cos(s) \cdot ds = 0$$

$$\int_{t-6\cdot\pi}^{t} 2b_1 \cdot \cos(s) \cdot ds = 0$$

The solution for the homogeneous equation of $y_1(t)$ is

$$y_0(t) = exp\{2 \cdot b_1 \cdot (\sin(t_0) - \sin(t))\}$$

so that the fundamental solution is

$$X(t,s) = exp\{2 \cdot b_1 \cdot (\sin(s) - \sin(t))\}$$

The integral in Eq. (18) is

$$\int_{0+}^{2\pi} X(t,s) \cdot \{-c_0 - (c_1 - c_2) \cdot \cos(s)\} \cdot y_2(s) \cdot ds$$

3.1.1 Case 1.A

If $c_1 = c_2$, the only contribution to this integral in the $a_1 - a_2$ term is from c_0. Substituting in Eq. (18), one gets

$$y_1(t) = y_0(t) + \int_{t_0}^{t} X(t,s) \cdot \{-c_0\} \cdot exp\{-c_0 \cdot (s - t_0) - 2 \cdot c_1 \cdot \sin(s)\} \cdot ds \cdot y_2(0) \tag{23}$$

The integral term $J \equiv y_1(t) - y_0(t)$ is equal (for $t_0 = 0$) to

$$J = -c_0 \cdot exp\{-2 \cdot b_1 \cdot \sin(t)\} \cdot \int_{t_0}^{t} exp\{-c_0 \cdot s + \sin(s) \cdot (2b_1 - 2c_1)\} \cdot ds \cdot y_2(t_0) \tag{24}$$

The result of the integral is a non-periodic function, so calculating the integral between the limits: t_k and $t_k + T$ will not give zero. In the special case $b_1 = c_1$, the integral term is much simpler, but still the result is not periodic.Thus, the function $y_1(t)$ is not periodic, unlike $y_0(t)$. Then the original variables $x_1(t)$ and $x_2(t)$ are a combination of a periodic part (that of $y_2(t)$ and $y_0(t)$) and a non-periodic part (J). If one adds an impulse to $y_1(t)$:

$$I_{1(k)} = \{y_1(t_k) - y_0(t_k)\} - \{y_1(t_k + T) - y_0(t_k + T)\},$$

this will correct the value of the function only for a single time point. Due to the dependence on the time delay, the behavior of the function for the next time interval will in general be different from that in the previous interval, so $y_1(t)$ will remain non-periodic. Therefore, in both cases, $b_1 \neq c_1$ and $b_1 = c_1$, the solution diverges for $t \to \infty$.

3.1.2 Case 1.B

Now assume $c_1 \neq c_2$ and $c_0 = 0$. Now the integral term is equal to

$$
\begin{aligned}
J &= \int_{t_0}^{t} X(t,s) \cdot \{-(c_1 - c_2) \cdot \cos(s)\} \cdot exp\{-(c_1 + c_2) \cdot \sin(s)\} \cdot \mathrm{d}s \cdot y_2(0) \\
&= -(c_1 - c_2) \cdot exp\{-2 \cdot b_1 \cdot \sin(t)\} \cdot \\
&\quad \int_{t_0}^{t} \cos(s) exp\{\sin(s) \cdot (2b_1 - (c_1 + c_2))\} \cdot \mathrm{d}s \cdot y_2(t_0)
\end{aligned}
\tag{25}
$$

If $2b_1 \neq c_1 + c_2$, then the expression above is equal to

$$
\begin{aligned}
J = &-(c_1 - c_2) \cdot exp\{-2 \cdot b_1 \cdot \sin(t)\} \cdot \\
&\frac{1}{(2b_1 - (c_1 + c_2)))} \cdot \\
&\{exp\{(2 \cdot b_1 - (c_1 + c_2)) \cdot \sin(t)\} - exp\{(2 \cdot b_1 - (c_1 + c_2)) \cdot \sin(t_0)\}\} \cdot y_2(t_0)
\end{aligned}
$$

and this is a periodic function, so that also $y_1(t)$ is periodic. If $2b_1 = c_1 + c_2$, then the expression is

$$
\begin{aligned}
J &= -(c_1 - c_2) \cdot exp\{-2 \cdot b_1 \cdot \sin(t)\} \cdot \int_{t_0}^{t} \cos(s) \cdot \mathrm{d}s \cdot y_2(t_0) \\
&= -(c_1 - c_2) \cdot exp\{-2 \cdot b_1 \cdot \sin(t)\} \cdot (\sin(t) - \sin(t_0))
\end{aligned}
$$

which is also periodic. Thus, regardless of the value of b_1, the solution is periodic, both for $x_1(t)$ and for $x_2(t)$.

3.2 Example 2

With the time delay: $r = \frac{\pi}{2}$, consider the following functions:

$$
h(t) = -\sin(t) \cdot exp\{2b_1 \cdot (\sin(t) - \cos(t))\}
$$

and $a_1(t), a_2(t)$ as in the previous example. Then $y_2(t)$ is the same as above, and for $y_1(t)$, we define the function $w(t) = exp\{2b_1 \cdot (\cos(t) - \sin(t))\}$.

Then Eqs. (16) and (17) become

$$
-\int_{t}^{t+2\cdot\pi} 2b_1 \cdot \sin(s) \cdot \mathrm{d}s = 0
$$

$$
-\int_{t-\frac{\pi}{2}}^{t} 2b_1 \cdot \sin(s) \cdot \mathrm{d}s = 2b_1 \cdot (\cos(t) - \sin(t))
$$

Now the solution for the homogenous equation of Eq. (15) is

$$y_0(t) = exp\{2 \cdot b_1 \cdot (\cos(t_0) - \cos(t))\}$$

so that the fundamental solution is

$$X(t,s) = exp\{2 \cdot b_1 \cdot (\cos(s) - \cos(t))\}$$

3.2.1 Case 2.A

If $c_1 = c_2$, the only contribution to this integral is from c_0.

Substituting in Eq. (17), one gets

$$y_1(t) = y_0(t) + \int_{t_0}^{t} X(t,s) \cdot \{-c_0\} \cdot exp\{-c_0 \cdot (t - t_0) - 2 \cdot c_1 \cdot \sin(s)\} \cdot \mathrm{d}s \cdot y_2(0) \tag{26}$$

The integral term $J \equiv y_1(t) - y_0(t)$ is equal to

$$J = -c_0 \cdot exp\{-c_0 \cdot (t - t_0) - 2 \cdot b_1 \cdot \cos(t)\} \cdot \int_{t_0}^{t} exp\{(2b_1 \cdot \cos(s) - 2c_1 \cdot \sin(s))\} \cdot \mathrm{d}s \cdot y_2(t_0) \tag{27}$$

The result of the integral is a non-periodic function, so calculating the integral between the limits: t_k and $t_k + T$ will not give zero. In fact, for this case,

$$\int_{0}^{2\pi} exp\{(2b_1 \cdot \cos(s) - 2c_1 \cdot \sin(s))\} \cdot \mathrm{d}s = 2\pi I_0(\sqrt{4 \cdot (b_1)^2 + 4 \cdot (c_1)^2})$$

where $I_0(x)$ is the modified Bessel function of order zero.

Thus, the function $y_1(t)$ is not periodic, unlike $y_0(t)$. Then the original variables $x_1(t)$ and $x_2(t)$ are a combination of a periodic part (that of $y_2(t)$ and $y_0(t)$) and a non-periodic part (J).

3.2.2 Case 2.B

Now assume $c_1 \neq c_2$ and $c_0 = 0$. Now the integral term is equal to

$$\begin{aligned} J &= \int_{t_0}^{t} X(t,s) \cdot \{-(c_1 - c_2) \cdot \cos(s)\} \cdot exp\{-(c_1 + c_2) \cdot \sin(s)\} \cdot \mathrm{d}s \cdot y_2(0) \\ &= -(c_1 - c_2) \cdot exp\{-2 \cdot b_1 \cdot \cos(t)\} \cdot \\ &\quad \int_{t_0}^{t} \cos(s) exp\{(2b_1 \cdot \cos(s) - (c_1 + c_2) \cdot \sin(s))\} \cdot \mathrm{d}s \cdot y_2(t_0) \end{aligned} \tag{28}$$

The result of the integration is not a periodic function. Also, for the special case $(c_1 + c_2) = 0$, the result is not periodic, and in that case,

$$\int_0^{2\pi} \cos(s) \cdot exp\{2b_1 \cdot \cos(s)\} \cdot ds = 2\pi I_1(2b_1)$$

where $I_1(x)$ is the modified Bessel function of order one. Thus, regardless of the value of b_1, the solution is not periodic, both for $x_1(t)$ and for $x_2(t)$. The solutions diverge for $t \to \infty$.

3.3 Example 3

With the time delay: $r = \frac{\pi}{2}$ consider the following functions:

$$h(t) = -\sin(t) \cdot exp\{2b_1 \cdot (\sin(t) - \cos(t))\}$$

and

$a_1(t) = c_0 + c_1 \cdot \sin(t)$, $a_2(t) = c_2 \cdot \sin(t)$ where c_0, c_1, c_2 are constants. Then for $0 < t$,

$$y_2(t) = exp\{-c_0 \cdot (t - t_0) + (c_1 + c_2) \cdot \cos(t)\} \cdot y_2\left(0^+\right) \tag{29}$$

and for $y_1(t)$, we define the function $w(t) = exp\{2b_1 \cdot (\cos(t) - \sin(t))\}$.

Then Eqs. (16) and (17) are the same as in Example 2 above,

and also the solution for the homogenous equation of Eq. (15) and consequently the fundamental solution are the same as in Example 2 above.

3.3.1 Case 3.A

If $c_1 = c_2$, the only contribution to this integral is from c_0.

Substituting in Eq. (17), one gets

$$y_1(t) = y_0(t) + \int_{t_0}^{t} X(t, s) \cdot \{-c_0\} \cdot exp\{-c_0 \cdot (s - t_0) + 2 \cdot c_1 \cdot \cos(s)\} \cdot ds \cdot y_2(0) \tag{30}$$

The integral term $J \equiv y_1(t) - y_0(t)$ is equal to

$$J = -c_0 \cdot exp\{c_0 \cdot (t_0) - 2 \cdot b_1 \cdot \cos(t)\} \cdot \int_{t_0}^{t} exp\{c_0 \cdot s + (2b_1 \cdot \cos(s) + 2c_1 \cdot \cos(s))\} \cdot ds \cdot y_2(t_0) \tag{31}$$

Thus, the function $y_1(t)$ is not periodic, unlike $y_0(t)$.

3.3.2 Case 3.B

Now assume $c_1 \neq c_2$ and $c_0 = 0$. Now the integral term is equal to

$$\begin{aligned} J &= \int_{t_0}^{t} X(t,s) \cdot \{-(c_1 - c_2) \cdot \sin(s)\} \cdot exp\{-(c_1 + c_2) \cdot \cos(s)\} \cdot ds \cdot y_2(0) \\ &= -(c_1 - c_2) \cdot exp\{-2 \cdot b_1 \cdot \cos(t)\} \cdot \\ &\int_{t_0}^{t} \sin(s) exp\{(2b_1 \cdot \cos(s) + (c_1 + c_2) \cdot \cos(s))\} \cdot ds \cdot y_2(t_0) \end{aligned} \tag{32}$$

If $2b_1 + c_1 + c_2 \neq 0$, then the expression above is equal to

$$\begin{aligned} J = &+ (c_1 - c_2) \cdot exp\{-2 \cdot b_1 \cdot \sin(t)\} \cdot \\ &\frac{1}{(2b_1 + (c_1 + c_2)))} \cdot \\ &\{exp\{(2 \cdot b_1 + (c_1 + c_2)) \cdot \cos(t)\} - exp\{(2 \cdot b_1 + (c_1 + c_2)) \cdot \cos(t_0)\}\} \cdot y_2(t_0) \end{aligned}$$

and this is a periodic function, so that also $y_1(t)$ is periodic. If $2b_1 = c_1 + c_2$, then the expression is

$$\begin{aligned} J &= -(c_1 - c_2) \cdot exp\{-2 \cdot b_1 \cdot \cos(t)\} \cdot \int_{t_0}^{t} \sin(s) \cdot ds \cdot y_2(t_0) \\ &= +(c_1 - c_2) \cdot exp\{-2 \cdot b_1 \cdot \cos(t)\} \cdot (\cos(t) - \cos(t_0)) \end{aligned}$$

which is also periodic. Thus, regardless of the value of b_1, the solution is periodic, both for $x_1(t)$ and for $x_2(t)$.

References

1. Hale, J.K., Verdyun-Lunel, S.M.: Introduction to Functional Differential Equations. Springer, New York (1993)
2. Graef, J.R., Kong, L.: Periodic solutions of first order functional differential equations. Appl. Math. Lett. **24**, 1981–1985 (2011)
3. Li, X., Bohner, M., Wang, C.-K.: Impulsive differential equations: periodic solutions and applications. Automatica **52**, 173–178 (2015)
4. Olach, R.: Positive periodic solutions of delay differential equations. Appl. Math. Lett. **26**, 1141–1145 (2013)
5. Faria, T., Oliviera, J.J.: Existence of positive periodic solutions for scalar delay differential equations with and without impulses. J. Dyn. Diff. Equat. **31**, 1223–1245 (2019)

6. Federson, M., Györi, I., Mesquita, J.G., Táboas, P.: A delay differential equations with an impulsive self-support condition. J. Dyn. Diff. Equ. (2019). https://doi.org/10.1007/s10884-019-09750-5
7. Agarwal, R.P., Berezansky, L., Braverman, E., Domoshnitsky, A.: Nonoscillation Theory of Functional Differential Equations with Applications. Springer, New York (2012)
8. Berezansky, L., Braverman, E.: On exponential stability of a linear delay differential equation with an oscillating coefficient. Appl. Math. Lett. **22**, 1833–1837 (2009)